P9-DTK-218

BUILDING TRADES ESTIMATING

Featuring

TIMBERLINE®

Precision Estimating
Extended Edition

AMERICAN TECHNICAL PUBLISHERS, INC.
HOMEWOOD, ILLINOIS 60430-4600

Leonard P. Toenjes

Acknowledgments

The publisher expresses appreciation for technical information and assistance provided by the Timberline Software Corporation and the American Society of Professional Estimators, St. Louis Metro Chapter Number 19 in the preparation of this textbook. The following companies and organizations provided additional technical information:

Andersen Windows, Inc.

Barclay & Associates

California Redwood Association

Cleaver-Brooks

General Electric Company

Georgia-Pacific

Greenlee Textron Inc.

Henny Penny Corporation

Hilti, Inc.

Increte Systems

Porter-Cable Corp.

RACO, Hubbell Electrical Products

American Technical Publishers, Inc. Editorial Staff

Technical Editor: Peter A. Zurlis
Copy Editor/Layout: Nanette E. Wargo
Illustrators: William J. Sinclair
             Christopher T. Proctor

1 2 3 4 5 6 7 8 9 – 00 – 9 8 7 6 5 4 3 2

Printed in the United States of America

ISBN 0-8269-0541-2

Contents

Introduction

Building Trades Estimating is a comprehensive introduction to building trades materials, methods, and estimating practices. The textbook contains 15 chapters covering estimating systems, printreading skills, sitework, doors and windows, mechanical systems, and specialty systems. The textbook features traditional (ledger sheet) and electronic (spreadsheet and estimating software) estimating methods. The textbook is designed for commercial contractors and estimators, and is based on the CSI MasterFormat™ using Timberline® Precision Estimating – Extended Edition software. Included with each textbook is a trial version of Timberline® Precision Estimating – Extended Edition software with a 6 month use period and a 50 line item limit. An introduction to using Timberline® Precision Estimating – Extended Edition is given at the end of Chapter 2. Individual topics are repeated at the end of each chapter throughout the book.

Each chapter contains Review Questions and Activities. The Review Questions test comprehension of chapter content. The Activities test estimating processes and consist of one or more of the following:

- Prints
- Cost Data
- Quantity Sheets
- Estimate Summary Sheets

The ledger sheet activity information (Prints, Quantity Sheets, and Estimate Summary Sheets) is provided at the end of each chapter in the textbook. The ledger sheet activities can be worked without the use of a computer. The spreadsheet and estimating software activity information (Prints, Quantity Sheets, and Estimate Summary Sheets) is provided on the CD-ROM. The spreadsheet and estimating software activities require the use of a computer. The spreadsheet and estimating software Prints require the use of Adobe® Acrobat® Reader. Adobe® Acrobat® Reader provides the user with the ability to enlarge images for greater clarity and other navigational functions. The spreadsheet and estimating software Quantity Sheets and Estimate Summary Sheets are Microsoft® Excel files. Microsoft® Excel 5.0 version or higher is required to open the spreadsheet and estimating software Quantity Sheets and Estimate Summary Sheets. Hard copies may be printed as required. The spreadsheet and estimating software activity information (Prints, Quantity Sheets, and Estimate Summary Sheets) is numbered based on the corresponding textbook chapter. Copies of the Prints, Quantity Sheets, and Estimate Summary Sheets can be made for instructional purposes without violation of copyright by printing directly to a printer.

Estimating procedures, Quantity Sheets, and Estimate Summary Sheets vary based on the instructor and/or training program. The Quantity Sheets and Estimate Summary Sheets provided with this textbook are designed for use with the Activities in each chapter of the textbook. Users are encouraged to modify the Quantity Sheets and/or Estimate Summary Sheets for specific instructional needs. The cost data provided for the Activities and used to develop material, labor, and equipment prices is for instructional purposes only. These values are approximate and vary based on the construction location.

A comprehensive Appendix, Glossary, and Index are also included. The Glossary consists of common estimating and building trades terms. The Instructor's Guide includes all answers to Review Questions and Activities. The CD-ROM includes Timberline® Precision Estimating – Extended Edition trial software, Prints, Quantity Sheets, and Estimate Summary Sheets, additional References, Glossary, FAQ link, and links to the American Tech and Timberline Software Corporation websites.

The CD-ROM also contains an electronic *Getting Started with Estimating* booklet. The *Getting Started with Estimating* booklet contains information concerning the operation of Precision Estimating –Extended Edition. To print this document insert the CD-ROM into the appropriate drive on the computer, click the **Start** button, and select **Run**.

If Adobe® Acrobat® Reader is already installed, enter d:\Documents\PEWinExtendedGettingStarted.pdf (where d: is the CD-ROM drive) and click **OK**. This will launch Adobe® Acrobat® Reader and open the file. Click **File** and **Print**.

If Adobe® Acrobat® Reader is not installed, enter d:\ATP\runacrobat.exe.lnk (where d: is the CD-ROM drive) and click **OK**. This will launch Adobe® Acrobat® Reader from the CD-ROM. When the program opens, click **File** and **Open** from the pull-down menu. Locate file d:\Documents\PEWinExtendedGettingStarted.pdf (where d: is the CD-ROM drive) and double-click to open. Click **File** and **Print**.

Before removing the CD-ROM, please note that the textbook cannot be returned if the CD-ROM seal is broken. To complete the Activities in *Building Trades Estimating*, the following must be performed:

1. Install Timberline® Precision Estimating –Extended Edition trial software following the instructions listed on the last page of this textbook. Timberline® Precision Estimating - Extended Edition trial software must be running to work the Estimating Software Activities.

2. With the CD-ROM installed in the appropriate drive, click the **Start** button and select **Run**.

- If Adobe® Acrobat® Reader is already installed:

  Enter d:\ATP\1start.pdf (where d: is the CD-ROM drive) and click **OK**. This will launch Adobe® Acrobat® Reader and open the file. Click START to bring up the homepage.

- If Adobe® Acrobat® Reader is not installed:

  Enter d:\ATP\runacrobat.exe.lnk (where d: is the CD-ROM drive) and click **OK**. This will launch Adobe® Acrobat® Reader from the CD-ROM. When the program opens, click **File** and **Open** from the pull-down menu. Locate file d:\ATP\1start.pdf (where d: is the CD-ROM drive) and double-click to open. Click START to bring up the homepage. The CD-ROM homepage contains buttons for navigation to files on the CD-ROM and Internet web sites.

3. Follow the instructions listed in the Activities at the end of each chapter. For example, to work Activity 3-2 click ACTIVITIES on the CD-ROM homepage to bring up the Activities menu. Click **Print 3-2** to open and view the print. A hard copy may be printed if required. Click BACK to return to the Activities menu. Click **Quantity Sheet 3-2** to open the Quantity Sheet in Microsoft® Excel. The Save As function should be used for storing files in a designated directory on the computer hard drive or removable media. Any combination of the Quantity Sheets, Estimate Summary Sheets, Prints, and Estimating Software may be displayed at the same time. To display several windows at the same time, open all windows to be displayed. Using the mouse, right-click in a blank area on the taskbar. The taskbar is the bar on the desktop that has the **Start** button on it. Click Tile Horizontally or Tile Vertically. To restore the windows to their original state, right-click the mouse in a blank area on the taskbar. Click Undo Tile.

Estimating Practices

Common components of good estimating practice should be followed to complete the estimating and takeoff processes successfully. Estimators should take a systematic approach to putting an estimate together, ensuring accuracy, comprehensiveness, consistency, and timeliness by using an organized, standardized system for each bid. The estimator must work with the most current set of plans and specifications, including all addenda.

ESTIMATING PROCESS

Estimating is the computation of construction costs of a project. The estimating process begins with a review of all contract documents. This includes the plan drawings, specification books including the general specifications, any addenda, and agreement forms. See Figure 1-1. An *addendum* is a change to the originally-issued contract documents. These documents are reviewed to determine if a particular construction project contains items that can be built by a particular construction company. The review also assists the estimator in developing a conceptual idea of the scope of the project and the estimated project budget.

Prebid Meeting

A *bid* is an offer to perform a construction project at a stated price. A *prebid meeting* is a meeting in which all interested parties in a construction project review the project, question the architect or owner concerning methods for accomplishing the work, and share information necessary to understand the entire scope of the work. At a prebid meeting, the estimator meets with

owners and possibly the occupants of the structure to be built. A *contractor* is an individual or company responsible for performing construction work in accordance with the terms of a contract. A *general contractor* is the contractor who has overall responsibility for a construction project. A *subcontractor* is a contractor who works under the direct control of the general contractor. General contractor and subcontractor representatives are also in attendance, providing an opportunity for contacts for future bidding information. In some cases, attendance at a prebid meeting may be required in order to submit a bid.

Figure 1-1. Printreading skills are the basis of successful estimating.

The architect and engineer are also normally in attendance at the prebid meeting to field questions concerning the project. The estimator may contact the architect in advance of the prebid meeting to acquire the names of all plan holders to obtain information concerning other potential bidders, including subcontractors and suppliers who may have assisted in compilation of the bidding documents. Any problems with the bidding documents that may result from the number of final copies required, alternate proposals required, subcontractor and supplier listings required, or unit pricing requirements may also be discussed at this time.

Bid Proposal

Bids and bid proposals include prices and costs for completion of a construction project. Bids and bid proposals are presented to the owner and architect in an understandable format. Each firm seeking to be awarded the construction contract submits information in an organized manner including quantity, cost, and overall pricing information.

A bid is generated that includes each section of the specifications and all items that are estimated and priced. Subcontractors are contacted for bids on any portion of the work that is not taken off and estimated by the estimator. *Takeoff* is the practice of reviewing contract documents to determine quantities of materials that are included in a bid. Bids are requested from subcontractors through personal contact or written request. This request notes the work to be performed, the project schedule, and the time at which bids are due.

Representatives from a construction company may also visit the job site to determine working and building conditions. Additional costs that may impact the final bid can be observed by a job site visit.

The estimator begins to take off and price each of the specification sections. For large projects, the takeoff process may be divided among several estimators, each with their own specialty. For example, individual estimators may specialize in electrical, mechanical, or earthwork. Items are entered in the columns of the master bid sheet or estimating software. See Figure 1-2.

Labor, material, and equipment costs are determined after quantity takeoff is completed. Labor, material, and equipment cost determination may be part of the quantity takeoff process in estimating software. Labor costs are highly variable and must be carefully determined based on worker skill and productivity. Material costs

are easier to determine based on supplier bids. Equipment expenses include depreciation, maintenance, and insurance costs.

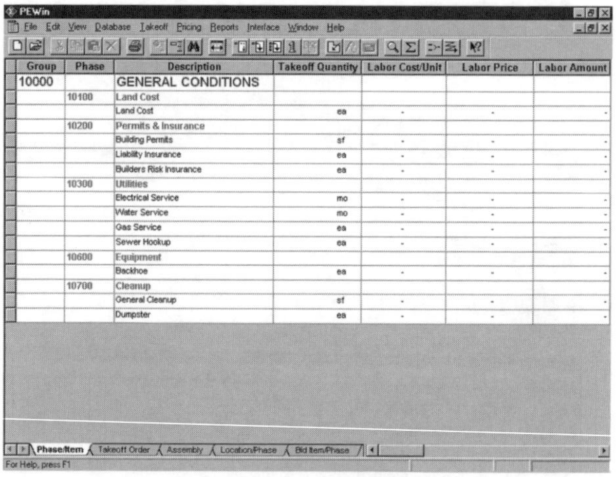

Figure 1-2. Estimating software may be used to compile bid information.

Subcontractor Quotation. Some portions of the work may be assigned to subcontractors for bidding and performance of the work. The scope of work to be bid and performed by a subcontractor must be clearly defined. A scope of work letter may be provided to the subcontractors who are bidding on portions of the work to ensure a complete and accurate subcontractor bid. See Figure 1-3. As subcontractor bids are received, they are entered into the quoted bid column of a ledger sheet or the proper cell in spreadsheets or estimating software.

A rough estimate should be generated by the estimator for comparison with the subcontractor bid. This ensures accuracy and provides a double check of the costs, especially when using unfamiliar subcontractors. An estimator may not always use the lowest subcontractor bid. Care is taken to use the lowest qualified bid that can complete the job successfully.

Proper and comprehensive requests for subcontractor and supplier quotations are required. Estimators should maintain a current mailing list and telephone listing of subcontractors and suppliers. This aids in prompt notification to subcontractors and suppliers. All notifications should include complete information about the project being bid including the job name, architect, bid date and time, information concerning plan availability, and the name and telephone number of the estimator. Estimators should make follow-up calls to all notified subcontractors and suppliers to determine if they will be providing information. Suppliers may be a source of information for the estimator concerning subcontractors who may be submitting bids for the project.

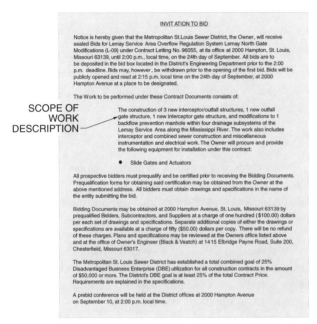

SCOPE OF
WORK
DESCRIPTION

Figure 1-3. A scope of work letter provides complete information to subcontractors bidding on portions of the work.

Systematic Approach

A systematic approach to estimating incorporates standard practices and procedures to ensure an accurate and comprehensive estimate. Successful estimating begins with an organized approach to the takeoff of quantities. The estimator must fully review and understand all portions of the project pertaining to the estimate being prepared. In addition, the estimator should use a consistent process when preparing all costs and quantities and work with the most current set of plans and specifications, including all addenda.

Accuracy. Accuracy in an estimate includes precise calculation of all labor, material, and equipment quantities and costs. Mathematical and printreading skills are essential for accurate estimates. Accurate totals are required as items are counted or calculated from the plans and specifications. Estimators commonly mark up plans and specifications to note items that have been counted to ensure that all necessary items are counted and no items are counted twice. See Figure 1-4. Accurate interpretation of all print symbols, abbreviations, and dimensions is also required for takeoff and estimating.

Figure 1-4. Prints are marked by the estimator to indicate quantities that have been taken off.

Comprehensiveness. Comprehensiveness in estimating ensures that all portions of the proposed work are included in the estimate. Care should be taken in reviewing the specifications and drawings. Some items may be indicated in the specifications and not on the drawings, or indicated on the drawings and not in the specifications. Marking up plans and specifications as a bid is generated ensures that all items are counted and no items are overlooked. A complete review of the specifications and addenda for all items that may affect the estimate is required. Items may be described in various portions of the plans and specifications. Estimators should not rely on a single source of product or material information in a set of plans and specifications.

Consistency. Consistent estimating practices include common procedures used by a company or individual that allow quick reference and ease of finding all items in the estimate. Patterns and habits are developed for creating good estimates. While individual methodologies vary, standard steps or procedures are followed for reviewing plans and specifications and developing each bid. A standard procedure is to review the specifications, review plan drawings, set up the bidding sheets or database (either ledger-based or computer-based), perform the quantity takeoff, price materials and labor, review the final bid, and double check for accuracy. After this is completed, all items are priced and checked and all profit, overhead, and taxes included, the job is ready for final review and a definitive bid.

Estimators should keep track of last minute items that have previously created problems on the final bid day and anticipate possible actions and solutions.

Timeliness. A timely estimating process requires that the most up-to-date set of plans and specifications are used for the bidding process. After the initial plans are released, an architect or owner may make changes to the project in the form of addenda. An addendum is a change to the originally issued contract documents. An addendum may include changes to the specifications and print drawings. Estimators must monitor the planning process and maintain awareness of any addenda up to the time of submission of the bid. Lack of knowledge of addenda can create a situation where a bid is developed based on outdated project information. An estimator may contact the architect a day prior to the bid submission day to ensure that all addenda have been taken into account in preparing the bid. A full consideration of the many variables associated with estimating is required from the beginning of the estimating process. Estimators analyze the entire project, paying special attention to unusual or highly-specialized items. Work on estimating or obtaining subcontractor or supplier bids should begin early in the process to minimize delays in completion of the final bid.

Estimators should ask questions and obtain timely answers concerning any items in the bid documents that require clarification. Lack of timely information at the time of submitting a bid may put the estimator in the position of making a judgment about the project based on incomplete information. Timely collection of information is required when preparing a complete, accurate bid.

VARIABLES

The estimator begins the bidding process with a set of plans, specifications, and addenda. Estimating costs, including preliminary and conceptual costs, design development, budgeting, and the construction bid, can be affected by the completeness of the architectural plans and specifications.

In addition to the architectural plans and specifications, the estimator may visit the job site to directly observe job conditions. Items such as existing structures, access to and from the job site, trees, shrubs, and surrounding buildings and neighborhoods may affect quantities and costs of the job. The estimator must

be aware of labor and material availability, the construction design for the project, and various government requirements in the construction area.

No two projects or estimates are the same. Conditions change involving the weather, job site employees, architectural drawings, subcontractors, suppliers, local government agencies, and a variety of other items. Estimators using bidding information from standardized industry information, company historical data, or electronic databases must keep the variability of each job in mind.

Project Location

The location of a project can introduce many variables into the estimating process. Access to the job site can greatly affect material hauling costs. A job site storage area for delivery and drop off of materials may or may not be available. Some materials required by the architectural design may not be available without high costs in the geographic location of the project. Labor markets are always changing based on the levels of employment in the local construction industry and other industries.

Overhead expenses also vary based on job site location. Overhead expenses include taxes, insurance, permits, and other location-specific expenses. A variety of building code requirements and legislative rules apply to construction projects and are different for various structures depending on the location.

Material Availability and Pricing. Material availability is the ability to obtain all items for a project from suppliers. Engineering developments in the design and properties of new materials create an almost endless variety of alternatives for structural engineers and architects. The exchange of information between engineers, product designers and manufacturers, building owners, architects, and estimators is required to ensure that the proper materials are designed and installed into the structure. Catalogs, web sites, and CD-ROMs that contain the latest developments in construction materials are available. See Figure 1-5. Planners of modern structures determine the proper uses of different materials, such as concrete, steel, wood, metal, plastic, glass, and fiberglass.

> ▲ *Construction costs are higher in some cities than in others. Cost adjustment factors may be added to or subtracted from standard cost values to account for the construction costs in specific cities.*

Figure 1-5. Catalogs and CD-ROMs are available that contain the latest developments in construction materials.

Estimators must take product availability into account because this affects price and job site scheduling. The use of scarce material in a particular geographic location may drive prices higher than normal. Lengthy delivery times for scarce material may also affect the project completion date. Proper planning for material pricing and ordering is essential to avoid costly penalties on jobs having penalties for time overruns. Material prices should be checked at the project location prior to pricing.

Estimators examine the bid documents to determine if the use of alternate materials is allowed. Some owners and architects may require cost breakdowns to be included in the final bid proposal. Material pricing is tracked more carefully and divided into individual categories when cost breakdowns are required. Additional staffing may be required at the time of the final bid preparation if cost breakdowns are required.

Labor Availability. Labor availability is the ability to hire the required skilled labor to build a specific project. The development of new construction methods has greatly changed the skills required by all involved in the building process and the procedures for building construction. Automation and prefabrication are playing an increasing role in the building process. New tools and procedures have increased the skill levels needed by tradesworkers on a job site. The availability of skilled labor in an area must be assessed when pricing a job and developing a bid. While standard labor rates are available from a variety of industry sources, an estimator must be familiar with the labor market in the project construction area. The availability and variability of union labor or open shop labor can have a significant affect on labor rates and productivity.

Labor rates and job scheduling can be affected by labor availability. The construction industry commonly faces variability of labor supply in various locations. Estimators should be aware of labor availability and its potential affect on labor costs and project scheduling.

Subcontractor Arrangements. Owners and architects may require a listing of subcontractors and suppliers to be submitted at the time of the final bid. Some owners may provide a listing of preapproved subcontractors and suppliers as the only firms that may submit bid proposals. Estimators should check the general conditions and specifications to ensure only approved subcontractors and suppliers are solicited for bid proposals.

Overhead Expenses. *Overhead* is the cost of doing business that is not related to a specific job. Overhead project location variables include local sales taxes, use taxes, other specialized local taxes, permit fees, insurance, security, and office staffing. The sales tax for materials should be determined to ensure proper material pricing. For example, some materials on construction projects for not-for-profit entities may be exempt from local sales tax depending on local and state legislation. See Figure 1-6. Fees for permits must be determined from each municipality and government agency responsible for inspections. Insurance rates for a project vary depending on the contractor experience, the security of the area surrounding the project, the type of building project, and other insurance industry cost trends. The estimator must carefully assess the insurance costs for each project. Location of a project in a dangerous or unsafe area may require the addition of security patrols, alarms, or security systems. Office staffing costs also vary depending on project location and the labor market for these support services.

Building Codes. Individual municipalities and government agencies have adopted various building codes. One of the most common codes is the Building Officials and Code Administrators International (BOCA) Code. A variety of other codes and regulatory agencies' rules are used for various portions of the building process such as electrical installation and fire protection. Even though some standardized building codes exist, local building and zoning commissions normally make local modifications to meet the needs of their residents. Codes and legal requirements may vary from city to city, state to state, and country to country, sometimes in an overlapping manner for various agencies. Estimators must remain up-to-date on the latest codes and legislation to meet all requirements.

MATERIAL SALES TAX EXEMPTION —

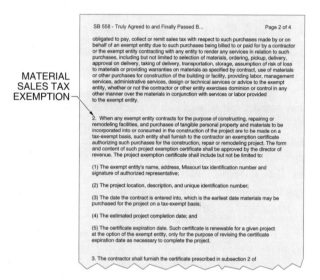

Figure 1-6. Estimators require knowledge of legislation that may affect product pricing.

Environmental Factors. Climatic requirements and the protection of the surrounding environment are two variables estimators must make allowances for in the bidding process. Specifications may describe certain climatic requirements that must exist for the placement of concrete, masonry, plaster, or other materials that may be affected by extremely hot, cold, wet, or dry conditions. See Figure 1-7. Some protection from the elements may be required to allow the project to stay on schedule and work to continue during unusual weather.

Protection of the environment surrounding a construction project may also be noted in the specifications. A job site visit is used to observe the conditions that exist and how they could affect the job progress and scheduling. Additional work may be required to protect certain environmental elements that could shift work schedules.

Government Factors. A variety of local, state, and national organizations have developed legislative and industry requirements that affect construction site progress and scheduling. Local building codes or ordinances may contain specific limits on various job site activities. State codes may allow work to proceed only when certain conditions exist related to environmental

factors. For example, many state Departments of Transportation allow certain types of pavement to be placed only when temperatures are within high and low limits.

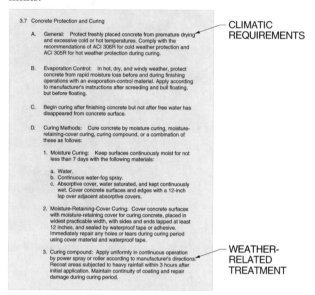

Figure 1-7. Estimators must consider climatic requirements that may affect work scheduling and material placement.

The U.S. government has several agencies that affect construction site activities including the Occupational Safety and Health Administration, the Environmental Protection Agency, and the Office of Federal Contract Compliance Programs. The *Occupational Safety and Health Administration (OSHA)* is a federal government agency established under the Occupational Safety and Health Act of 1970, which requires all employers to provide a safe environment for their employees. OSHA reviews and monitors safety conditions on construction jobs. Estimators must ensure that any costs associated with OSHA regulation requirements are included in the bid. The *Environmental Protection Agency (EPA)* is a federal government agency established in 1970 to control and abate pollution in the areas of air, water, solid waste, pesticides, radiation, and toxic substances. The EPA monitors the effects of a project on the environment. Estimators must include costs of environmental protection in bids as mandated by the EPA. The *Office of Federal Contract Compliance Programs (OFCCP)* is a federal agency whose mandate is to promote affirmative action and equal employment opportunity on behalf of minorities, women, the disabled, and Vietnam veterans. The OFCCP monitors federally funded projects to ensure various civil rights legislation is followed. Costs may be incurred that are associated with recordkeeping, recruitment, and training. These costs must be included in the overhead costs of the project being bid.

Construction Systems

A *construction system* is a method used in the design and construction of a structure, including materials, construction sequence, structural design, and finish materials. The construction system used in a structure has a direct effect on labor costs, material quantities, and equipment requirements. After consideration of all the factors involved, the owner, architect, and engineer choose the construction system that best fits a particular project. Construction systems include frame, structural steel, concrete, and masonry systems. Each has specific applications and advantages. Most structures combine several of the systems to achieve their overall design and purpose.

Frame. *Frame construction* is construction in which a structure is built primarily of wood structural members. In frame construction, a series of small components, including joists, studs, plates, runners, braces, trimmers, cripples, and rafters are joined together to form a rigid frame structure. These components may be wood or metal. Frame construction is used in all types of buildings, but is primarily used in small structures such as dwellings and small commercial buildings. Frame construction is commonly used for interior partitions. See Figure 1-8. Openings between framing members allow for installation of mechanical and electrical systems. After frames are built and mechanical and electrical systems and insulation are installed, the frames are covered with gypsum drywall, masonry veneer, or wood or metal siding. Flexibility of design and economy are advantages in using frame construction.

Structural Steel. *Structural steel construction* is construction in which a series of horizontal beams and trusses and vertical columns are joined to create large structures with open areas. In structural steel construction, lightweight steel members are used to build industrial buildings and storage structures, or large beam and truss assemblies are used to build skyscrapers and bridges. Structural steel construction enables great flexibility of design. See Figure 1-9. Many of the largest buildings in the world are built of structural steel construction.

Special consideration is given to rigging requirements, working at heights, lifting heavy members into place,

and safety for all on the job site. Structural steel buildings have a number of bracing systems which give them great strength against exterior stresses such as wind, earthquakes, and other imposed loads. Many different types of cladding can be attached to the exterior of these structures. *Cladding* is wall surface material attached to a structural steel frame to span between supporting members and provide closure to the structure. Cladding material includes glass, metal, masonry, and precast concrete.

Figure 1-8. Frame construction is based on wood or metal components joined into frames that are strengthened by rigid sheathing or bracing.

Structural steel construction is commonly used for long span roof truss systems and bridges. Careful engineering and design is performed to ensure that the proper steel shapes, sizes, and connections are used to meet all load requirements.

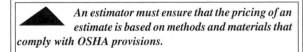

An estimator must ensure that the pricing of an estimate is based on methods and materials that comply with OSHA provisions.

Figure 1-9. Structural steel construction is used for many commercial building projects.

Concrete. *Concrete construction* is construction in which concrete and reinforcing steel are placed in forming materials to create a high-strength finished structure. Concrete construction may consist of poured-in-place or precast concrete. *Poured-in-place concrete* is plastic (uncured) concrete that is poured (placed) in wood or metal forming material that is set to a specific shape and acts as a mold for the concrete. Reinforcing steel is set in the forms and concrete is placed into the forms around the reinforcing steel. The concrete provides compressive strength and the reinforcing steel provides tensile strength. This mixture creates a long-lasting structure. See Figure 1-10. The forms are removed after the concrete has reached a specified degree of set.

Precast concrete is concrete components that are formed, placed, and cured to a specific strength at a location other than their final installed location. Precast concrete members include beams, columns, pipes, walls, flooring sections, and exterior cladding. Consideration must be given to transporting and lifting the large, heavy members into place. Tilt-up construction is a variation of precast construction. *Tilt-up construction* is a method of concrete construction in which concrete members are cast horizontally at a location close to their final position and tilted in place after removal of the forms. A *bond breaker* is a chemical compound applied to the surface of concrete forming material or the slab that comes in contact with the concrete as it is placed. The bond breaker prevents bonding of the newly-placed concrete with forming materials or existing slabs. Reinforcing and lifting hardware is installed in the wall system prior to concrete placement. After the concrete has been placed, finished, and cured, the walls are tilted up into place.

Figure 1-10. Concrete construction combines concrete and reinforcing steel for compressive and tensile strength.

Masonry. *Masonry construction* is construction in which masonry members such as clay bricks or tiles, concrete bricks or blocks, and natural or artificial stone are set in mortar. *Mortar* is a bonding mixture consisting of lime, cement, sand, and water. Masonry members may be reinforced with various steel shapes. Structures of large size and weight-bearing capacity can be built with masonry members. Masonry members are fire-resistant and often used as fire breaks between adjoining areas of a structure. A *fire break* is a space or fire-resistant materials between structures or groups of structures to prevent fire from spreading to adjacent areas. Masonry construction uses brick and block members that are available in a variety of sizes and shapes. See Figure 1-11. Brick and block members are made from clay, concrete, and glass. Stone is also used for structural and decorative applications. Considerations must be made to provide openings for piping and ductwork for mechanical and electrical systems during masonry installation. Masonry is also commonly used as a veneer material for other construction systems, such as frame, structural steel, and concrete. When used as a veneer material, ties are used to anchor the masonry veneer to the structure.

Estimating Practices

An *estimating practice* is the system used to integrate all parts of the estimating process in a cohesive, consistent, reliable manner to ensure an accurate final bid. The four estimating practices are detail takeoff, crew-based, design build (schematic), and model estimating. The system used depends on company practice, the project being estimated, and estimator preference. These estimating practices may be used with a ledger sheet, spreadsheet, or estimating software.

In some cases, there may be allowances in the bidding documents for value engineering. *Value engineering* is a process in which construction personnel such as project managers, estimators, and engineers employed by the firm developing the bid are allowed to suggest changes to the contract documents. The changes are intended to deliver the same needs to the owner as presented in the original document but in a less expensive or more productive method. Value engineering allows architects and estimators to work together and design components of a structure that may be more efficient and economical than components designed by architects or estimators alone. Design features may be changed based on estimator input in value engineering.

> ▲ *An estimator should ensure that the prices quoted by a supplier are for materials that satisfy the requirements of the specifications.*

Figure 1-11. Concrete block used in masonry construction is available in a variety of sizes and shapes.

Detail Takeoff. *Detail takeoff estimating* is an estimating practice in which each individual construction component is taken off. Some projects require detail takeoff. In detail takeoff estimating, all costs and quantities are determined on an individual basis. For example, a takeoff for a section of concrete sidewalk may include individual entries and calculations for the cubic yards of concrete, forming materials, subsurface preparation, excavation labor, excavation equipment costs, concrete placement labor costs, concrete finishing labor costs, form removal labor, and final grading.

Crew-based. *Crew-based estimating* is an estimating practice in which a range of quantities and costs are included into a single calculation based on a specific quantity of construction put in place. These costs may include labor, material, and equipment. For example, a single takeoff entry for a section of concrete sidewalk based on the number of square feet of sidewalk to be put in place may be tied directly to a standard, basic crew cost that includes all labor, material, and equipment for the operation. Crew-based estimating is commonly based on company historical data of total costs per unit of construction put in place.

Design Build. *Design build (schematic) estimating* is an estimating practice in which the development of an estimate is based on the building function or the functional area of a building. Each building function or functional area is priced based on company historical records of costs per square foot according to the function. Where possible, design build estimating may be priced by individual building systems such as exterior walls, roofs, or structural members. This pricing requires some general quantity takeoff, but is not as detailed as detail takeoff or crew-based estimating.

Model. *Model estimating* is an estimating practice in which an estimate is produced by inputting fundamental structure parameters such as building function, size, height, location, and construction system. Company historical records based on a combination of these parameters produces a general price estimate. This method requires no detail or crew takeoffs to produce a general price.

Labor

One of the highest risk variables in construction project estimating is the cost of construction labor. Labor cost per hour, availability, and skill level may vary greatly depending on the project location, time of year, and work conditions. Estimators must take these items into account when creating labor pricing for various items.

Working Conditions and Wage Rates. Working conditions at a job site affect the rate of work performed. Little working space, limited storage, and difficult delivery situations normally lead to lower rates of work. For example, a small job site may increase congestion of workers and materials, thereby lowering work rates and increasing labor costs. An extremely large job site may also lower work rates and increase labor costs by the increased time necessary to reach various areas of the structure. This can apply to vertical and horizontal distances.

Wage rates refer to costs of worker wages and benefits. Working conditions and wage rates vary significantly from project to project. Factors affecting working conditions and wage rates include project location, material storage space, the trade performing a given construction task, the existence of a building trade union, and the contract agreements that may or may not exist. For example, the cost per hour of worker wages and benefits for a craftworker covered by a union labor agreement varies from workers who are not covered by such an agreement. Jurisdictional claims by various trade unions must also be taken into account in preparing labor costs. Labor costs for various material installations vary depending on the labor trade performing the installation and wage and benefit rates.

Training. Labor training includes all skill development programs including vocational education programs, apprenticeships, and journey-level upgrading programs. The quality of work performed by craftworkers can vary greatly from area to area and from trade to trade. An estimator should contact industry sources in the construction project area to determine if there are significant quality concerns due to untrained construction labor. See Figure 1-12. Additional work hours may be required for project completion if training is insufficient to produce quality construction in a minimum amount of time. Additional costs may also be incurred due to rework, repair, and patching of inadequate work to bring it up to the standards required by the owner and architect.

Scheduling

The scheduling of a project affects the estimated costs for materials, labor, and overhead. Material costs may be higher or lower at various times of the year or various market cycles. For example, concrete costs may be higher during winter months in cold climates due to required admixtures in the concrete that inhibit freezing and help the concrete develop a quick set. Jobs that require a fast track scheduling may demand overtime or shift work for craftworkers. Overhead costs may be affected by the time necessary to complete a project, the number of craftworkers required for completion, and penalties for not completing a project on a specific schedule.

Figure 1-12. Highly-trained labor can reduce construction costs and time and improve the quality of the work.

Liquidated damages are penalties assessed against the contractor or subcontractor based on failure to complete work within a specific time period. Estimators should check specifications for the inclusion of liquidated damages to ensure that the estimating process is integrated with the entire construction process. This includes a review of all portions of the specifications.

Bid Schedule and Location. The estimator should consult the general conditions of the specifications to determine the date, time, and location for submission of the final bid. The date and time should be scheduled to take into consideration the day of the week, potential holiday schedules, staffing availability on bid day, and conflicts with other jobs that may be bidding at the same time. Bids due early in the morning require scheduling considerations on the day prior to bidding. The location for the delivery of the bid is considered in order to leave an adequate amount of time to properly prepare the final bid documents and deliver them to the proper bid location.

REFERENCE DATA

Estimators use a vast library of standardized information in pricing and bidding of materials, equipment, and labor. Many private vendors collect market information and publish reference materials. These reference materials include costs that can be put together with the quantity takeoff for pricing. Associations such as the American Institute of Architects (AIA) and the Construction Specifications Institute (CSI) provide a variety of printed material for building standards and specifications. Contractors in business for several years often have their own historical reference data. Contractors analyze previously built projects to determine costs for labor, material, equipment, and overhead for various construction projects. Historical or third party reference data may be available and stored electronically or in print form.

Cost Indexes

A *cost index* is a compilation of a number of cost items from various sources combined into a common table for reference use. Estimators can consult a broad range of cost indexes to determine percentage changes in construction cost items. Cost indexes are divided into a broad range of categories including the construction system, labor, material, equipment, and geographic locations. Sources include public agencies such as the Bureau of Reclamation, the U.S. Department of Commerce, and many private sources specializing in cost indexes such as industrial buildings, chemical process plants, reinforced concrete buildings, and a variety of other specialties. Cost index information is provided in book format and on CD-ROM.

Printed References

Many printed references containing tables, charts, and various cost information are available for use by an estimator. The printed references are arranged to allow an estimator to quickly find labor, material, and equipment costs in relation to the quantity takeoff. See Figure 1-13. Component costs, prices according to square, cubic, or linear measure, and standard labor rates may be obtained from printed reference charts.

> ▲ *Construction cost manuals should be used only as a pricing guide because the prices represent average costs of a number of contractors and may not match the costs incurred by a particular construction company.*

SWITCH AND RECEPTACLE PLATES					
Material	Craft@Hr	Unit	Material	Labor	Total
Combination Decorator and Three Standard Switch Plates					
4 gang brown	E1@.20	Ea	4.88	5.73	10.61
4 gang ivory	E1@.20	Ea	4.88	5.73	10.61
4 gang white	E1@.20	Ea	4.88	5.73	10.61
Semi-Jumbo Switch Plates					
1 gang brown	E1@.05	Ea	.69	1.43	2.12
1 gang ivory	E1@.05	Ea	.69	1.43	2.12
1 gang white	E1@.05	Ea	.75	1.43	2.18
1 gang gray	E1@.05	Ea	.75	1.43	2.18
2 gang brown	E1@.10	Ea	1.51	2.86	4.37
2 gang ivory	E1@.10	Ea	1.51	2.86	4.37

PRE-ENGINEERED STEEL BUILDINGS*						
	Craft@Hr	Unit	Material	Labor	Equipment	Total
40′ × 100′ (4000 SF)						
14′ eave height	H5@.074	SF	6.37	2.98	.75	10.10
16′ eave height	H5@.081	SF	7.15	3.26	.83	11.24
20′ eave height	H5@.093	SF	8.08	3.75	.95	12.78
60′ × 100′ (6000 SF)						
14′ eave height	H5@.069	SF	6.15	2.78	.70	9.63
16′ eave height	H5@.071	SF	6.29	2.86	.72	9.87
14′ eave height	H5@.065	SF	5.12	2.62	.66	8.40
16′ eave height	H5@.069	SF	5.44	2.78	.70	8.92
20′ eave height	H5@.074	SF	6.04	2.98	.75	9.77
100′ × 200′ (20,000 SF)						
14′ eave height	H5@.063	SF	5.02	2.54	.64	8.20
16′ eave height	H5@.066	SF	5.25	2.66	.67	8.58
20′ eave height	H5@.071	SF	5.64	2.86	.72	9.22

* 26 gauge colored galvanized steel roof and siding with 4 in 12 (20 lb live load) roof. Cost per square foot of floor area. Costs do not include foundation or floor slab. Add delivery cost to site. Equipment is a 15 ton truck crane and a 2 ton truck.

Figure 1-13. Printed references contain charts that provide information concerning labor, material, and equipment costs for use when developing an estimate.

Component Costs. Items that are counted from a set of prints can be priced according to component costing. For example, a structure may contain 5 windows of a certain type. The estimator can locate this window in the reference material and find a cost per window including material and labor. Component costs are available for a variety of materials such as doors, windows, special fixtures, louvers, and other items that are listed on a plan.

Unit Prices. Unit (crew-based) pricing is the calculation of material and labor prices in a single step based on unit items. Labor rates are calculated based on standard material quantities and judgments made about the production levels of labor per material quantity unit. Items such as concrete flatwork, painting, gypsum drywall, and flooring are priced according to the square foot or square yard. For example, a standard concrete driveway estimate is developed by calculating the number of square feet of driveway to be placed and finished. This directly affects the labor costs for finishing. A labor cost for concrete finishing based on the number of square feet to be finished is found in printed reference charts. See Figure 1-14. Other materials are unit priced for labor and materials based on linear feet or other unit price measures.

Figure 1-14. Material and labor costs may be calculated based on the square feet of work to be performed.

> ▲ *The best source of cost information is company historical data which matches estimate prices to how a company actually performs its construction.*

Labor Rates. Many government organizations and trade associations provide labor rates for various geographic areas. Estimators can use these sources to determine costs per hour for various construction craftworkers. After the estimator has determined the number of work hours required for a particular unit price of material from printed reference tables, wage rates can be added to the calculations to determine total labor costs. For example, a printed reference table may indicate the amount of carpenter time necessary to hang a 3'-0" metal swinging door in a metal frame is 45 min. Another printed reference table is accessed to determine the wage and benefit costs for a carpenter. Multiplication of these two numbers provides the labor cost based on these printed reference tables.

Electronic References

As computers have become more widely used in the estimating process, reference tables and databases have become available on electronic media. Diskettes and CD-ROMs containing labor, material, and equipment costs can be purchased and integrated into various bidding software packages. Electronic databases can be transferred quickly into bidding spreadsheets to generate costs. See Figure 1-15. Experienced contractors may develop their own electronic databases for costs based on experiences on their job sites.

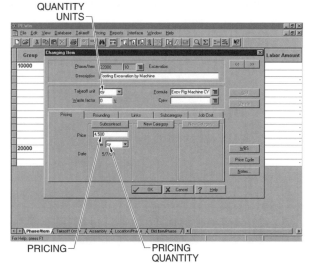

Figure 1-15. Estimating software uses electronic databases of cost information for quantity pricing.

Component Costs. Component costs are located by searching an electronic database for particular words or items. The costs can be copied into the estimate spreadsheet or may be entered automatically in more sophisticated software that connects the component cost database with the takeoff spreadsheet.

Unit Prices. Electronic databases contain similar information for unit pricing as printed references. When estimating with a computerized system, electronic information concerning unit pricing may be calculated automatically by entering the dimensions of the quantity of material to be bid and the type of material.

Assembly Costs. For estimating software, an assembly cost is the costing of a number of common construction materials combined into a unit assembly, such as a wall made up of a number of various materials and components. Common construction assemblies may be included in electronic databases. Items such as various wall assemblies, floor coverings, or ceiling finishes may be taken off and priced as assemblies. See Figure 1-16. For example, a wall built of 2″ × 6″ studs spaced 16″ on center (OC) and braced with let-in bracing and blocking is included as an assembly. The number of linear feet of this wall is determined by the estimator. The electronic database automatically determines the number of studs, plates, amount of blocking and bracing materials, and labor costs tied to each item. This greatly simplifies the bidding process for standard assemblies.

Labor Rates. A wide range of labor rate information is available in electronic database form. In addition to diskettes and CD-ROMs, various Internet access services are available to electronically download wage rates for different craftworkers in various locations. This adds accuracy to the estimate and may be tied directly to the estimate spreadsheet.

Equipment Rates. Costs for rental, maintenance, and operations of various construction equipment are included on electronic database references. Rates vary according to equipment availability, the volume of construction work in an area, and the equipment required. These rates can be entered into databases or spreadsheets where necessary.

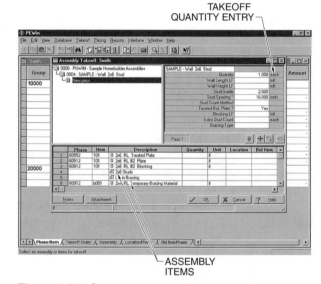

Figure 1-16. Common construction assemblies may be unit priced in estimating software.

Estimating

_____ 1. _____ is the computation of construction costs of a project.

_____ 2. An _____ is a change to the originally-issued contract documents.

   A. addition   C. edition
   B. addendum   D. neither A, B, nor C

_____ 3. A(n) _____ is an offer to accept a construction project at a stated price.

T   F 4. A prebid meeting is a meeting in which all interested parties in a construction project review the project, question the architect or owner concerning methods for accomplishing the work, and share information necessary to understand the entire scope of the work.

_____ 5. _____ is the cost of doing business that is not related to a specific job.

_____ 6. A construction _____ is a method used in the design and construction of a structure, including materials, construction sequence, structural design, and finish materials.

_____ 7. _____ construction is construction in which a structure is built primarily of wood structural members.

_____ 8. _____ construction is construction in which a series of horizontal beams and trusses and vertical columns are joined to create large structures with open areas.

_____ 9. _____ construction is construction in which concrete and reinforcing steel are placed in forming materials to create a high-strength finished structure.

_____ 10. _____ construction is construction in which masonry members are set in mortar.

T   F 11. Cladding is wall surface material attached to a structural steel frame to span between supporting members and provide closure to the structure.

T   F 12. Poured-in-place concrete is plastic concrete that is poured in wood or metal forming material set to a specific shape.

_____ 13. _____ concrete is concrete components that are formed, placed, and cured to a specific strength at a location other than the final installed location.

_____ 14. _____ is a bonding mixture consisting of lime, cement, sand, and water.

_____ 15. A(n) _____ is a space of fire-resistant materials between structures or groups of structures to prevent fire from spreading to adjacent areas.

_____ 16. A cost _____ is a compilation of a number of cost items from various sources combined into a common table for reference use.

_____ 17. An estimating _____ is the system used to integrate all parts of the estimating process in a cohesive, consistent, reliable manner to ensure an accurate final bid.

T   F 18. In tilt-up construction, concrete members are cast vertically at a location close to their final position.

_____ **19.** _____ is the practice of reviewing contract documents to determine quantities of materials that are included in a bid.

_____ **20.** _____ engineering is a process in which construction personnel are allowed to suggest changes to the contract documents.

Concrete Blocks

_____ **1.** Standard

_____ **2.** Solid

_____ **3.** Corner

_____ **4.** Corner sash

U.S. Government Agencies

_____ **1.** OSHA

_____ **2.** EPA

_____ **3.** OFCCP

A. Controls and abates pollution of air, water, solid waste, pesticides, radiation, and toxic substances

B. Requires employers to provide a safe environment for their employees

C. Promotes affirmative action and equal employment opportunities for minorities, women, disabled, and Vietnam veterans

Estimating Practices

Briefly describe each of the following.

1. Detail takeoff

2. Crew-based

3. Design build

4. Model

Activities
Estimating Practices

Name _____ Date _____

Activity 1-1 – Interpreting Specifications

_____ **1.** Concrete surfaces must be kept continuously moist for not less than _____ days.

_____ **2.** Comply with recommendations of _____ for cold weather protection during curing.

_____ **3.** The sides and ends of moisture-retaining cover must be lapped at least _____ ″.

_____ **4.** Concrete may be cured by _____ curing, moisture-retaining cover curing, curing compound, or a combination of these.

T   F   **5.** Continuous water-fog spray may be used for moisture curing.

_____ **6.** The sides and ends of a moisture-retaining cover must be sealed by _____ tape or adhesive.

_____ **7.** Comply with recommendations of _____ for hot weather protection during curing.

3.7 Concrete Protection and Curing

A. General: Protect freshly placed concrete from premature drying and excessive cold or hot temperatures. Comply with the recommendations of ACI 306R for cold weather protection and ACI 305R for hot weather protection during curing.

B. Evaporation Control: In hot, dry, and windy weather, protect concrete from rapid moisture loss before and during finishing operations with an evaporation-control material. Apply according to manufacturer's instructions after screeding and bull floating, but before floating.

C. Begin curing after finishing concrete but not after free water has disappeared from concrete surface.

D. Curing Methods: Cure concrete by moisture curing, moisture-retaining-cover curing, curing compound, or a combination of these as follows:

1. Moisture Curing: Keep surfaces continuously moist for not less than 7 days with the following materials:

a. Water.
b. Continuous water-fog spray.
c. Absorptive cover, water saturated, and kept continuously wet. Cover concrete surfaces and edges with a 12-inch lap over adjacent absorptive covers.

2. Moisture-Retaining-Cover Curing: Cover concrete surfaces with moisture-retaining cover for curing concrete, placed in widest practicable width, with sides and ends lapped at least 12 inches, and sealed by waterproof tape or adhesive. Immediately repair any holes or tears during curing period using cover material and waterproof tape.

3. Curing compound: Apply uniformly in continuous operation by power spray or roller according to manufacturer's directions. Recoat areas subjected to heavy rainfall within 3 hours after initial application. Maintain continuity of coating and repair damage during curing period.

Activity 1-2 – Interpreting Cost Data

_____   **1.** The material cost per square foot of floor area for a 40′ × 100′ steel building with a 14′ eave height is $_____.

_____   **2.** The labor cost per square foot of floor area for a 60′ × 100′ steel building with a 16′ eave height is $_____.

_____   **3.** The equipment cost per square foot of floor area for a 40′ × 100′ steel building with a 20′ eave height is $_____.

_____   **4.** The hours required per square foot of floor area for a 60′ × 100′ steel building with an eave height of 14′ is _____ hr.

T   F   **5.** Costs include a foundation or floor slab.

PRE-ENGINEERED STEEL BUILDINGS*						
	Craft@Hr	Unit	Material	Labor	Equipment	Total
40′ × 100′ (4000 SF)						
14′ eave height	H5@.074	SF	6.37	2.98	.75	10.10
16′ eave height	H5@.081	SF	7.15	3.26	.83	11.24
20′ eave height	H5@.093	SF	8.08	3.75	.95	12.78
60′ × 100′ (6000 SF)						
14′ eave height	H5@.069	SF	6.15	2.78	.70	9.63
16′ eave height	H5@.071	SF	6.29	2.86	.72	9.87

* 26 gauge colored galvanized steel roof and siding with 4 in 12 (20 lb live load) roof. Cost per square foot of floor area. Costs do not include foundation or floor slab. Add delivery cost to site. Equipment is a 15 ton truck crane and a 2 ton truck.

Activity 1-3 – Estimating Software Activity

_____   **1.** The stud spacing for the assembly is _____″.

_____   **2.** The phase number for the 2×6 RL treated plate is _____.

_____   **3.** The unit of measure for the 2×6 plates is _____.

T   F   **4.** The assembly is for a 2×6 stud wall.

_____   **5.** The item number for the 2×4 RL temporary bracing material is _____.

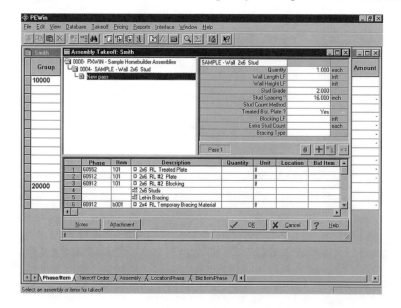

Estimating Methods

An estimating method is the approach used by an estimator to perform project analysis, takeoff, and pricing in a consistent and organized manner. Estimating is generally grouped into traditional and electronic estimating methods. In the traditional estimating method, quantities are calculated and entered onto ledger sheets for pricing. In the electronic estimating method, takeoff and pricing is accomplished using computer spreadsheets and integrated database systems. Electronic estimating methods include spreadsheet and estimating software use. Advantages of electronic estimating include time savings during the estimating process, ease of making last-minute changes that can be incorporated into the entire bid, and help in reducing the possibility of mathematical errors or missed bid items.

ESTIMATING METHODS

An *estimating method* is the approach used by an estimator to perform project analysis, takeoff, and pricing in a consistent and organized manner. Estimators use many different methods for quantity takeoff and pricing of construction projects. Estimating methods are based on a process that begins with a set of plans and printed specifications. A review of the specifications is made to determine the work contained in a particular construction project. The applicable portions of the printed specifications are highlighted and the plans are marked for takeoff. The quantities are calculated and entered into ledger sheets or software cells for pricing. All quantities are multiplied by the appropriate unit costs and totaled to determine a final price.

The architect and owner may include a listing of standard bid categories. The estimator reviews the listing of bid categories to organize takeoff and pricing. A decision is made by a contractor to bid on the project if the work to be done can be performed by the contractor. Estimating is generally grouped into traditional and electronic estimating methods.

TRADITIONAL ESTIMATING METHODS

The *traditional estimating method* is an estimating method in which quantities are calculated and entered onto ledger sheets for pricing. All quantities are multiplied by the appropriate unit costs. All results of this multiplication are totaled to determine a final cost.

Plan Markup

Plan markup is the color coding or marking off of items during the takeoff process. The marking of plans is a common practice for estimators. See Figure 2-1. For most construction projects, it is difficult to remember the sections of the project that have been taken off and quantities determined without a visual reminder. An estimator studies the plans and uses several colored markers to show different materials taken off and those that are recorded.

> ▲ *The most efficiently-produced estimates are developed by following a well-organized, methodical estimating procedure.*

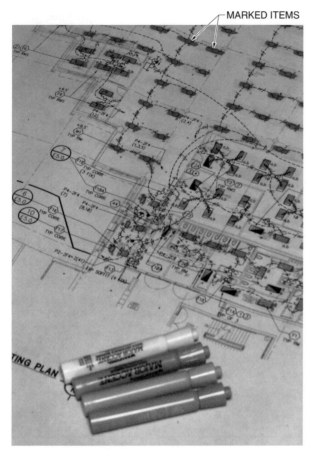

Figure 2-1. Marking up plans ensures all items to be taken off are accounted for by the estimator.

Accuracy. Printreading skills are required by an estimator to mark and take off only those portions of the plans that pertain to the work being bid. For example, an estimator working with a set of plans to determine the number of light standards in a parking lot may mark each light standard with a red X or a colored number to indicate it has been included in the total calculations. Mistaking bollards or drains for light standards would lead to an inaccurate bid. A *bollard* is a stone guard attached to a corner or a freestanding stone post to protect it against damage from vehicular traffic.

Comprehensiveness. After plans are marked and quantities listed on a ledger sheet, the estimator reviews all of the specifications and plans a final time to ensure all applicable items have been taken off. The final review should include a review of the entire project including the general conditions, detail drawings, and portions of the drawings that may not appear to apply to the work being bid. This ensures all items have been included in the plan markup and quantity takeoff.

In addition to materials, a variety of general conditions and overhead costs must be included in a comprehensive bid. Estimators should pay special attention to scheduling of construction on the job and any other items that may affect the overall bid.

Ledger Sheets

A *ledger sheet* is a grid consisting of rows and columns on a sheet of paper into which item descriptions, locations, code numbers, quantities, and costs are entered. As prints and specifications are marked up and quantities determined, the estimator enters these quantities on a ledger sheet. See Figure 2-2. Estimators should use standard categories on ledger sheets to ensure consistency and to present the information in a format familiar to others in the construction process. Skill in mathematical calculations speeds the final tabulation of the material quantities. A well-designed ledger sheet can be cross-referenced to other portions of the bid and used in other estimating calculations.

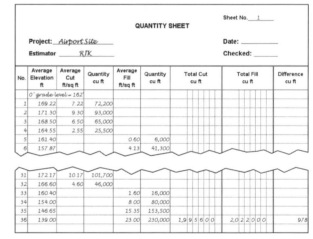

Figure 2-2. The traditional estimating method uses ledger sheets to track quantities taken off from prints.

Categories. The general conditions section of the specifications commonly contains a listing of the bid categories. Companies that bid on a project must submit their bid according to a format given by the architect and owner. See Figure 2-3. Individual ledger sheets may be developed for each type of work being taken off. For example, all items pertaining to pavement on a project, including curbing, sidewalks, concrete flatwork, asphalt paving, striping, parking blocks, and other incidental pavement items, may be placed on one ledger sheet. A numerical code is given to each item. For building construction, these codes are commonly based on the CSI MasterFormat™. See Figure 2-4.

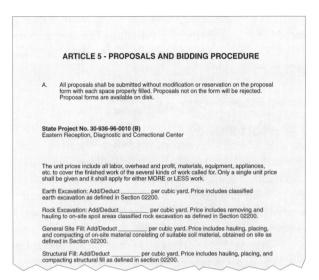

ARTICLE 5 - PROPOSALS AND BIDDING PROCEDURE

A.   All proposals shall be submitted without modification or reservation on the proposal form with each space properly filled. Proposals not on the form will be rejected. Proposal forms are available on disk.

State Project No. 30-936-96-0010 (B)
Eastern Reception, Diagnostic and Correctional Center

The unit prices include all labor, overhead and profit, materials, equipment, appliances, etc. to cover the finished work of the several kinds of work called for. Only a single unit price shall be given and it shall apply for either MORE or LESS work.

Earth Excavation: Add/Deduct _____ per cubic yard. Price includes classified earth excavation as defined in Section 02200.

Rock Excavation: Add/Deduct _____ per cubic yard. Price includes removing and hauling to on-site spoil areas classified rock excavation as defined in Section 02200.

General Site Fill: Add/Deduct _____ per cubic yard. Price includes hauling, placing, and compacting of on-site material consisting of suitable soil material, obtained on site as defined in Section 02200.

Structural Fill: Add/Deduct _____ per cubic yard. Price includes hauling, placing, and compacting structural fill as defined in section 02200.

Figure 2-3. The general conditions commonly contain standardized bid categories to provide consistency of bids from all bidding companies.

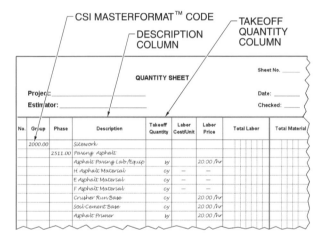

Figure 2-4. Use of standard CSI MasterFormat™ codes helps organize the bidding process.

The *CSI MasterFormat*™ is a master list of numbers and titles for organizing information about construction requirements, products, and activities into a standard sequence. The CSI MasterFormat™ code is entered in the left column of the ledger sheet. Other internal company codes may be entered in an adjoining column. A description column is created on the ledger sheet indicating the material taken off, such as gravel, concrete forms, carpet, etc. A takeoff quantity column is included to indicate the numerical quantity and the units used, such as square feet, linear feet, or cubic yards. Additional columns include material and labor pricing information and total cost for each item.

For road, bridge, and highway construction, numerical codes are developed by each state department of transportation. See Figure 2-5. Bids must be assembled according to these categories, which vary from state to state. Estimators can obtain these codes from each state department of transportation. The categories make it easier to find individual items and make necessary changes when project requirements change due to addenda or material and labor price changes.

PRELIMINARY QUANTITIES

J6P1265
RTE 40/61
CO   ST. CHARLES

201-10.10	CLEARING - METRIC	HECTARE	1.35
201-20.10	GRUBBING - METRIC	HECTARE	0.30
202-20.10	REMOVAL OF IMPROVEMENTS	LUMP SUM	1
203-10.05	CLASS A EXCAVATION - METRIC	CU METER	2,830
203-55.05	EMBANKMENT IN PLACE - METRIC	CU METER	209,833
203-60.05	COMPACTING EMBANKMENT	CU METER	2,460
206-30.05	CLASS 3 EXCAVATION - METRIC	CU METER	1,506
207-20.05	LINEAR GRADING CLASS 2 - METRIC	METER	45
301-20.05	MINERAL AGGREGATE (BITUMINOUS BASE) - METRIC	MEGAGRAM	2,422
301-60.19	ASPHALT CEMENT (BITUMINOUS BASE) PG 64-22 - METRIC	MEGAGRAM	128.3
304-05.05	TYPE 5 AGGREGATE FOR BASE (100 MM THICK) - METRIC	SQ METER	12,318
310-50.05	GRAVEL (A) OR CRUSHED STONE (B) - METRIC	CU METER	27
403-40.30	MINERAL AGGREGATE (ASPHALTIC CONCRETE) (TYPE I-C MIX) - METRIC	MEGAGRAM	737
403-81.24	ASPHALT CEMENT (ASPHALTIC CONCRETE) PG 64-2 (TYPE I-C MIX) - METRIC	MEGAGRAM	38.7
407-10.00	TACK COAT - METRIC	LITER	1,550
408-20.10	PRIME-LIQUID ASPHALT RC 70 OR MC 30 - METRIC	LITER	12,400

Figure 2-5. Standard estimating categories may be mandated by government agencies.

CSI MasterFormat™ Division 2 – Site Preparation includes subsurface investigation such as core drilling, groundwater monitoring, and seismic investigation.

Mathematical Requirements. Estimating requires proficiency with different mathematical calculations. Estimators must add quantities and linear totals, multiply various quantities by costs per unit and labor costs to determine pricing, and calculate areas and volumes to determine square feet, square yards, and various cubic measurements. Familiarity with common formulas speeds the process of bid preparation and ensures a high level of accuracy.

Cross-References. As an estimate is developed, columns from various ledger sheets may cross-reference each other. For example, there may be a certain type and number of reinforcing steel bars needed in foundation walls, concrete sidewalks, and concrete roof decks. Cross-reference across these various portions of the ledger sheet allows for a total number of reinforcing bars to be calculated and one order to be placed at one price rather than three.

ELECTRONIC ESTIMATING METHODS

The *electronic estimating method* is an estimating method in which takeoff and pricing is accomplished using computer spreadsheets and integrated database systems. Electronic estimating methods have had a great effect on the estimating process. Electronic estimating methods include spreadsheets and estimating software use. Spreadsheets have streamlined the calculation process by allowing estimators to enter standard calculations into individual cells. Estimating software allows for material pricing, labor costs, and quantity calculations to be determined through the use of standard item, crew, and assembly information.

Advantages of electronic estimating include time savings during the estimating process, ease of making last-minute changes that can be incorporated into the entire bid, and help in reducing the possibility of mathematical errors or missed bid items. In addition, electronic estimating enables standardization throughout a large construction company, integration with other company departments such as accounting, scheduling, and job costing operations, and flexibility of final bid reporting. Electronic estimating also allows estimators to cut and paste information from previous bids into a new bid. Some electronic estimating programs allow easy conversion from English to metric measure.

Many vendors supply estimating software programs which range from simple to complex. Some basic programs may operate in DOS. Other Windows-based programs range from spreadsheet programs that are customized to the ledger sheet format to more sophisticated electronic estimating programs that integrate the takeoff process through use of a digitizer and quantity and pricing calculations.

Standardized Formats

Electronic estimating programs vary from estimator-created spreadsheets to estimating software systems that include descriptions, material quantities, labor costs, and totals. Standard row and column formats are common in all electronic estimating programs. Features available depend on the sophistication of the software. Advanced features include database interactivity, the use of standard assemblies and standardized preset formulas for various estimating activities, digitizing to calculate quantities, interfacing with drawings available on computer aided drafting (CAD) systems or plans scanned into a computer memory, and the compilation of a variety of final reports and forms.

Spreadsheets. A *spreadsheet* is a computer program that uses cells organized into rows and columns that perform various mathematical calculations when formulas and numbers are entered. The foundation of most estimating software programs is the computer spreadsheet. See Figure 2-6. Spreadsheets are designed with a standardized row and column format for the entry of material quantities and descriptions, labor costs and prices, and other items required by the estimator. The quality of the final bid is determined by the calculation information entered into spreadsheet cells. Precautions should be taken by the estimator to verify that the spreadsheet is correctly adding the proper range of cells to arrive at the correct final results. Estimators may purchase software with pre-formatted cell calculations that can be copied into an existing spreadsheet program.

Database Interactivity. *Database interactivity* is the direct transfer of stored information by a software program from a database into a spreadsheet. One of the most time saving features of estimating software is the ability of sophisticated programs to connect quantity takeoff amounts with material and labor unit pricing information to quickly calculate costs. The estimator maintains a database as part of the estimating software program that stores standard costs and formulas for labor, materials, etc. See Figure 2-7. Proper layout, installation, and maintenance of the database is required for estimating software programs. Database codes must be organized to ensure consistency of estimates and ease in finding various materials and work processes.

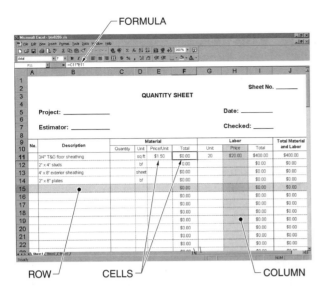

Figure 2-6. A spreadsheet uses cells organized into rows and columns that perform various mathematical calculations when formulas and numbers are entered.

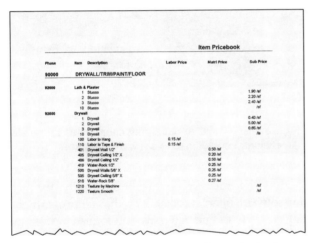

Figure 2-7. Estimating software uses electronic databases containing labor and material prices that interface with computer spreadsheets to facilitate the estimating process.

As each quantity for an item is entered, information connected to the item is selected from the database and entered onto the bid spreadsheet. Material unit quantity and labor cost information per unit is automatically transferred from the database. The estimator may make final adjustments in the spreadsheet calculations without changing or disturbing the database information.

Electronic database systems are commonly available with pre-loaded database information for estimating building or highway construction. These database items are customized by the estimator. Pre-loaded da-

tabases include items such as types and rates for labor, equipment, materials, and overhead costs.

For department of transportation work, state codes and preliminary quantities may be transferred electronically from bulletin board services. The state-maintained bulletin boards allow estimators to download codes, descriptions, and preliminary material quantities by logging onto the service, requesting the appropriate job information, and transferring this information into estimating software programs specifically designed to accommodate road and bridge building construction.

A major advantage of estimating software programs is consistency of data. Information concerning labor and material pricing and formulas is stored in the database for reuse without the need to reenter repetitive information. This produces consistent costs and formulas for all estimators on a large project or for companies having many estimators. Care should be taken by an estimator to ensure that the database pricing is applicable to the project being bid. Estimators should make any necessary adjustments to the database pricing for the specific project. Even when using a standard database, estimators must be aware that no two projects are exactly alike.

Assemblies. An *assembly* is a collection of items needed to complete a particular unit of work. Standard construction assemblies are entered into the database at a standard labor, material, and equipment quantity and cost per standard unit of measure. For example, a common wall assembly includes labor, material, and equipment based on the number of linear feet of wall. See Figure 2-8. The wall assembly includes quantities and costs for labor, materials, and equipment for framing, finish, and possibly door and window openings. The linear feet of the assembly is determined from the prints and entered into the estimating software spreadsheet. The quantities for each item in the assembly are calculated by the software based on the assembly information obtained from the database. Estimating software also allows for activity setups that can transfer information to a spreadsheet concerning labor crew costs, materials, equipment, and overhead based on the construction operation such as pipe laying or paving work.

> ▲ *For floor framing, the assembly function in estimating software includes the correct quantities of plywood, underlayment, nails, and bridging sets based on the square feet of floor.*

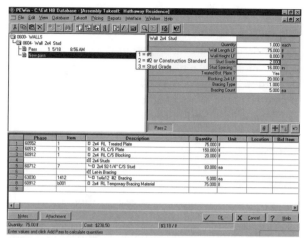

Figure 2-8. Standard assemblies enable an estimator using estimating software to calculate several items concurrently.

Final Reporting. Final reporting includes the compilation of all the various information required for the different members of the construction team. After all components of a bid are entered and a final estimate is determined, some estimating software programs allow the printing of the final bid information in various formats for suppliers, subcontractors, construction personnel, schedulers, and for presentation to the owner in an attempt to win the project. Estimating software has the flexibility to rearrange information for reports as needed by the various groups without the information being reentered. Other estimating software reporting features include subcontractor comparisons, cost reporting by construction activity, customization of bid formats as required by various owners such as state departments of transportation, and review of company job costs based on historical data.

Plan Markup

Electronic plan markup is becoming more widely used for estimating. Electronic plan markup can be done with a digitizer on paper prints or with an interface with CAD drawings. A *digitizer* is an electronic input device that converts analog data to digital form. A digitizer is used with an estimating software program to input drawing coordinates directly into the software. This reduces the need to enter quantities and dimensions using the computer keyboard. See Figure 2-9. The estimating software reads coordinates for various points on the prints and determines linear feet, square feet, and other measurements depending on the drawing scale. For CAD drawings, the takeoff is done on the computer screen. Care must be taken not to change or

manipulate various components of the drawings. Architects often insert material codes into the CAD drawings to allow for interfacing with estimating software. The CAD drawings are transferred into the estimating software with quantity calculations made during the transfer. Some CAD systems enable drawings that are scanned into the computer to be marked with various colors on the computer screen. As with the traditional estimating method, marking the components that have been entered into the estimating software is required to avoid duplication and ensure all required items are included in the bid.

DIGITIZER

Figure 2-9. A digitizer uses print coordinates to transfer dimensions into estimating software programs.

Estimating Software Internet Capability. Estimating software Internet capability is the ability of estimating software to link and transfer information from the world wide web into the estimating software. Prints can be scanned into computer memory and made available to estimators through the Internet. Electronic plan rooms exist in several locations throughout the country. By obtaining a password, members of electronic plan rooms can search through specifications and prints on a computer screen. When projects or items are found that can be bid by a company, the specifications and plans can be electronically retrieved for use with estimating software. In some cases, these plans can be marked up on the computer screen to facilitate quantity takeoff. For example, BuildNET (www.abuildnet.com) is a database of construction subjects that can be searched according to the interests of the builder. The site also contains information on computer software solutions, companies, and general information for the construction marketplace. In addition, e-builder™

(www.e-builder.net) is an Internet-based communication tool designed to enhance the exchange of information among construction project participants.

Integration with Total Job Site Management Systems

Use of estimating software allows for the quantity, labor, and overhead information to be used by others on the construction project team. Project managers can compare accounting information to the bid information to track costs readily throughout a construction project. Software programs that assist in scheduling of construction work can connect to the estimating software to track material quantities and labor use throughout the course of a project. The compatibility of estimating programs with CAD programs is affecting the future of the estimating process by enabling the automatic calculation of quantities. Architectural CAD programs that are compatible with estimating software enable material quantities to be calculated from the CAD drawing by the estimating software and entered into the software spreadsheet cells automatically.

Job Cost Analysis. *Job cost analysis* is the study of the final costs of building a project as compared to the original estimate. As a construction project proceeds, the owner, architect, and project managers for various contractors need to ensure that the costs of the project are staying within the estimated amounts. Integration of estimating software with job cost analysis programs can show areas where savings or cost overruns may have occurred. This integration allows for quick, timely information availability to assist in decisions concerning the remainder of the project. At the completion of the project, the information obtained concerning the initial estimate and the final job cost can be integrated into company historical cost records to improve the accuracy of future estimates.

Scheduling and Progress Tracking. Scheduling for construction includes information concerning material delivery times, work start and completion times, and subcontractor start and completion times. Several software programs are available for the specific purpose of scheduling construction projects and keeping track of various work phases. See Figure 2-10. Integration of estimating software with scheduling programs allows for ongoing analysis of areas where changes in the scheduling or the estimated materials and/or labor may be necessary. As with job cost information, this can be integrated into company historical data for future use.

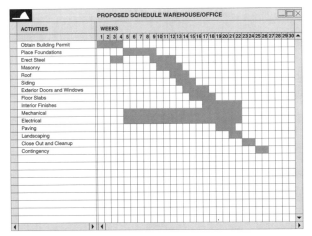

Figure 2-10. A job scheduling program can interface with estimating software to create an integrated project management system.

CAD Compatibility. *CAD compatibility* is the ability of an estimating software package to directly transfer quantity takeoff information from a CAD-generated set of plans. As computer systems become increasingly sophisticated, programs are being developed to directly integrate estimating software with CAD programs. This development allows for the computer software to automatically develop quantity takeoffs based on item codes embedded in CAD drawings. As labor unit prices and material costs are added to an integrated database, the ability exists for the use of a single program to design and produce the plan drawings and develop material and labor quantities.

Estimating software data can be integrated with job cost analysis programs to ensure the costs of a project are staying within the estimated amounts.

ESTIMATING SOFTWARE USE

Computerized estimating systems, such as Timberline Software Corporation's Precision Collection®, offer contractors a high level of estimating capability and many significant advantages over traditional spreadsheet packages. Benefits of using Timberline Software Corporation's Precision Estimating software include:

Estimating software may be integrated with CAD programs to perform material quantity takeoffs automatically from the CAD drawings.

- Pricing, formulas, takeoff items, etc. needed to create an estimate are stored permanently in a database that resides on the computer. This information must be entered only once in the database to be able to retrieve the information when building estimates on the spreadsheet.
- As the estimate is created on the spreadsheet, easy-to-use spreadsheet tools can be used that perform calculations and allow instant changes to the estimate as it is being developed.
- Existing estimates can be easily retrieved and used as the basis for new estimates. A library of basic estimates can be built, and minor modifications can be made to fit an existing estimate to a new project.
- Alternate versions of an estimate can be created to analyze and fine-tune the final estimate.
- A few menu selections enable a professional estimate to be created for clients, with add-ons allocated and hidden or with information summarized.
- Unlike typical memory-based spreadsheets, Precision Estimating Software's exclusive disk-based spreadsheet remains on the computer hard drive. This eliminates the requirement of saving the estimate and prevents any work done on an estimate from being lost.

The following lessons are a comprehensive procedure designed to introduce the common features of Timberline Software Corporation's Precision Estimating – Extended Edition trial software. Additional information concerning Timberline Software Corporation's Precision Estimating – Extended Edition trial software is provided in the Getting Started With Estimating booklet located on the CD.

PRECISION ESTIMATING – EXTENDED EDITION . . .

INSTALLING PRECISION ESTIMATING – EXTENDED EDITION

Implementing a computerized estimating system does require an investment in computer hardware, software, and training. However, this investment is recovered quickly as estimates become more complete, accurate, and timely. With the software handling data, calculations, and presentation of results, the estimator is free to concentrate on the creative aspects of estimating.

Precision Estimating – Extended Edition is designed to work on Microsoft® Windows 95, 98, and Windows NT™. Requirements include at least a 486 or Pentium™ computer with 16 MB of memory, 80 MB of available hard disk space, a VGA monitor, a CD drive, and a mouse.

About the Demo Version

The demo is a fully functional version of Precision Extended, however, a 180-day time lock and a 50-item limit has been placed on the program.

Installing the Software

The software must be installed using the Setup command. The application files cannot be copied directly onto the computer.

1. Insert the CD-ROM into the appropriate drive on the computer.

2. Select Run from the Start menu.

. . . PRECISION ESTIMATING – EXTENDED EDITION . . .

3. Enter the letter of the source drive followed by a colon (:) and the setup command. For example: d:setup. Click OK.

4. Follow the on-screen instructions. When prompted for the application components to be installed, make sure that both the Precision Estimating and Sample Databases boxes are checked.

5. Accept the default destination folder (c:\Program Files\Timberline\Precision).

6. Once the setup has finished, select Yes to restart the computer.

Starting the Demo

1. From the Start menu, choose Programs, Precision Estimating, Precision Estimating – Extended Edition.

2. The first time Precision Extended is started, a Demo Activation window appears indicating that 180 days remain to use the software. Click OK.

3. The next window to appear states "You must have a database open to work with Precision Estimating. Would you like to open a database now?" Select Yes.

4. Once the database window appears, double-click on Sample Ext Commercial G.C. and then double-click on the pei.dat file. *Note:* You are now connected to the Sample Ext Commercial G.C. database and ready to create an estimate.

CREATING AN ESTIMATE AND STORING ESTIMATE INFORMATION

In Precision estimating software, a new estimate is created by specifying a folder and filename for the estimate. Estimating information is stored by inputting estimate information in the Estimate Information window.

Creating an Estimate

1. Click the New Estimate button (⬜) to display the New Estimate window.

2. Enter a filename (Stanton Industries) for the new estimate at the Estimate File Name tab.

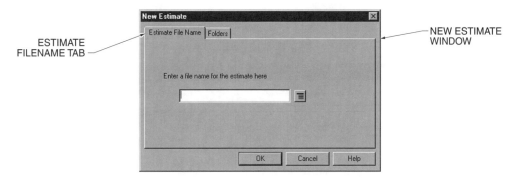

ESTIMATE FILENAME TAB

NEW ESTIMATE WINDOW

3. Click the Folders tab to display the Folders window. The Folders tab lets the estimator specify the folder where the estimate is stored and the database used with the estimate.

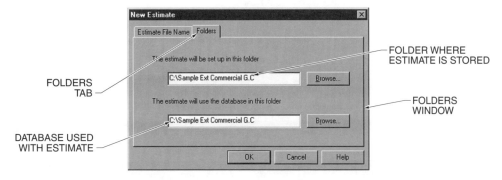

FOLDERS TAB

DATABASE USED WITH ESTIMATE

FOLDER WHERE ESTIMATE IS STORED

FOLDERS WINDOW

4. Accept the defaults given (c:\Program Files\Timberline\Precision\Sample Ext Commercial G.C.) and click OK to open the Estimate Information window.

. . . PRECISION ESTIMATING – EXTENDED EDITION . . .

Storing Estimate Information

The Estimate Information window consists of eight tabs where the estimator can enter general information about the job being estimated. The first tab is the Main window. To move to other tabs, click the appropriate tab. The only required field is the project name. The project name appears in the heading of every printed page of the estimate report. The project name also appears on the report cover page when printed. All other fields are optional. Close the Estimate Information window and the spreadsheet displays. *Note:* Choose Estimate Information from the Takeoff menu if any information in the Estimate Information window must be changed later.

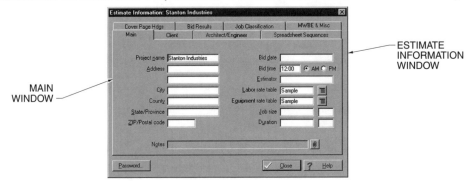

DRAG AND DROP QUICK TAKEOFF

Quick takeoff is a Precision estimating software function that enables the estimator to enter quantities directly into the estimating software spreadsheet. Quick takeoff is the easiest way to place items into an estimate. The results are shown immediately because work is being done directly in the spreadsheet. When using the quick takeoff function, the estimating software performs the calculation using the cost information and formulas stored in the database.

1. Open the Quick Takeoff window by clicking the Quick Takeoff (🖳) button.

2. Select the items to be added to the spreadsheet by double-clicking through each level of the database (group, phase, and item). Double-click Group 1000.00 General Requirements, Phase 1101.00 Personnel: Supervision, and Item 10 Superintendent.

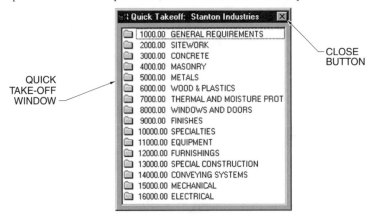

3. Close the Quick Takeoff window by clicking the close (✖) button in the upper right-hand corner of the Quick Takeoff window.

4. Enter 48 in the Takeoff Quantity cell for the superintendent item and press ENTER.

QUICK TAKEOFF ENTER DIMENSIONS FUNCTION

In Precision estimating software quick takeoff, the enter dimensions function acts as a switch that controls whether the enter dimensions window opens when selected items are dragged to the spreadsheet. A check mark next to the function causes the enter dimensions window to open whenever an item is taken off. Leaving the function unmarked loads the selected items directly into the spreadsheet. Quantities can then be entered in the takeoff quantity column of the spreadsheet.

1. Open the Quick Takeoff window by clicking the Quick Takeoff button.

2. Right-click in the Quick Takeoff window.

3. A check mark is displayed next to the function name, indicating that the Enter Dimensions function is turned ON.

. . . PRECISION ESTIMATING – EXTENDED EDITION . . .

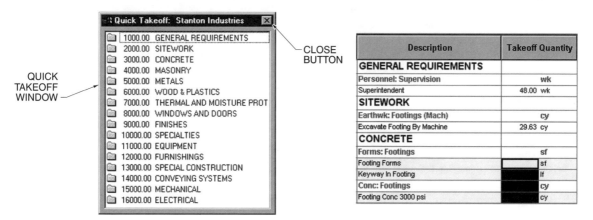

4. Double-click on 2000.00 Sitework and 2220.25 Earthwork: Footings (Mach). Choose item 10, excavate footing by machine, by double-clicking on the item or dragging and dropping the item to the spreadsheet.

5. Item 10 has an associated formula. The Enter Dimensions function prompts the user with the appropriate variables based on this formula. In the Enter Dimensions window, enter 1 for Quantity, 50 for Length, 4 for Width, and 4 for Depth. Click OK to send the item to the spreadsheet.

6. Next, disable the Enter Dimensions function by right-clicking in the Quick Takeoff window and choosing Enter Dimensions from the shortcut menu. The check mark no longer displays next to the function name.

MULTIPLE ITEM QUICK TAKEOFF

The Precision estimating software quick takeoff function can be used to take off multiple items and enter dimensions for all of the items at one time.

1. Open the Quick Takeoff window.

2. Take off items 3111.00 10 Footing Forms, 3111.00 50 Keyway in Footing, and 3306.00 c 30 Footing Conc 3000 psi. Hold down the CTRL key and click on item 10 Footing Forms, item 50 Keyway in Footing, and item c 30 Footing Conc 3000 psi.

Note: In addition to taking off items by double-clicking each item or dragging and dropping the items to the spreadsheet, time is saved by holding down the CTRL key and clicking on multiple items to be taken off. Once all items have been tagged, drag and drop them to the spreadsheet.

3. Once all items have been added to the spreadsheet, the dimensions can be entered for all items at once. Position the cursor in the takeoff quantity cell for the Footing Forms and while keeping the mouse button pressed, drag the cursor down to the last item taken off (Footing Concrete). Once these cells have been highlighted, right mouse click anywhere in the block of cells and select Enter Dimensions from the shortcut menu. Enter 50 for Length, 3 for Width, and 4 for depth of concrete.

4. Click the Add to Quantity button.

Note: Each item selected has an associated formula. The enter dimensions function prompts the estimator with the appropriate variables needed to calculate the formula.

. . . PRECISION ESTIMATING – EXTENDED EDITION . . .

CALCULATOR USE

Precision estimating software contains a calculator into which values and operators are entered to calculate various building material quantities. The calculator contains a single field that prefills with the value from the field where the cursor was placed before the calculator was opened.

1. Perform a quick takeoff for item 4221.10 rw 1 Blk 12″ Standard Face Reg Wt masonry blocks by clicking the Quick Takeoff button and double-clicking through the group, phase, and item categories.

2. Close the Quick Takeoff window by clicking the close button.

3. Calculate the number of blocks needed for a 50′ × 8′ block wall by clicking on the takeoff quantity cell for the 12″ block and opening the calculator by clicking the calculator (🖩) button from the toolbar menu. Click the clear button to delete the contents of the calculator if required.

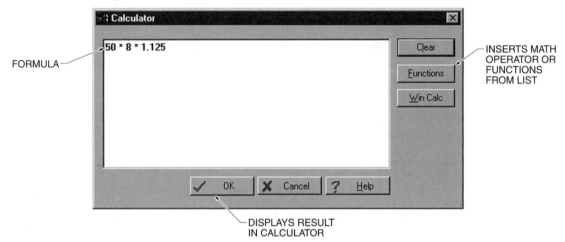

4. Develop the formula. Use the calculation: wall length × wall width × # of blocks per sq ft. Enter 50 * 8 * 1.125 directly into the window or click the Functions button to insert a math operator or functions from the list.

    Note: The takeoff unit for this item is "each" so the formula developed must calculate the total number of blocks.

5. Click ENTER to display the result in the calculator.

6. Press OK to insert the result into the quantity cell and close the calculator.

    Note: The value generated by a formula or the calculator must always be in the units specified as the takeoff unit for the item. For example, if the takeoff unit is in cubic yards, the calculation must also return the value in cubic yards.

LISTS

Lists are available throughout Precision estimating software to help make selections, provide ways to find information, and to sort the contents in alternate order.

Expanding and Collapsing Lists

1. Open the Quick Takeoff window.

2. Using the mouse, double-click the group to expand from the group level to the phase level. Double-click the phase to expand from the phase level to the item level. Double-click the phase to collapse back to the phase level. Double-click the Group to collapse back to the item level.

    Using the shortcut menu, right-click in the list to display the shortcut menu to expand or collapse the entire list. Select Collapse All or Expand All.

    Using the keyboard, highlight the group and press the Right Arrow key to expand the list from the group level to the phase level. Highlight the phase and press the Right Arrow key to expand the list from the phase level to the item level. Highlight the Phase and press the Left Arrow key to collapse the list back to the phase level. Highlight the phase and press the Left Arrow key to collapse the list back to the item level.

Sorting Lists

    The Item list can be sorted by description or phase/item. To select the sort order, right-click in the item list. A shortcut menu displays. Select Description or Phase/Item. The list is immediately re-sorted.

Finding Items

    In any list, the quickest way to get to a specific entry item is by typing its first few characters. The entry item displays at the top of the window, and the list moves to the closest match at the same level. If there is no match, the cursor moves to the end of the list. To find items in a list, choose the sort option that is most useful for the search. If necessary, use the Left Arrow key to collapse the list or the Right Arrow key to expand the list. Using correct capitalization, enter the first few numbers or characters of the entry.

. . . PRECISION ESTIMATING – EXTENDED EDITION . . .

ASSEMBLY TAKEOFF

An *assembly* is a collection of items needed to complete a particular unit of work. Assemblies enable the estimator to take off multiple items with a single operation. Assemblies also enable the estimator to obtain a cost per unit for a group of items.

Assemblies may be database assemblies or one-time assemblies. A *database assembly* is an assembly that is stored in the database and can be used in any estimate based on that database. A *one-time assembly* is an assembly that is stored with the estimate and cannot be used in other estimates.

1. Open the Assembly Takeoff window by clicking the Assembly Takeoff button. The Assembly Takeoff window displays the assembly list pane, the dimension pane, and the item grid pane. The assembly list pane contains the assemblies listed in the database. The dimension pane lists the dimensions/variables from the formulas assigned to the items. The item grid pane displays the items in the selected assembly in the assembly list pane.

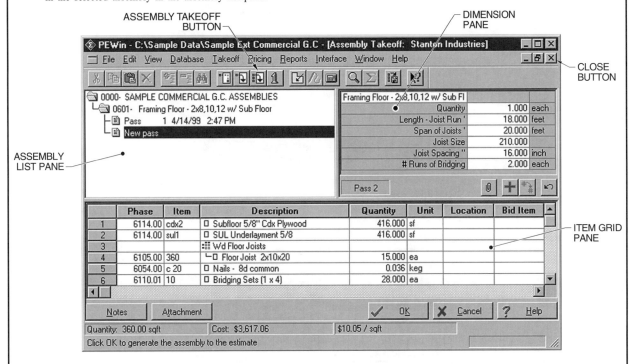

2. Double-click the 0000 – Sample Commercial G.C. Assemblies and then double-click assembly 0601 – Framing Floor – 2 × 8, 10, 12 w/Sub Floor. The items in the assembly display in the item grid and the variables from the formulas assigned to those items display in the dimension pane. *Note:* Some quantities in the dimension pane prefill from the last time the assembly was used.

3. Enter the dimensions for the assembly. Enter 1 for the Quantity, 18 for Length – Joist Run, 20 for Span of Joists, 210 for Joist Size, 16″ for Joist Spacing, and 2 for # Runs of Bridging. Select ENTER. *Note:* If there were two identical joist runs, a quantity of 2 would have been entered for this assembly. Also note that after the last dimension is input and ENTER is pressed, takeoff quantities are automatically generated to the item grid.

4. Click OK to send the items to the spreadsheet.

5. Close the Assembly Takeoff window by clicking the close button in the upper right-hand corner of the Assembly Takeoff window. *Note:* The items taken off using this assembly are viewed by sorting the spreadsheet in assembly order by clicking the assembly tab at the bottom of the spreadsheet. The estimate is closed by clicking the close button in the upper right-hand corner of the spreadsheet screen.

ASSEMBLY ITEM ADDITION

In Precision estimating software, adding, deleting, or substituting items in an assembly may be necessary for unique applications. Any changes made to an assembly apply to the current estimate only and do not affect the assembly in the database.

1. Create a new estimate. Name the estimate Parker Ind – Phase 2.

2. Open the Assembly Takeoff window by clicking the Assembly Takeoff button. Double-click 0000 – Sample Commercial G.C. Assemblies and then double-click Assembly 0601 – Framing Floor – 2 × 8, 10, 12 w/Sub Floor. Enter the dimensions for the assembly. Enter 18 for Length – Joist Run, 20 for Span of Joists, 210 for Joist Size, 16 for Joist Spacing, and 2 for # Runs of Bridging.

... PRECISION ESTIMATING – EXTENDED EDITION ...

3. Click the Add Pass (➕) button.

4. Display the item list by right-clicking in the assembly list pane and choosing List Items from the shortcut menu.

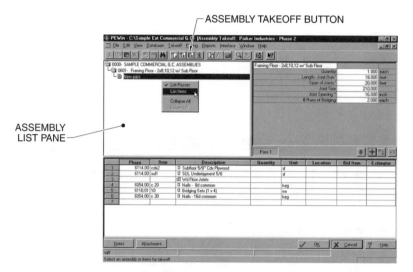

ASSEMBLY TAKEOFF BUTTON

ASSEMBLY LIST PANE

5. Add an item to the assembly by drilling down through 7000.00 Thermal and Moisture Protection, 7211.20 Insulation: Sound Blanket, and by selecting 30 Sound Blanket 16″ × 3-⁵/₈″.

6. Input the variables for the assembly addition by right-clicking the item quantity field in the item grid for the sound blanket and selecting SF L×H from the shortcut menu. Enter 1 for Quantity, 50 for Length, and 10 for Height for the dimensions for the floor in the SF L×H window.

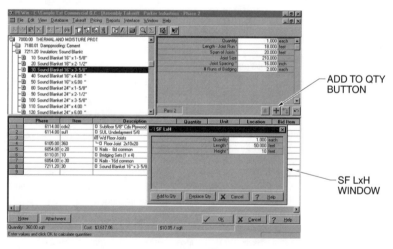

ADD TO QTY BUTTON

SF LxH WINDOW

7. Send the items to the spreadsheet by clicking the Add to Qty button and clicking OK.

8. Close the Assembly Takeoff window. *Note:* Sound Blanket 16″ × 3-⁵/₈″ has been added to the framing assembly.

CREATING AN ASSEMBLY ON THE FLY

An assembly is created in Precision estimating software by taking off several items and saving these items as an assembly.

1. Open the Item Takeoff window by clicking the Item Takeoff button (🔲).

2. Take off items 8110.01 10 Metal Doors, 8110.01 20 Metal Frames, 8710.01 160 Door Closers, 8710.01 170 Weatherstripping, and 8710.01 190 Kickplates.

3. Enter 7 in the Quantity field for Metal Doors and press ENTER. Highlight the quantity field again and while keeping the mouse button pressed, drag the cursor down to the last item taken off (Kickplates). Once these cells have been highlighted, right-click anywhere in the highlighted block of cells and select Fill Down from the shortcut menu.

... PRECISION ESTIMATING – EXTENDED EDITION ...

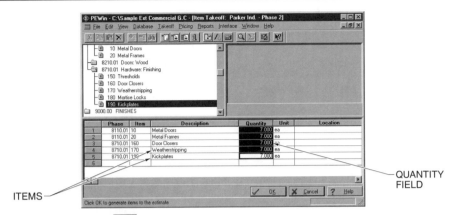

ITEMS

QUANTITY FIELD

4. Click the Save as Assembly button ([icon]).

5. Enter 10 as the assembly number in the Assembly field. Each assembly must have a unique code. Enter Metal Door Assembly in the Description field.

6. Click on One-time ([radio: Database / One-time]) to save the assembly only with this estimate. The new assembly is not saved to the database.

7. Click OK to save the assembly.

8. Click Close to return to the Item Takeoff window.

9. Click OK and Close to send this assembly to the spreadsheet.

   Note: Click on the Assembly Takeoff button to see the estimate name added to the Assembly list. Double-click on Parker Ind. – Phase 2 to see the new assembly.

ASSEMBLY TAKEOFF ITEM SUBSTITUTION

Precision estimating software enables items to be substituted in an assembly.

1. Take off assembly 0932s Wall – S Studs 25 Ga – $^5/_8$ GWB (simple) by selecting assembly 0932s in the Assembly Takeoff window. *Note:* Any changes made to the assembly for this estimate will not affect the original assembly.

2. In the Takeoff grid, highlight any field in the row containing item 9253.30   40 GWB $^5/_8''$ All Size Regular.

3. Right-click in the highlighted area.

4. Choose Substitute Item from the shortcut menu. The Item List opens at the highlighted item.

5. Double-click item 9253.30 30 GWB ½″ All Size Regular to place it in the Takeoff grid and substitute it for the original item 9253.30 40 GWB $^5/_8''$ All Size Regular.

6. Take off the wall by entering 50 for Length and 8 for Height of the Wall.

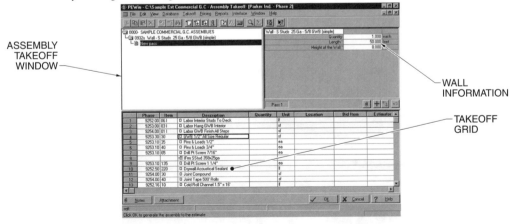

ASSEMBLY TAKEOFF WINDOW

WALL INFORMATION

TAKEOFF GRID

7. Hit ENTER or click Add Pass to generate item quantities to the Takeoff grid. Click OK and Close to send these items and their quantities to the spreadsheet.

. . . PRECISION ESTIMATING – EXTENDED EDITION . . .

TAKEOFF USING ONE-TIME ITEMS

One-time items in Precision estimating software enable an estimator to take off items that are not currently in the database. Details can be added through the Detail Window and, if desired, one-time items for a particular estimate can be automatically saved to the database.

1. Open the One-time Item window by clicking the One-time Item ([1]) button in the toolbar.

2. Right-click in the Phase field and select Edit Phase from the shortcut menu. Click the Add button to create the Phase record. Input 15100 for Phase and Copper Pipe for description. Click OK and then click close.

3. Create the item record by filling the One-time Item fields. Input Copper Pipe 1″ for Description, 125 for Quantity, and lnft for Takeoff Unit.

4. Select the categories by clicking and placing a check in the Labor and Material Category boxes.

ONE-TIME ITEM WINDOW

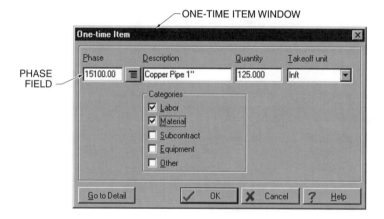

PHASE FIELD

5. Click the Go to Detail button and enter .50 for Labor Unit Price and .98 for the Material Unit Price for the item.

6. Exit the Detail window by clicking OK and then the Close ([X]) button. The one-time item displays in the spreadsheet.

BUILDING DATABASES USING ONE-TIME ITEMS

One-time items may be taken off and automatically added to the database in Precision estimating software.

1. Click the One-time Item button ([1]) on the toolbar to open the One-time Item window.

2. Create the Phase record by right-clicking in the Phase field and selecting Edit Phase from the shortcut menu. Click the Add button.

3. Input 16100 in the Phase field and Subcontractors in the Description field. Click OK and close the window.

4. Complete the One-time Item fields. Enter 16100 for Phase, Electrical Rough-In for Description, 1 for Quantity, and lsum for Takeoff unit.

5. Uncheck the Labor and Material Categories and place a check mark in the Subcontract Category box and click OK. The one-time item displays on the spreadsheet.

PHASE FIELD

ONE-TIME ITEM WINDOW

. . . PRECISION ESTIMATING – EXTENDED EDITION . . .

6. Highlight the Electrical Rough-in description and press CTRL-S to automatically jump to the Sub Amount field. Input 15000 and hit ENTER.

7. Hold down the CTRL key and hit the Left Arrow key to automatically jump to the far left column on the spreadsheet.

8. Save the one-time items associated with this estimate to the database by clicking Save One-time Items to Database from the Pricing menu. *Note:* This window presents each one-time item in the estimate to be saved or rejected. As each one-time item appears, the Item List goes to the phase specified for that item. The information shown can be accepted, new item codes entered, item details edited, different phases specified, or the item skipped altogether.

9. Precision Estimating opens the Save One-time Items to Database window. Enter 10 in the Item Field and click OK.

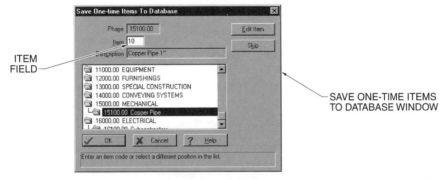

10. Scroll down and highlight Phase 16100, enter Item 10, and select OK. Once finished, click OK to return to the spreadsheet.

   Note: When Precision Estimating finds a one-time item, it opens the Save One-time Items to Database window. Options include: entering the desired item code in the Item field, clicking OK to accept if an item code has been proposed, or clicking on a different item in the item list to have Precision Estimating suggest a different item code. Additional item information can be added, such as the formula, waste factor, and prices by clicking the Edit Item button and filling in the desired fields in the Adding Item window.

FINALIZING ESTIMATE

In Precision estimating software, finalizing an estimate is done after takeoff is complete.

1. Ensure the estimate is maximized by pressing the maximize button (▢) in the upper right-hand corner of the estimate.

2. Hide columns by right-clicking on the column heading and selecting Hide from the shortcut menu. Hide all columns on the spreadsheet except Group, Phase, Description, Labor Amount, Material Price, Material Amount, Sub Amount, and Total Amount. *Note:* Several columns can be hidden at once by holding down the left mouse button and dragging the highlight across the column headings.

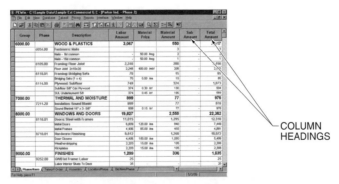

3. Make price changes by placing the cursor in the material price cell, inputting the change, and pressing ENTER. For example, a reduction in the price of metal doors is input by placing the cursor in the Material Price cell for Metal Doors, changing the price from $120/ea to $110/ea, and pressing ENTER. The Material Amount and the Total Amount columns are adjusted automatically.

4. Collapse the level of detail shown on the spreadsheet by pressing the collapse button (⊟) on the toolbar. The collapse button may be pressed again for a more summarized view. Expand the estimate by pressing the expand button (⊞) twice to show full detail.

5. Show estimate totals by category (labor, material, etc.) by pressing the totals button (Σ) on the toolbar.

. . . PRECISION ESTIMATING – EXTENDED EDITION

6. Create an addon for profit and overhead by pressing the edit addon button (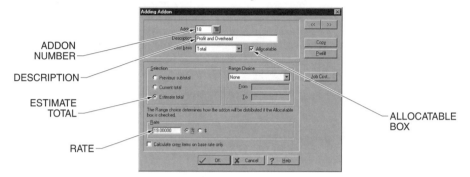) in the totals window. The addon window opens. Click the add button and set up the addon by inputting 18 in the Addon field and Profit and Overhead for Description. Check the allocatable box and Estimate Total under selection. Input a rate of 19%.

ADDON NUMBER

DESCRIPTION

ESTIMATE TOTAL

RATE

ALLOCATABLE BOX

7. Click OK and Close to exit the addon window.

8. Click the insert addon button () in the totals window to apply this addon to the estimate. Select addon 10 and 18 by double-clicking on each and Close to exit the insert addon window. The addon amounts are calculated automatically.

9. Close the totals window by clicking OK and Close. *Note:* Each addon has been proportionally spread across all estimate items. The addon and grand total columns can be displayed by right-clicking the Total Amount Column heading and selecting Show Hidden Columns from the shortcut menu.

10. Generate a user-defined spreadsheet report by selecting spreadsheet from the reports menu.

REPORTS MENU

SPREADSHEET SELECTION

11. Click Report Options and Prefill from Spreadsheet buttons. Uncheck the boxes marked Print Horizontal Gridlines, Print Vertical Gridlines, and Print Cover Page and place a check in the Minimize Overline Columns box. Select Preview.

REPORT OPTIONS BUTTON

PREFILL FROM SPREADSHEET BUTTON

PRINT PREVIEW WINDOW

PREVIEW BUTTON

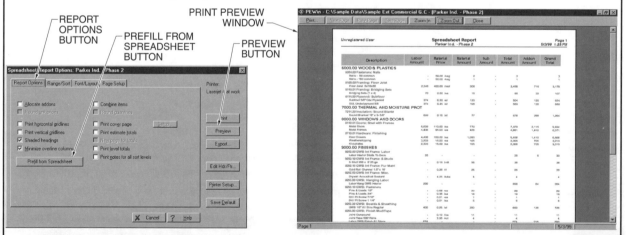

12. Click twice anywhere on the report to zoom in. Click Close in the Print Preview window and click Cancel and Close in the report options screen to get back to the spreadsheet.

Review Questions
Estimating Methods

Name _____ Date _____ Chapter

Estimating

_____ **1.** The _____ estimating method is an estimating method in which quantities are calculated and entered onto ledger sheets for pricing.

_____ **2.** Plan _____ is the color coding or marking off of items during the takeoff process.

_____ **3.** A(n) _____ sheet is a grid consisting of rows and columns on a sheet of paper into which item descriptions, locations, code numbers, quantities, and costs are entered.

_____ **4.** A(n) _____ is a computer program that uses cells organized into rows and columns that perform various mathematical calculations when formulas and numbers are entered.

_____ **5.** The _____ estimating method is an estimating method in which takeoff and pricing is accomplished using computer spreadsheets and integrated database systems.

_____ **6.** A(n) _____ is a collection of items needed to complete a particular unit of work.

_____ **7.** A(n) _____ is an electronic input device that converts analog data to digital form.

T   F   **8.** Estimating methods are based on a process that begins with a set of plans and printed specifications.

_____ **9.** The _____ MasterFormat™ is a master list of numbers and titles for organizing information about construction requirements, products, and activities into a standard sequence.

_____ **10.** Database _____ is the direct transfer of stored information by a software program from a database into a spreadsheet.

T   F   **11.** An estimating method is the approach used by an estimator to perform project analysis, takeoff, and pricing in a consistent and organized manner.

_____ **12.** Job cost _____ is the study of the final costs of building a project as compared to the original estimate.

_____ **13.** CAD _____ is the ability of an estimating software package to directly transfer quantity takeoff information from a CAD-generated set of plans.

T   F   **14.** Timberline Software Corporation's Precision Estimating – Extended Edition is designed to work in Microsoft® Windows® 95, 98, or NT operating systems.

T   F   **15.** The foundation of most estimating software programs is the bid document.

T   F   **16.** Consistency of data is a major advantage of estimating software programs.

T   F   **17.** At the end of the installation process, a shortcut may be added to the Start Menu for the Precision Estimating group, enabling quick and easy access to the software.

Job Scheduling

T   F   **1.** Obtaining the job permit is the first activity performed.

T   F   **2.** The roof is completed before the floor slabs are poured.

_____ **3.** _____ and _____ work require the most time.

_____ **4.** Placing the foundations requires _____ weeks.

_____ **5.** The Warehouse/Office takes _____ weeks to complete.

T   F   **6.** Landscaping is started after paving is completed.

T   F   **7.** No other activities are scheduled during cleanup.

_____ **8.** _____ weeks are required to erect the steel.

T   F   **9.** Interior finishes are completed before electrical.

T   F   **10.** Masonry and siding are started at the same time.

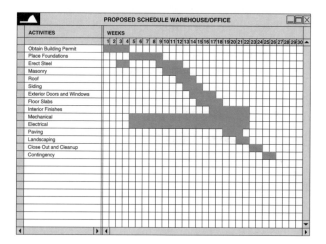

Spreadsheet

_____ **1.** Row

_____ **2.** Column

_____ **3.** Cell

_____ **4.** Formula

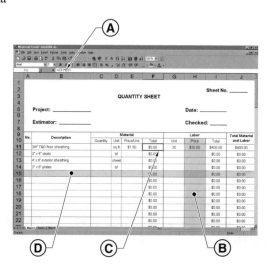

Activities
Estimating Methods

Name _____ Date _____

Activity 2-1 – Field Conversion

Use the field conversion method to convert the inch dimensions to decimal parts of a foot. See Appendix.

_____ 1. $8^5/_8''$ equals _____'.

_____ 2. $4^3/_8''$ equals _____'.

_____ 3. $5^1/_4''$ equals _____'.

_____ 4. $3^3/_8''$ equals _____'.

_____ 5. $1^1/_8''$ equals _____'.

_____ 6. $^7/_8''$ equals _____'.

_____ 7. $9^5/_8''$ equals _____'.

_____ 8. $7^1/_2''$ equals _____'.

_____ 9. $11^5/_8''$ equals _____'.

_____ 10. $2^3/_4''$ equals _____'.

Activity 2-2 – Unit Conversion

COST DATA			
Material	**Unit**	**Material Unit Cost***	**Labor Unit Cost***
Joist hangers, heavy duty 12 gauge, galvanized			
2″ × 4″	100	863.50	137.50
2″ × 6″	100	918.50	145.50
2″ × 8″	100	979	145.50
2″ × 10″	100	1045	145.50
2″ × 12″	100	1182.50	145.50

* in $

_____ 1. The total material unit cost for fifty-two 2″ × 4″ joist hangers is $_____.

_____ 2. The total labor unit cost for 266 2″ × 6″ joist hangers is $_____.

_____ 3. The material unit cost for each 2″ × 8″ joist hanger is $_____.

_____ 4. The total material and labor cost for a project requiring 124 2″ × 10″ joist hangers is $_____.

_____ 5. The total material unit cost for ninety-six 2″ × 12″ joist hangers is $_____.

COST DATA			
Material	**Unit**	**Material Unit Cost***	**Labor Unit Cost***
Armored cable, 600 V, Cu (BX)			
No. 14 gauge, 2-conductor	100 lf	49.50	116.60
No. 12 gauge, 2-conductor	100 lf	50.05	122.10
No. 10 gauge, 2-conductor	100 lf	88	140.80
No. 8 gauge, 3-conductor	100 lf	189.20	215.60

* in $

_____ 6. The total material and labor cost for 1025 lf of No. 14 gauge, 2-conductor armored cable is $_____.

_____ 7. The total material unit cost for 365 lf of No. 12 gauge, 2-conductor armored cable is $_____.

_____ 8. The total labor unit cost of 837 lf of No. 10 gauge, 2-conductor armored cable is $_____.

_____ 9. The cost per linear foot for No. 8 gauge, 3-conductor armored cable is $_____.

_____ 10. The total material and labor cost for 482 lf of No. 12 gauge, 2-conductor armored cable including 12% for overload and profit is $_____.

COST DATA				
Material	**Unit**	**Material Unit Cost***	**Labor Unit Cost***	**Equipment Unit Cost***
Concrete walls, 8″ thick				
8′ high	cu yd	107.25	130.90	16.39
14′ high	cu yd	136.40	218.90	27.50
Concrete walls, 12″ thick				
8′ high	cu yd	98.45	92.95	11.72
14′ high	cu yd	109.45	149.60	18.81

* in $

_____ 11. The total material unit cost for a 12″ thick, 8′ high, and 27′ long concrete wall is $_____.

_____ 12. The total labor unit cost for a 12″ thick, 14′ high, and 54′ long concrete wall is $_____.

_____ 13. The total material, labor, and equipment cost for an 8″ thick, 8′ high, and 66′ long concrete wall is $_____.

_____ 14. The total equipment unit cost for an 8″ thick, 14′ high, and 42′ long concrete wall is $_____.

_____ 15. The total material, labor, and equipment cost for an 8″ thick, 8′ high, and 135′ long concrete wall including 8% for overhead and profit is $_____.

Printreading Skills

Specifications are written information included with a set of prints clarifying and supplying additional data. Specifications, along with the prints, describe the entire building process and project. The CSI MasterFormat™ is a master list of numbers and titles for organizing information about construction requirements, products, and activities into a standard sequence. The CSI MasterFormat™ contains 16 divisions that define the broad areas on construction. An estimator must be able to read and interpret architectural and shop drawings. Familiarity with various lines, drawing scales, symbols, and abbreviations is required to develop an accurate bid.

SPECIFICATIONS

Construction of residential, commercial, and industrial buildings requires a set of prints and specifications. *Specifications* are written information included with a set of prints that clarify and supply additional data. Specifications provide additional details that could not be shown on the prints or that require further description. Specifications, along with the prints, describe the entire building process and project. Specifications contain information related to legal issues, building materials, construction procedures, and quality control issues of construction.

Specifications are an organized presentation of bidding information, contract requirements, and all phases of the construction process. Standard forms, language, and formats provide direction for all involved to ensure clear and accurate communication. Different construction projects and architects require the specifications to be used in a variety of ways. The specifications must be completely reviewed during the bidding process to fully understand the project and to develop an accurate bid.

Estimators should note areas where the specifications are in conflict or inconsistent with the prints. For example, the foundation plan may indicate that the basement slab is to be placed directly on the soil below the slab. The specifications may state that the slab is to be placed on 4″ of crushed gravel. The estimator or contractor must obtain written permission or clarification from the architect and/or owner when a conflict exists between the prints and the specifications.

An architect is responsible for the detailed design and coordination of a construction project. The architect is the central person in a group of design professionals including civil engineers, structural engineers, mechanical engineers, and electrical engineers. Meetings take place between the architect and the owner or developer to determine the needs of the owner or developer. After deliberation and conferences, the architect prepares preliminary sketches of the project and submits them to the owner or developer for approval. The owner or developer suggests changes and returns the final requirements to the architect for the final work on the plans.

The needs of the structure, the materials used in the building of the structure, and the methods by which these materials are placed and erected must be clearly specified for a project to be erected from a set of plans. All parties concerned with the project must understand the architect's concept of the structure. The architect or developer submits the specifications and working drawings to several contractors for bidding. The specifications are the most important initial document for the contractor.

Specifications are divided into sections that cover the requirements of each trade involved in building a structure. Some portions of the specifications have more sections than others. A basic set of specifications is usually required to carry the average project to completion. Estimators must be familiar with the most pertinent parts of the specifications to carefully analyze every section, paragraph, and line. The misreading of one word or paragraph may distort the entire meaning and requirements of a particular section of the specifications and can lead to a faulty bid or a dangerous deviation from the plans. Care must be taken, because in some cases various sections of the specifications can give conflicting information.

General Conditions

The general conditions section of the specifications includes overhead expenses that cover the entire construction project. A construction project involves many people, agencies, and trades. The construction process also involves city, county, state, and federal agencies, banks, bonding companies, and insurance companies. The concerns of all the parties involved must be stated in writing so that every entity bidding for the job is aware of the requirements. This gives each business involved in the bidding process a fair and equal chance to arrive at the best and most complete qualified bid.

The general conditions also state the method used for bidding. The general conditions indicate whether the project is for a fixed bid, cost plus, time and material, or another project delivery method such as design build. The general conditions may also contain the standardized bidding form that all contractors must use to submit a bid.

> ▲ *Supplementary conditions include modifications to the general conditions for requirements unique to a specific project. Supplementary conditions include anti-pollution measures, health and safety criteria, etc.*

CSI MasterFormat™

The *Construction Specifications Institute (CSI)* is an organization that developed standardized construction specifications. The CSI, in cooperation with the American Institute of Architects (AIA), the Associated General Contractors of America (AGC), the Associated Specialty Contractors (ASC), and other industry groups, developed the *CSI MasterFormat™ for Construction Specifications* and *The Uniform System for Construction Specifications, Data Filing, and Cost Accounting*. These specification standards apply mainly to projects in the United States and Canada. Estimators working on projects outside the United States and Canada must review the specifications much more carefully.

The *CSI MasterFormat™* is a master list of numbers and titles for organizing information about construction requirements, products, and activities into a standard sequence. The CSI MasterFormat™ contains 16 divisions that define the broad areas of construction. See Figure 3-1. The 16 divisions are numbered 1 through 16. Each division is designed to give complete written information about individual construction requirements for building and material needs. Each division is divided into subclassifications. For example, Division 16 – Electrical has subclassifications such as transmission and distribution, lighting, and communications. Each subclassification has a reference number for ease of identification. For example, the reference number for transmission and distribution is 16300.

Division 1 – General Requirements. Division 1 of the specifications includes the forms that must be submitted by contractors for payment and for documentation of various project requirements pertaining to materials, change orders, and substitutions. Health and safety, quality control, and contract closeout are also described.

The first section of Division 1 gives a description of the overall construction work. This is done with an index of all the plan drawings and several paragraphs giving a general overview of the project. See Figure 3-2. General descriptions of alternates are provided, along with the bidding procedures for building projects that have several alternate additional construction items. Procedures for filing for approval and payment of change orders and unit pricing are listed. Division 1 also includes temporary facilities and controls, and administrative and product requirements.

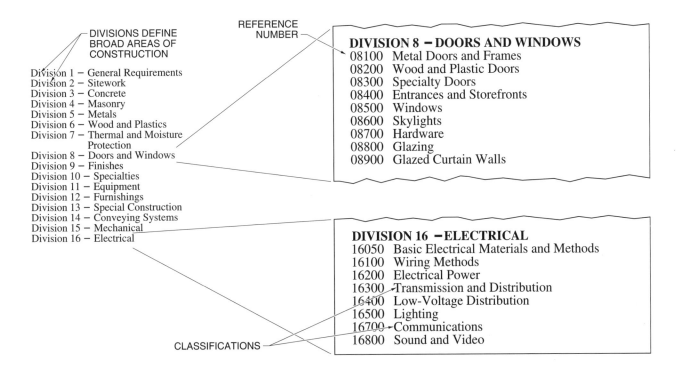

Figure 3-1. The CSI MasterFormat™ contains 16 divisions for the categorization of construction activities.

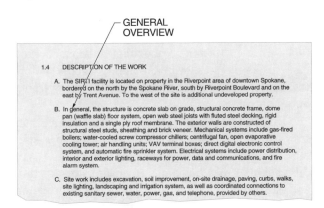

Figure 3-2. A general overview describing the size and scope of the construction project is included in Division 1 of the specifications.

The time in which the construction project must be completed is given in Division 1 of the specifications. Depending on the project, there may be a bonus for early completion, sharing of potential construction cost savings, or penalties assessed if construction is not completed during a certain time. Procedures for completion of the project are listed. These include the final cleaning and preparation of the project, submission of all documents such as contract drawings, specifications, addenda, change orders, shop drawings, warranties, and operation and maintenance data.

Division 2 – Site Construction. Division 2 of the specifications includes information concerning items located below ground, such as foundations, tunnels, pipes, and piers, as well as items on the ground, such as landscaping, fencing, and paving. Responsibilities for subsurface exploration, excavation, compaction, and disposal of excavated materials are part of Division 2.

Division 2 of the specifications details the contractor responsibilities for testing and handling of the construction site soils, fill, and backfill materials. Bearing capacities for the subsurface must be achieved and measured according to industry standards such as the American Society for Testing and Materials (ASTM). Removal and use of surface soils and necessary soil improvements are given. Specific soil and slope finish information may be provided. See Figure 3-3. Information is also given in Division 2 for drainage piping, utility holes, and inlets.

Paving materials and standards for their use and installation are described in Division 2. The specifications include the weather conditions under which paving may or may not be installed, in addition to slopes and smoothness requirements.

Soil mixtures, trees and shrubs, grasses, finish grading, fertilizers, and mulches are detailed in Division 2.

Installation, protection, and maintenance procedures for plant materials are also given. Piping and connection information is included when irrigation systems are installed.

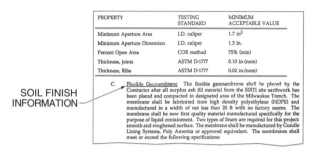

PROPERTY	TESTING STANDARD	MINIMUM ACCEPTABLE VALUE
Minimum Aperture Area	I.D. caliper	1.7 in²
Minimum Aperture Dimension	I.D. caliper	1.3 in.
Percent Open Area	COE method	75% (min)
Thickness, joints	ASTM D-1777	0.10 in.(nom)
Thickness, Ribs	ASTM D-1777	0.02 in.(nom)

SOIL FINISH INFORMATION

C. Flexible Geomembrane The flexible geomembrane shall be placed by the Contractor after all surplus ash fill material from the SIRTI site earthwork has been placed and compacted in designated area of the Milwaukee Trench. The membrane shall be fabricated from high density polyethylene (HDPE) and manufactured in a width of not less than 20 ft with no factory seams. The membrane shall be new first quality material manufactured specifically for the purpose of liquid containment. Two types of liners are required for this project: smooth and roughened surface. The membrane shall be manufactured by Gundle Lining Systems, Poly America or approved equivalent. The membranes shall meet or exceed the following specifications:

Figure 3-3. Division 2 of the specifications includes soil finish information.

Division 3 – Concrete. Division 3 of the specifications contains information concerning concrete, including materials, procedures for placement, curing, and finishing, formwork construction and removal, reinforcing methods, and other related information. Precast concrete members are also described in Division 3.

The primary material described in Division 3 is concrete. Concrete material information includes the cement to be used, aggregate, admixtures, and the quality of water. Reinforcing steel is described according to ASTM references. Accessories to be placed in concrete such as chairs for supporting reinforcing steel, dowels, and anchors may be described according to manufacturer references. See Figure 3-4. Materials for forms, form release agents, grout, joint fillers, water stops, and curing compounds are also described.

Masonry requirements are given in Division 4 of the specifications.

REINFORCING STEEL DESCRIBED ACCORDING TO ASTM REFERENCES

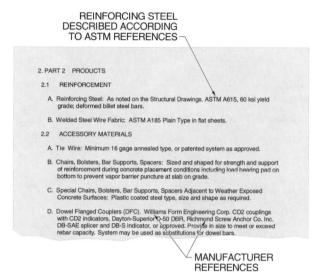

2. PART 2 PRODUCTS

2.1   REINFORCEMENT

A. Reinforcing Steel: As noted on the Structural Drawings. ASTM A615, 60 ksi yield grade; deformed billet steel bars.

B. Welded Steel Wire Fabric: ASTM A185 Plain Type in flat sheets.

2.2   ACCESSORY MATERIALS

A. Tie Wire: Minimum 16 gage annealed type, or patented system as approved.

B. Chairs, Bolsters, Bar Supports, Spacers: Sized and shaped for strength and support of reinforcement during concrete placement conditions including load bearing pad on bottom to prevent vapor barrier puncture at slab on grade.

C. Special Chairs, Bolsters, Bar Supports, Spacers Adjacent to Weather Exposed Concrete Surfaces: Plastic coated steel type, size and shape as required.

D. Dowel Flanged Couplers (DFC). Williams Form Engineering Corp. CD2 couplings with CD2 indicators, Dayton-Superior D-50 DBR, Richmond Screw Anchor Co. Inc. DB-SAE splicer and DB-S indicator, or approved. Provide in size to meet or exceed rebar capacity. System may be used as substitutions for dowel bars.

MANUFACTURER REFERENCES

Figure 3-4. Specifications include manufacturer references to define types and qualities of materials and hardware.

Division 4 – Masonry. Division 4 of the specifications addresses components of masonry construction, including masonry units, mortars, reinforcement, and accessories. Different mortars are used in various applications, such as load-bearing masonry walls, nonload-bearing masonry walls, and tuckpointing. Mortar components detailed include cement, aggregates, water, bonding agents, coloring, and admixtures (plasticizers and water repellents). See Figure 3-5.

Sizes and colors of face brick and concrete masonry units may be described according to a specific manufacturer. When stone is supplied, a specific supplier may be named to ensure stone quality and uniformity. Other masonry material information given includes metal ties and anchors, flashing, and control joints.

Division 5 – Metals. Division 5 of the specifications includes metal used on a construction project. Metal used on a construction project includes structural steel members such as columns, beams, and joists; steel decking for floors, walls, and roofs; light-gauge metal framing members; metal stairs; and ornamental metals such as handrails, ladders, and expansion joints. Cold-formed metal framing members are included in Section 05400 but are normally supplied with other interior members such as gypsum drywall. Flashing and sheet metal are described in Division 7, Section 07600. Metal piping and conduit specifications are described in Division 15 and Division 16.

PORTLAND CEMENT/LIME MORTARS			
Type/Description	Portland Cement*	Hydrated Lime or Lime Putty*	Sand*
M – Mortar of high compressive strength (minimum 2500 psi) after curing 28 days and with greater durability than some other types. Used for masonry below ground and in contact with the earth, such as foundations, retaining walls, and access holes. Type M withstands severe frost action and high lateral loads.	1	¼	3
S – Mortar with a fairly high compressive strength (minimum 1800 psi) after curing 28 days. Used in reinforced masonry and for standard masonry where maximum flexural strength is required. Also used when mortar is the sole bonding agent between facing and backing units.	1	½	4¼
N – Mortar with a medium compressive strength (minimum 750 psi) after curing 28 days. Used for exposed masonry above ground and where high compressive strength or lateral masonry strengths are required.	1	1	6
O – Mortar with a low compressive strength (minimum 350 psi) after curing 28 days. Used for general interior walls. May be used for load-bearing walls of solid masonry if axial compressive stress does not exceed 100 psi and wall is not exposed to weathering or freezing.	1	2	9

* proportion by volume

Figure 3-5. Division 4 of the specifications addresses components of masonry construction including masonry units, mortars, reinforcement, and accessories.

Steel material specifications commonly refer to ASTM standards for structural steel shapes, coatings, and connectors such as bolts, nuts, and washers. See Figure 3-6. The shape, diameter, metal, and pipe schedule specifications are given for railings and other ornamental iron.

Division 6 – Wood and Plastics. Division 6 of the specifications includes rough wood framing, finish woodworking, and plastic materials such as plastic laminate. Division 6 also provides information concerning heavy timber framing members.

As with other materials, various industry group standards are used as references for lumber products. Lumber product industry groups include the National Forest Products Association (NFPA), the Western Wood Products Association (WWPA), and the American Plywood Association (APA). Woodwork quality standards are provided by the Architectural Woodworking Institute (AWI). See Figure 3-7. Wood-related materials described include lumber and lumber treatments such as fireproofing, softwood and hardwood lumber and plywood, finish woods such as oak and maple, and fasteners such as nails, bolts, and lag screws. Finish cabinetry (casework) information includes plastic laminate grades and various hardware. When manufactured wood casework is used instead of shop-built cabinets and millwork, the manufactured wood casework is described in Division 12.

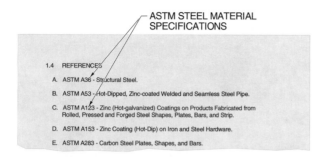

ASTM STEEL MATERIAL
SPECIFICATIONS

1.4 REFERENCES

A. ASTM A36 - Structural Steel.

B. ASTM A53 - Hot-Dipped, Zinc-coated Welded and Seamless Steel Pipe.

C. ASTM A123 - Zinc (Hot-galvanized) Coatings on Products Fabricated from Rolled, Pressed and Forged Steel Shapes, Plates, Bars, and Strip.

D. ASTM A153 - Zinc Coating (Hot-Dip) on Iron and Steel Hardware.

E. ASTM A283 - Carbon Steel Plates, Shapes, and Bars.

Figure 3-6. Steel material specifications commonly refer to ASTM standards for structural steel shapes, coatings, and connectors.

Fireproofing materials stop the passage of fire between floors of a structure.

Division 7 – Thermal and Moisture Protection. Division 7 of the specifications covers a wide range of different construction products including asphalt roofing, rubberized roofing, mastics, waterproof coatings, vapor barriers, sheet metal flashing, insulation materials, fireproofing materials, and joint sealants. These products are used to stop moisture movement or to provide thermal insulation to a structure.

Due to the specialized nature of many thermal and moisture protection products, Division 7 of the specifications relies heavily on manufacturer names and product numbers. See Figure 3-8. The specification information is performance-based where common materials are used. For example, specifications for expanded polystyrene insulation board include the required board density, thermal resistance, and compressive strength. Schedules of applications for various materials may be included to assist in locating the placement of each material mentioned in the specifications.

Division 8 – Doors and Windows. Division 8 of the specifications contains information concerning doors and windows. Division 8 also contains schedules for doors, windows, and their necessary hardware. See Figure 3-9.

FINISH REQUIREMENTS
RELATED TO AWI STANDARDS

WOODWORK QUALITY STANDARDS			
Joint Tolerance*	Premium Grade	Custom Grade	Economy Grade
Maximum Gap Between Exposed Components	$\frac{1}{64}$	$\frac{1}{32}$	$\frac{1}{16}$
Maximum Length of Gap in Exposed Components	3	5	8
Maximum Gap between Semi-Exposed Components	$\frac{1}{32}$	$\frac{1}{16}$	$\frac{1}{8}$
Maximum Length of Gap in Semi-Exposed Components	6	8	12

* in in.

Note: No gap may occur within 48′ of another gap.

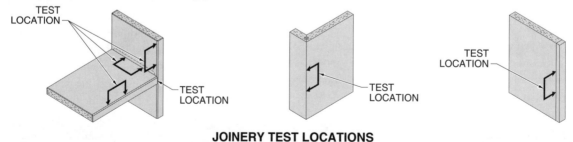

JOINERY TEST LOCATIONS

Figure 3-7. The Architectural Woodworking Institute provides woodwork quality standards.

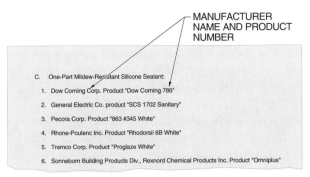

Figure 3-8. Thermal and moisture protection products in Division 7 of the specifications are designated by manufacturer names and product numbers.

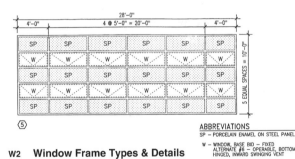

DF1 Door & Frame Schedule
Spokane Intercollegiate Research & Technology Institute

Door No.	Type	Finished Opening Size	Thk.	Material	Glazing	U.L. Rating	Frame Type	Details Head	Hinge Jamb	Strike Jamb
100A	FG	6'-0" x 8'-0"	-	Alum.	2-I-2	-	100AA	23/A11.9	23/A11.9	-
100B	FG	6'-0" x 8'-0"	-	Alum.	T-1	-	100BA	23/A11.9	23/A11.9	-
100AA	5	9'-0" x 10'-0"	4½"	Alum.	2-I-2	-	-	17/A11.9	18/A11.9	-
100BA	5	10'-10" x 10'-0"	4½"	Alum.	T-1	-	-	19/A11.9	20/A11.9	-
101	N	6'-0" x 7'-0"	-	H.S.	W	1½ HR	2	3	4	4
102	F	3'-6" x 7'-0"	-	H.S.	-	20 M	1	28/A11.9	9/A11.9	8/A11.9
103	N	3'-6" x 7'-0"	-	H.S.	W	20 M	1	1	2	2

W2 Window Frame Types & Details

ABBREVIATIONS
SP — PORCELAIN ENAMEL ON STEEL PANEL

W — WINDOW, BASE BID — FIXED
ALTERNATE #6 — OPERABLE, BOTTOM
HINGED, INWARD SWINGING VENT

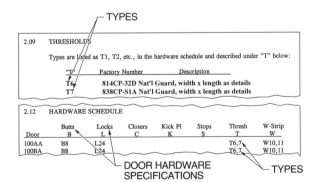

Figure 3-9. Division 8 of the specifications provides door, window, and hardware schedules and detail information that can be cross-referenced to the prints.

Door information is included for swinging metal doors and frames, swinging wood doors and frames, access doors, overhead doors and grills, glass doors, and sliding doors. Additional information concerning door frames, louvers, glass lights, and astragals is given in Division 8. An *astragal* is a molding used to cover the opening between a pair of doors to provide a weather seal. A schedule of door information relating each door to a numbered opening on the architectural drawings is also included.

Types and finishes of frames, glazing, and weatherstripping comprise a large portion of the window specifications. Windows may be specified by manufacturer product codes. Other window performance requirements given include ability to withstand wind pressures, deflection, air leakage, thermal performance, and water leakage. Drawings in the specifications may give additional details for large or complicated window installations. The specifications may also include a schedule for placement of various glazing materials such as glass types and glazing compounds.

A variety of hardware components are described in the specifications with information concerning the type and location for installation. Manufacturer names, designs, sizes, finishes, and functions are given in the hardware specifications. The hardware schedule for each door includes hinges, locks, door closers, door pulls, push plates, kickplates, stops, bolts, coordinators, thresholds, weather stripping, and door seals.

Metal door and frame information is given in Division 8 of the specifications.

Division 9 – Finishes. Division 9 of the specifications details specific information concerning applications of finish materials in each building area. Buildings have different usage areas that require different floor, wall, and ceiling finishes. Various tile, wood products, gypsum products, plaster, cement materials, paint, and special treatments are necessary to meet the different usage requirements.

Finish materials for each area of a large building may be noted on a room finish schedule. See Figure 3-10. The finish materials included in Division 9 of the specifications include metal lath and plaster, gypsum products, metal stud framing and accessories, ceramic floor and wall tile, resilient flooring, carpeting, wood flooring, suspended ceiling systems, special wall and ceiling coverings, and paint materials including stains, varnishes, and exterior and interior paint. Manufacturers of various finish materials are commonly named.

Division 10 – Specialties. Division 10 of the specifications contains a list of special items that may or may not be part of a construction project. In a manner similar to other divisions of the specifications, there is extensive reliance on manufacturer names and products. Specialty items are items commonly not bid by a general contractor, but more likely used by subcontractors and specialty contractors. Examples of specialty items include toilet compartments, fireplaces, flagpoles, awnings, shelving, laundry accessories, and access flooring. See Figure 3-11.

Division 11 – Equipment. Various structures require equipment to be built into the initial project. Projection screens, dishwashers, ovens, water treatment machinery, and a variety of industrial equipment may be installed during construction. Industry-specific equipment such as library, bank, detention, and church equipment is included in Division 11 of the specifications. The equipment is part of the initial construction bid. The general contractor and subcontractors are responsible for obtaining and installing the specialized equipment prior to completion of the project according to the information provided in Division 11. Equipment specifications rely on specific manufacturer model specifications.

Division 12 – Furnishings. Division 12 of the specifications provides information concerning metal and wood cabinetry and countertops including materials, finishes, hardware, fabrication, and installation. See Figure 3-12. Other information commonly contained in Division 12 includes installed seating such as for a theater or classroom as well as window blinds and curtains.

Division 13 – Special Construction. Division 13 of the specifications provides information concerning highly-specialized construction. Information from the manufacturer for each product or material is required. Common areas of Division 13 include pre-engineered metal buildings, seismic controls, and building automation systems that control such items as lighting, fire protection, and heating and cooling systems.

ROOM FINISH SCHEDULE

Room No.	Name	Floor Mat.	Floor Fin.	Floor Col.	Base Mat.	Base Fin.	Base Col.	Walls North Mat.	North Fin.	North Col.	South Mat.	South Fin.	South Col.	East Mat.	East Fin.	East Col.	West Mat.	West Fin.	West Col.	Ceiling Mat.	Ceiling Fin.	Ceiling Col.	Ceiling Ht.	Notes
000	LOBBY	14	F3		21	----		33	F8	P1	33	F8	P1	32	F8	P1	33	F8	P1	41	----		11'-½"	
000A	STAIR 1	10	F1		20 23	F17 F17		33	F8	P2	33	F8	P2	33	F8	P2	33	F8	P2	41	----		11'-½"	
001	CORRIDOR	14	F3		21	----		32	F8	P1	33	F8	P1	32	F8	P1	33	F8	P1	41	----		11'-½"	
009	MEN'S TOILET	13	F2		22	F2		32	F11 F6	P4	32	F11 F6	P4	32	F11 F6	P4	32	F11 F6	P4	42	F8	P4	8'-0"	
009A	VESTIBULE	13	F2		22	F2		32	F8		32	F8		32	F8		32	F8		42	F8	P4	8'-0"	
010	WOMEN'S TOILET	13	F2		22	F2		32	F11 P6	P4	32	F11 P6	P4	32	F11 P6	P4	32	F11 P6	P4	42	F8	P4	8'-0"	

Materials

Floor Materials

10	Concrete (CONC)
11	Slate Pavers (SP)
12	Not Used
13	Ceramic Mosaic Tile (CMT)
14	Vinyl Composition Tile (VCT)

Base Materials

20	Hardwood (Maple)
21	Vinyl Cove
22	Ceramic Mosaic Tile

Wall Materials

31	Concrete
32	Gypsum Wallboard (Single Layer) on Metal Studs
33	Gypsum Wallboard (Single Layer) on Metal Furring

Ceiling Materials

41	Exposed Concrete
42	Gypsum Wallboard

Finish

F1	Concrete Hardener/Sealer
F2	Clean & Seal
F3	Clean, Wax, & Buff
F4	Tackable Wall Covering
F5	Vinyl Fabric Wall Covering
F6	Ceramic Mosaic Tile
F7	Paint - Masonry System
F8	Paint - GWB System
F9	Traffic Membrane
F10	Rubber Treads, Risers, Sheet Rubber Nonslip Tile @ Landing
F11	High Build Glazed Coating

Color

P1	Columbia #5391W
P2	Columbia #5393M
P3	Columbia #5394D
P4	Columbia #5390W

Figure 3-10. Room finish schedules in Division 9 of the specifications contain room finish information.

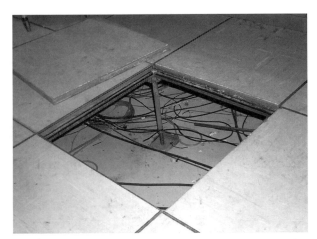

Figure 3-11. Division 10 of the specifications includes specialized items such as access flooring.

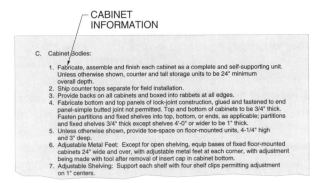

Figure 3-12. Division 12 of the specifications includes cabinet, countertop, installed seating, and window treatment information.

Division 14 – Conveying Systems. Division 14 of the specifications describes the various conveying systems used in different buildings. Machinery for moving people, materials, and equipment is included in most large construction projects. Elevators are essential in large office buildings. Retail stores often use elevators and escalators to move customers. Banks use pneumatic tubes to shuttle containers from a drive-up window to the teller area. In manufacturing and power plants, conveyors and monorail systems move materials and products from one area to another.

Elevators are the most common conveying system defined in Division 14. Elevator specifications include the rated net capacity, rated speed, travel distance, number of stops, car size, etc. See Figure 3-13. A variety of government and industry organizations such as the American National Standards Institute (ANSI), the Americans with Disabilities Act (ADA), the National

Electrical Code (NEC®), and Underwriters Laboratories Inc. (UL®), provide regulations and standards that apply to elevator construction.

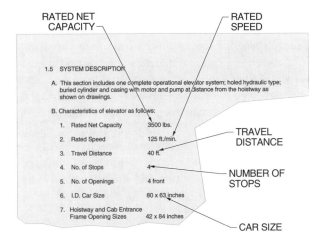

Figure 3-13. Division 14 of the specifications provides information concerning conveying systems such as elevators.

Many different conveying systems are designed for commercial and industrial use. Overhead lifts and monorail systems are normally electrically-controlled. Overhead lifts and monorail systems are described in the specifications according to their lifting capacity, speed of lift and movement, and height of lift required. Conveying systems move items as varied as cans, coal, and pieces of mail. Conveyor type, speed, and capacity are part of Division 14 specifications.

Division 15 – Mechanical. Division 15 of the specifications describes various mechanical systems such as heating, ventilating, and air conditioning equipment, fixtures, piping, instrumentation, and controls. Architects, mechanical engineers, electrical engineers, estimators, and various general contractors and subcontractors must coordinate their work to ensure proper operation of mechanical systems. Basic material and method information such as pipe, pipe hangers, and pipe supports, duct and pipe connectors, insulation, gauges, flow control and measurement devices, electric motors and starters, and fuel tanks are given. Many different control systems are installed on mechanical systems that are defined in Division 15.

Estimators must ensure that care is taken when evaluating subcontractor bids concerning Division 15. Plumbing subcontractors and mechanical subcontractors may have overlapping areas or uncovered areas depending on their assessment of the specifications and the scope of their bids. Estimators must carefully assemble and completely analyze the scope of subcontractor bids in this area.

Several subclassifications of Division 15 apply to plumbing work. Water supply and treatment information includes connections to available water supplies and fire protection service pipes, valves, meters, and hydrants. Wastewater disposal and treatment information is given for waste pipes and fittings and storm water drains. Pipes and valves are specified by pipe, size, material, and ability to withstand certain pounds per square inch (psi) of pressure. See Figure 3-14. Piping and fittings for natural gas, compressed air systems, and vacuum systems are included in Division 15 of the specifications. Tanks described include water heaters and storage tanks. Plumbing fixture information is given for sinks, toilets, drinking fountains, and faucets.

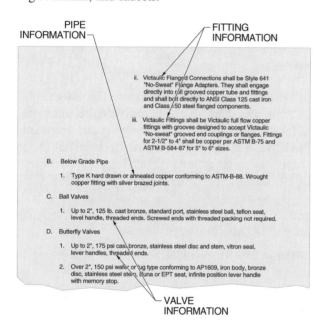

Figure 3-14. Plumbing specifications give pipe, valve, and fitting information.

Heating, ventilating, and air conditioning systems generate heat from sources such as boilers or natural gas heaters, provide cooling with systems containing refrigerants, chillers, and compressors, and distribute air throughout large areas with various air handlers, motors, fans, and duct systems. See Figure 3-15. In hot and cold water systems, specifications are given for piping, circulating pumps, and heating and cooling transfer equipment. Small heating and cooling systems such as unit heaters and air conditioners for special conditions are also specified. Other air handling situations described include the filtering, removal, and exhausting of fumes and smoke.

Figure 3-15. Ductwork carries heated or cooled air throughout a building from a forced-air HVAC system.

Agencies and codes that regulate fire protection equipment and develop standards include the National Fire Protection Association (NFPA), Uniform Mechanical Code, Uniform Building Code (UBC), Occupational Safety and Health Administration (OSHA), and local fire authorities. All fire protection systems are designed and installed in accordance with these agencies and codes. Specifications describe piping, sprinkler heads, check valves, and fire department connections for wet pipe systems. Dry pipe systems include similar information as wet pipe systems in addition to air compressor and air pressure requirements.

Division 16 – Electrical. Division 16 of the specifications provides wiring, equipment, and finish information for electrical systems. Division 16 information includes descriptions of electrical site work, raceways and conduit, panelboards, lighting, communication and telephone systems, and heating and cooling systems. The NEC® is updated every three years and is the code on which many of the specifications rely. For example, subsection 700-9(d)(2) requires that equipment for feeder circuits in occupancies for greater than 1000 people or for buildings above 75′ with assembly, educational, residential, detention and correctional, business, or mercantile occupancies, shall be located in spaces fully protected by approved automatic fire suppression systems or spaces with a 1-hr fire resistance rating. See Figure 3-16.

> ▲ *The correct size pipe must be used for plumbing applications. Undersized pipe can cause backed-up sewer pipes, unacceptably low water volume from faucets or showers, and inadequate volume for outdoor sprinklers.*

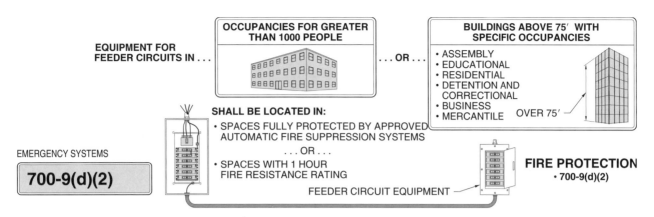

Figure 3-16. Many electrical specifications rely on information from the NEC®.

Electrical wiring information in Division 16 of the specifications includes conductors and cables, raceways and boxes, and wiring devices and connections. Wire size, insulation, and installation information in addition to lighting and sign information is also included. See Figure 3-17. The specifications contain quality assurance requirements and acceptable manufacturer names and product numbers for each of these areas.

Department of Transportation

A different numerical and description classification system is used for road, highway, bridge, and other heavy construction performed by each state department of transportation. Each state uses a code developed internally by their state department of transportation (DOT). See Figure 3-18. Proposals for bids are commonly distributed with each category listed. A preliminary quantity may be provided along with each code. These quantities indicate the preliminary estimating calculations made by DOT engineers for a specific project. Estimators must use these codes to submit a bid to a DOT in the required format. Some electronic estimating systems designed specifically for heavy/highway construction can be preloaded with the codes for a particular state or series of states.

ARCHITECTURAL PRACTICES

An estimator must be able to read and interpret architectural and shop drawings. Familiarity with various lines, drawing scales, symbols, and abbreviations is required to develop an accurate bid.

All individuals involved in the construction process require a common source of information that is legally

and functionally reliable. All members of the building team must read and interpret prints with skill and accuracy to ensure the project is in accordance with building codes, the needs of the owner(s), and the design of the architect(s). The drawings that make up a large set of prints are separated into different divisions to help individuals readily find information. The use of symbols and abbreviations and the ability to interpret portions of the prints across the divisions are necessary skills in the construction and estimating processes.

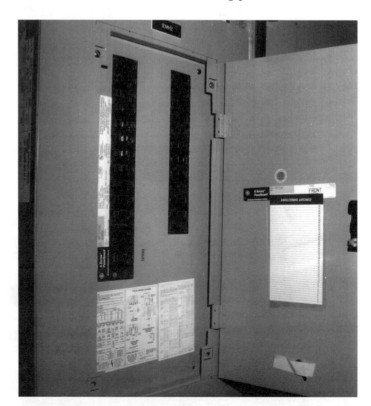

Conduit and panelboard information is included in Division 16 of the specifications.

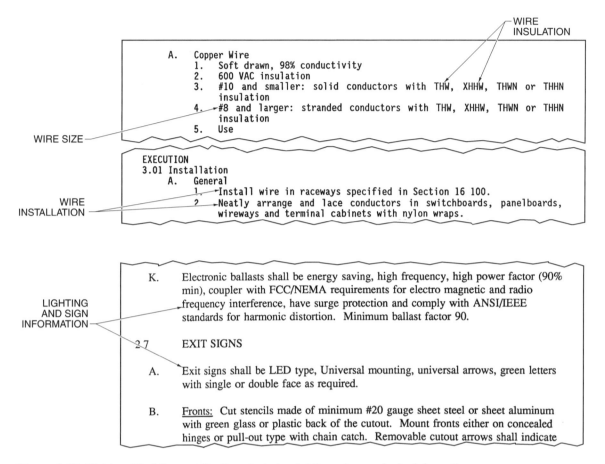

WIRE INSULATION

WIRE SIZE

A.   Copper Wire
    1.   Soft drawn, 98% conductivity
    2.   600 VAC insulation
    3.   #10 and smaller: solid conductors with THW, XHHW, THWN or THHN insulation
    4.   #8 and larger: stranded conductors with THW, XHHW, THWN or THHN insulation
    5.   Use

WIRE INSTALLATION

EXECUTION
3.01 Installation
    A.   General
        1.   Install wire in raceways specified in Section 16 100.
        2.   Neatly arrange and lace conductors in switchboards, panelboards, wireways and terminal cabinets with nylon wraps.

LIGHTING AND SIGN INFORMATION

K.   Electronic ballasts shall be energy saving, high frequency, high power factor (90% min), coupler with FCC/NEMA requirements for electro magnetic and radio frequency interference, have surge protection and comply with ANSI/IEEE standards for harmonic distortion. Minimum ballast factor 90.

2.7   EXIT SIGNS

A.   Exit signs shall be LED type, Universal mounting, universal arrows, green letters with single or double face as required.

B.   Fronts: Cut stencils made of minimum #20 gauge sheet steel or sheet aluminum with green glass or plastic back of the cutout. Mount fronts either on concealed hinges or pull-out type with chain catch. Removable cutout arrows shall indicate

Figure 3-17. Division 16 of the specifications includes lighting, sign, and wire information.

MISSOURI DEPARTMENT OF TRANSPORTATION	
CONSTRUCTION CLASSIFICATION CODES	
CODE	DESCRIPTION
201-10.10	CLEARING - METRIC
201-20.10	GRUBBING - METRIC
202-20.10	REMOVAL OF IMPROVEMENTS
203-10.05	CLASS A EXCAVATION - METRIC
203-55.05	EMBANKMENT IN PLACE - METRIC
203-60.05	COMPACTING EMBANKMENT
206-30.05	CLASS 3 EXCAVATION - METRIC
207-20.05	LINEAR GRADING CLASS 2 - METRIC
301-20.05	MINERAL AGGREGATE (BITUMINOUS BASE) - METRIC
301-60.19	ASPHALT CEMENT (BITUMINOUS BASE) PG 64-22 - METRIC
304-05.05	TYPE 5 AGGREGATE FOR BASE (100 MM THICK) - METRIC
310-50.05	GRAVEL (A) OR CRUSHED STONE (B) - METRIC
403-40.30	MINERAL AGGREGATE (ASPHALTIC CONCRETE) (TYPE I-C MIX) - METRIC
403-81.24	ASPHALT CEMENT

Figure 3-18. Each state department of transportation uses its own construction classification codes for bid documents.

One of the most necessary and difficult skills to develop when creating an estimate from a large set of prints is the ability to visualize an entire project and the relationship of all components from a set of prints and specifications. In addition to an overall view, it is often necessary to obtain information from several different drawings and the specifications to fully understand a single building element. The ability to pull information from a variety of sources and combine this information into a common understanding is the most important requirement in estimating takeoff for a large building project.

Dimensions

Architects use a variety of notation systems to give dimensional information concerning a construction project. A combination of various types of lines and symbols note beginning and ending points for linear dimensions. Dimensions for site work, building work, finish work, and details all use different systems for quantities and drawing scales.

Lines. A variety of lines are used to depict print components. Each line serves a specific function in a set of prints. The type of line used depends on the location of the item shown and the dimensioning system. The basic line types are object, leader, hidden, cutting plane, section, break, dimension, center, and extension lines. See Figure 3-19. These lines are used to show different electrical and plumbing lines. Line types and their use are standard among architects.

An *object line* is a line that defines the visible shape of an object. Object lines are solid and without breaks. Object lines are the most common lines shown on a print. A *leader line* is a line that connects a dimension, note, or specification with a particular feature of the drawn object. Leader lines have an arrow at the end where they contact the edge of an object. Leader lines are drawn at any angle required to make the connection between the written description and the object. A *hidden line* is a line that represents shapes that cannot be seen. Hidden lines are used wherever there is a distinct change in the surface of an object. Hidden lines are drawn as dashed lines.

A *cutting plane line* is a line that shows where an object is imagined to be cut in order to view internal features. A cutting plane line is a thick line with a 90° arrow on each end. The direction of the arrows indicate the direction in which the detail is viewed. A *section line* is a line that identifies the internal features cut

by a cutting plane line. Section lines are drawn at an angle to object lines. Specific configurations of section lines identify the particular type of material. A *break line* is a line that shows internal features or avoids showing continuous features. Break lines may be drawn as straight lines with a zig-zag in the middle or as a freehand, staggered line. Break lines are used to discontinue a drawing where no further length of an object need be shown.

A *dimension line* is a line that is used with dimensions to show size or location. Dimension lines are thinner that object lines. Dimension lines are commonly broken for the placement of numerals giving the measurement. A *centerline* is a line that locates the center of an object. Centerlines are thin, dark lines broken by short dashes. Centerlines are used to locate windows, doors, openings, and other fixtures. An *extension line* is a line that extends from surface features and terminates dimension lines. Extension lines are drawn at 90° to the object and dimension lines. Extension lines do not touch the object line. Dimension, center, and extension lines provide a system for measurement determination in the English or metric system of measurement. See Appendix.

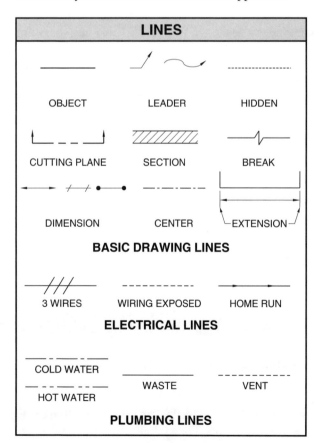

Figure 3-19. Architects use a variety of lines to depict print components.

English. The English system of measurement uses the fractional system of feet, inches, and fractions of an inch. The English system is currently the most common system of measurement used in building construction in the United States. Elevations above sea level and distances for linear and angular measure are also given in feet and tenths and hundredths of a foot. See Appendix.

Metric. The metric system of measurement uses measurements based on the meter (m). One meter is equal to 39.37″. Other common metric units of measure include the kilometer (1000 m), centimeter (.01 m), and millimeter (.001 m). The metric system is being used more widely in many construction projects. Areas are based on square meters.

Scales. Different drawing scales are used for various types of drawings. Architectural drawings use scales ranging in size from $\frac{1}{32}″ = 1′$ to $3″ = 1′$ and are measured with an architect's scale. An architect's scale is used for building plans. Shop and detail drawings may be drawn to actual size for small detailed items. Large site drawings use scales of 10, 20, 30, or more feet to an inch and are measured with an engineer's scale. An engineer's scale measures distances up to $1″ = 1000′$. Metric scales divide measurements into ratios such as 1:25. This notation denotes a drawing that is $\frac{1}{25}$th actual size. Other metric ratios include 1:10, 1:20, 1:50, 1:100, 1:150, and 1:200. See Figure 3-20.

Print Divisions

Prints are separated into different divisions to help individuals readily find information. Print divisions are denoted by a capital letter prefix. Architectural prints are denoted by a capital A followed by a page number. In a similar manner, structural prints are denoted by an S, mechanical prints by an M, electrical prints by an E, and civil prints by a C. See Figure 3-21. Architects may use other divisions depending on the nature of the building project. For example, site mechanical plans may be identified by the designation MPE, denoting mechanical, plumbing, and electrical on one drawing. Print numbering begins with the number 1 within each division. For example, the architectural prints may run from page A1 to page A65. Print divisions may be subdivided when several pages apply to the same elements. For example, the structural prints begin with page S1. Subdivisions may be given as S1.1, S1.2, and S1.3.

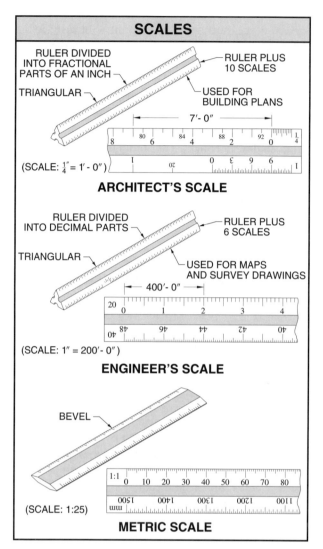

Figure 3-20. A variety of measurement scales are used by architects, engineers, and estimators.

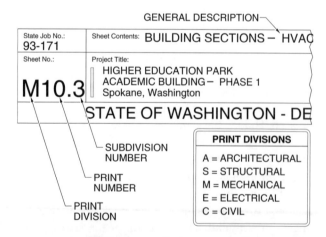

Figure 3-21. Prints are separated into divisions for ease of information retrieval.

Architectural. Architectural prints include general building information, floor plans, elevation views, section drawings, and detail drawings. Architectural prints are commonly the largest division of prints for a construction project.

Structural. Structural prints provide information about sizes, styles, and placement for foundations, beams, columns, joists, and other framing and load-bearing members. Framing and load-bearing members may be built of wood, masonry, reinforced concrete, or structural steel.

Mechanical. Mechanical prints include plumbing, heating, ventilating, and air conditioning information including ductwork, piping, and equipment placement and sizes. Information about piping for fire protection systems may also be part of the mechanical prints or may be provided in a separate division.

Electrical. Electrical prints indicate the capacity and placement of power plant systems, lighting, cable trays, conduit and panel schedules, finish fixtures, wiring, switches, and any other electrical installations. Electrical prints commonly contain numerical and alphabetic codes that are cross-referenced to schedules in the specifications for many types of electrical controls and fixtures.

Civil. Civil prints include overall site layout, grading, elevations, and topographical information. Other information includes site drainage, paving designs, and parking layout. Landscaping may also be a part of the civil prints or provided separately.

Drawings

Different drawings are used on prints. Drawings used on prints commonly include perspective, oblique, isometric, and orthographic projection. See Figure 3-22. A *perspective drawing* is the representation of an object as it appears when viewed from a given point. Perspective drawings illustrate various objects in a pictorial manner. Perspective drawings are commonly used for showing the exterior of a structure.

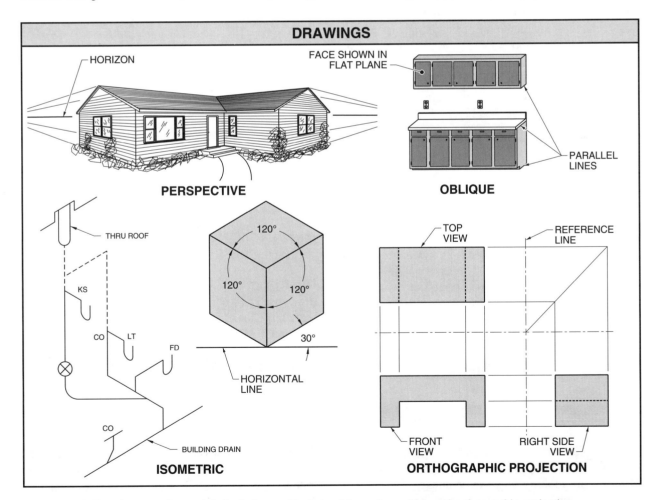

Figure 3-22. Drawings used on prints include perspective, oblique, isometric, and orthographic projection.

An *oblique drawing* is a drawing in which one face of an object is shown in a flat plane with the receding lines projecting back from the face. To show a three-dimensional appearance, parallel lines are drawn from the corners and edges of the flat plane at any angle other that 90°. Oblique drawings are often used to show kitchen and bathroom cabinets.

An *isometric drawing* is a drawing in which all horizontal lines are drawn at a 30° angle from horizontal and all vertical lines are drawn at a 90° angle from horizontal. Isometric drawings are used in some types of detail drawings.

An *orthographic projection drawing* is a drawing in which each face of an object is projected onto flat planes at 90° to one another. No allowances are made for depth or distance of view. Orthographic projection drawings are used for elevation drawings, plan views, and detail drawings.

Elevation Drawings. An *elevation drawing* is an orthographic view of a vertical surface without allowance for perspective. Elevation drawings show the vertical surfaces of a structure or object along with material and dimension information. Materials are noted with architectural notes and symbols. Dimensions are given with extension and dimension lines. Elevation drawings are commonly used to show interior and exterior wall surfaces and details. See Figure 3-23.

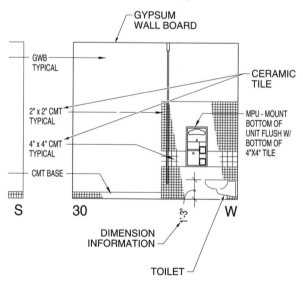

Figure 3-23. Elevation drawings show the appearance of a surface projected onto a flat plane.

Plan Views. A *plan view* is a drawing of an object as it appears looking down from above. Plan views use orthographic projection to show horizontal di-

mensions of a structure by means of a cutting plane approximately 5' above each floor line. See Figure 3-24. A *cutting plane* is a line that cuts through a part of a structure on a drawing. The plan view cutting plane slices through walls, cabinets, doors, and all other items it encounters. Items above the cutting plane are noted by the architect and drawn with hidden lines. Items hidden from view and objects below the floor level may also be shown with hidden lines on plan views.

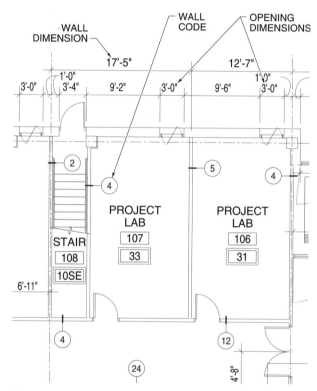

Figure 3-24. Plan views are commonly used to show floor plan information.

Detail Drawings. A *detail drawing* is a drawing showing a small part of a plan, elevation, or section view at an enlarged scale. Detail drawings provide information about specialized construction items that are not clearly shown in another part of the drawing. See Figure 3-25. Cutting planes may be used for detail drawings. The cutting plane refers to a separate elevation, sectional view, or detail drawing given for that area. Items encountered at the cutting plane are projected onto a flat surface, as on an orthographic projection. When cutting planes are used for detail drawings, they are referred to as section views. Section views illustrate a portion of an object that requires architectural detail.

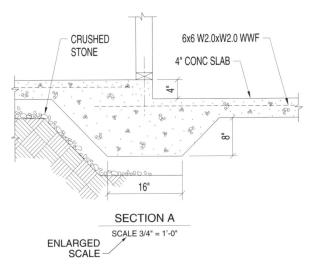

SECTION A
SCALE 3/4" = 1'-0"

ENLARGED
SCALE

Figure 3-25. Detail drawings provide large scale information about objects that are difficult to illustrate.

Symbols and Abbreviations

A *symbol* is a pictorial representation of a structural or material component used on prints. An *abbreviation* is a letter or group of letters representing a term or phrase. Interpretation of prints requires an understanding of symbols and abbreviations. Determination of the proper building material is based on accurate reading of symbols and abbreviations. See Figure 3-26. See Appendix.

Symbols. A common set of symbols is used on architectural prints. The symbols denote materials, building objects, and various dimensions and locations. Architects on large building projects commonly provide their own illustrations for symbols used on a particular project. The illustrations include elements that are repeated often and specialty items for one specific job. The symbols shown on a plan by an architect supersede any other common interpretations for the symbols.

Abbreviations. In a manner similar to symbols, architects commonly include a listing of abbreviations for a set of prints. The list is not complete, but provides a partial list of common and specific specialty terms. Periods are commonly used with abbreviations to avoid confusion when the abbreviation spells a word.

ESTIMATING SOFTWARE – CREATING AN ESTIMATE AND STORING ESTIMATE INFORMATION

In Precision estimating software, a new estimate is created by specifying a folder and filename for the estimate. Estimating information is stored by inputting estimate information in the Estimate Information window. See Figure 3-27.

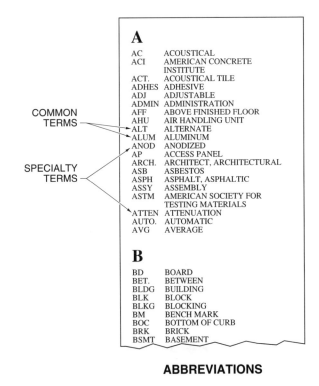

ABBREVIATIONS

Figure 3-26. An understanding of symbols and abbreviations is required for the interpretation of prints.

PRECISION ESTIMATING SOFTWARE — CREATING AN ESTIMATE AND STORING ESTIMATE INFORMATION

Creating an Estimate

1. Click the New Estimate button (🗋) to display the New Estimate window.

2. Enter a filename (Stanton Industries) for the new estimate at the Estimate File Name tab.

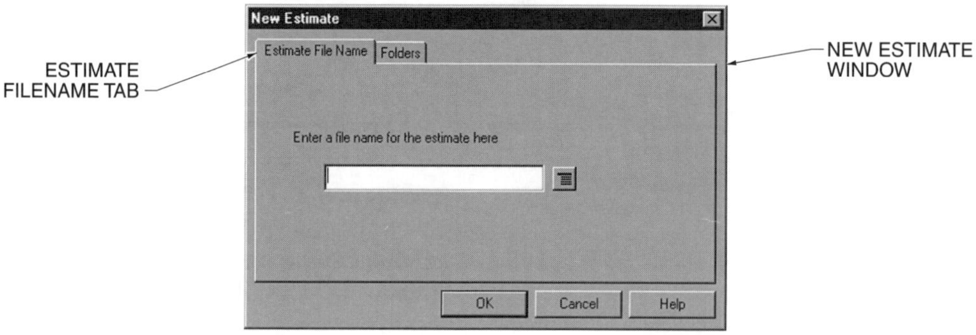

ESTIMATE FILENAME TAB

NEW ESTIMATE WINDOW

3. Click the Folders tab to display the Folders window. The Folders tab lets the estimator specify the folder where the estimate is stored and the database used with the estimate.

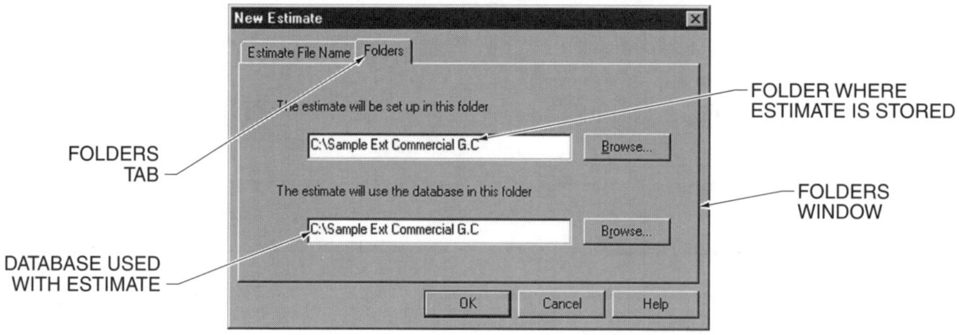

FOLDERS TAB

FOLDER WHERE ESTIMATE IS STORED

FOLDERS WINDOW

DATABASE USED WITH ESTIMATE

4. Accept the defaults given (C:\Program Files\Timberline\Precision\Sample Ext Commercial G.C.) and click OK to open the Estimate Information window.

Storing Estimate Information

The Estimate Information window consists of eight tabs where the estimator can enter general information about the job being estimated. The first tab is the Main window. To move to other tabs, click the appropriate tab. The only required field is the project name. The project name appears in the heading of every printed page of the estimate report. The project name also appears on the report cover page when printed. All other fields are optional. Close the Estimate Information window and the spreadsheet displays. *Note:* Choose Estimate Information from the Takeoff menu if any information in the Estimate Information window must be changed later.

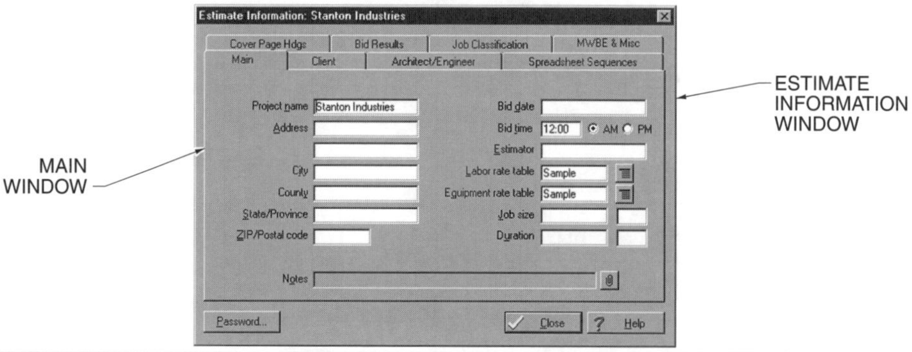

MAIN WINDOW

ESTIMATE INFORMATION WINDOW

Figure 3-27. In estimating software, a new estimate is created by specifying a folder and filename for the estimate.

Estimating

_____ 1. _____ are written information included with a set of prints that clarify and supply additional information.

_____ 2. The CSI MasterFormat™ contains _____ divisions that define the broad areas of construction.

T   F   3. Division 1 covers General Requirements.

_____ 4. Division 2 covers _____.

    A. Site Construction   C. Concrete

    B. Masonry   D. Metals

_____ 5. Division _____ covers Special Construction.

    A. 7   C. 13

    B. 8   D. 14

_____ 6. Division _____ covers Electrical.

_____ 7. Division _____ covers Concrete.

T   F   8. Division 7 covers Wood and Plastic.

_____ 9. Division 11 covers _____.

_____ 10. Division _____ covers Doors and Windows.

_____ 11. Division 15 covers _____.

    A. Metals   C. Masonry

    B. Furnishings   D. Mechanical

T   F   12. Conveying systems are not covered in the CSI MasterFormat™.

T   F   13. A different numerical and description classification system is used for road, highway, bridge, and other heavy construction performed by each state DOT.

_____ 14. The _____ system of measurement uses the fractional system of feet, inches, and fractions of an inch.

_____ 15. Prints are separated into different divisions denoting their type by _____.

    A. sequential numbers   C. icons

    B. capital letter prefixes   D. neither A, B, nor C

_____ 16. A(n) _____ view is a drawing of an object as it appears looking down from above.

_____ 17. A(n) _____ drawing is a drawing showing a small part of a plan, elevation, or section view at an enlarged scale.

_____ **18.** A(n) _____ is a letter or group of letters representing a term or phrase.

_____ **19.** A(n) _____ is a pictorial representation of a structural or material component used on prints.

_____ **20.** In estimating software, a new estimate is created by specifying a folder and _____ for the estimate.

Material Symbols

_____ **1.** Wood panel   _____ **11.** Steel

_____ **2.** Plywood   _____ **12.** Small scale metal

_____ **3.** Glass   _____ **13.** Blocking

_____ **4.** Concrete   _____ **14.** Rough wood member

_____ **5.** Gravel   _____ **15.** Lath and plaster

_____ **6.** Rock   _____ **16.** Acoustical tile

_____ **7.** Earth   _____ **17.** Ceramic tile

_____ **8.** Cut stone   _____ **18.** Structural clay tile

_____ **9.** Brick   _____ **19.** Blanket insulation

_____ **10.** Rubble   _____ **20.** Concrete block

Basic Drawing Lines

_____ **1.** Object

_____ **2.** Hidden

_____ **3.** Break

_____ **4.** Section

_____ **5.** Cutting plane

_____ **6.** Leader

_____ **7.** Dimension

_____ **8.** Extension

_____ **9.** Center

Electrical Lines

_____ **1.** 3 wires

_____ **2.** Wiring exposed

_____ **3.** Home run

(A)   (B)   (C)

Plumbing Lines

_____ **1.** Waste

_____ **2.** Vent

_____ **3.** Cold water

_____ **4.** Hot water

(A)   (B)

(C)   (D)

Drawings

_____ **1.** Isometric

_____ **2.** Oblique

_____ **3.** Perspective

_____ **4.** Orthographic projection

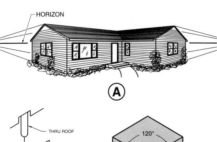

(A)

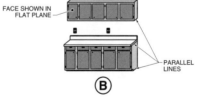

(B)

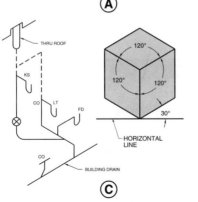

(C)

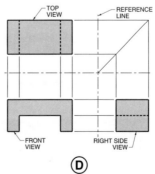

(D)

Printreading

_____   **1.** Earth

_____   **2.** Gravel

_____   **3.** Concrete

_____   **4.** Rough wood member

_____   **5.** Dimension

_____   **6.** Leader

_____   **7.** Break line

_____   **8.** Center line

_____   **9.** Extension line

_____   **10.** Object line

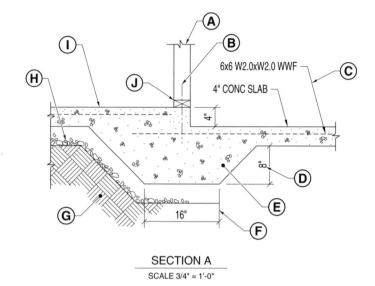

6x6 W2.0xW2.0 WWF

4" CONC SLAB

16"

SECTION A
SCALE 3/4" = 1'-0"

Scales

_____   **1.** Architect's

_____   **2.** Engineer's

_____   **3.** Metric

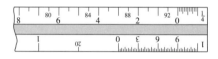

Ⓐ

Ⓑ

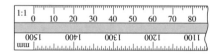

Ⓒ

Activities
Printreading Skills

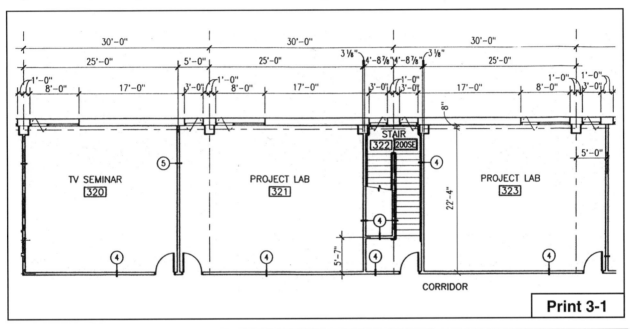

Name _____ Date _____ **Chapter 3**

Activity 3-1 – Ledger Sheet Activity

Refer to Print 3-1 and Quantity Sheet No. 3-1. Take off the square feet of floor area for TV Seminar Room 320, Project Lab 321, and Project Lab 323. Calculate the total square footage of the three rooms.

Print 3-1

QUANTITY SHEET

Sheet No. __3-1__

Project: _____

Estimator: _____

Date: _____

Checked: _____

No.	Description	Dimensions				Unit		Unit		Unit		Unit
		L	W									
	TV seminar room 320											
	Project lab 321											
	Project lab 323											
	Total square footage											

Activity 3-2 – Spreadsheet Activity

Refer to Print 3-2 and Quantity Sheet No. 3-2 on the CD-ROM. Determine the square footage of Microcomputer Classroom 326. Convert all measurements to decimal parts of a foot.

Activity 3-3 – Estimating Software Activity

Create a new estimate and name it Activity 3-3. Input *Roberts Industries* for client name and *Johnson Group* for architect. Click the **Quick Takeoff** button and take off a job sign by double-clicking **1000.00 General Requirements**, double-clicking **1591.00 Office Equipment**, and double-clicking **1591.10 Job Sign**. Close the Quick Takeoff window. Enter 1 for the Takeoff Quantity. Print a standard estimate report by selecting **Reports** then **Standard Estimate**. Click **Print** to print the report.

General Conditions

The general conditions describe the rights and responsibilities of all parties involved in a construction project. The general conditions include items that affect a project but do not appear on print drawings. Construction estimating items include material, labor, equipment, overhead, and profit. Material quantities are calculated based on the type of each material required. Labor pricing is determined based on the number of labor hours required and the cost of labor per hour. Equipment used on a construction project includes power equipment such as earth-moving, hauling, and lifting equipment, temporary job site equipment, and communication equipment. Overhead costs are expenses that are not attributable to any one specific operation. Profit is the monetary benefit realized by a construction firm at the completion of the building process.

GENERAL CONDITIONS

General conditions are the written agreements describing building components and various construction procedures included in the specifications for a construction project. The general conditions spell out the rights and responsibilities of all parties involved in a construction project. The relationships between the various parties to the construction contract are also given in the general conditions.

Many items that affect the project but do not appear on print drawings are included in the general conditions. The general conditions and specifications include construction cost items as diverse as weather protection requirements for concrete, modification procedures, and bonding and insurance requirements. Estimators must take all portions of the general conditions and specifications into account when calculating overall job costs for items that may otherwise be missed in the final price of a project.

MATERIAL QUANTITY TAKEOFF

An estimator calculates the quantities of the materials required after the determination that the plans and specifications, including the general conditions, contain items pertinent to the construction company providing the work. Items to be calculated are located in the specifications and on the print drawings. The quantity for each item is calculated at each location and entered on the ledger sheet, spreadsheet, or estimating software.

Measurement Systems

Items are calculated in square feet, cubic feet, linear feet, square yards, cubic yards, acres, metric quantities such as meters or hectares, or in individual units. The proper takeoff quantity measurement must be used for each material. Estimators check the specifications and plans to determine whether quantities are in the English or metric system of measurement. For standardized ledger sheets, spreadsheets, and estimating software systems, quantities for each material may be listed on the sheet to assist the estimator in using the proper measurement system for each takeoff item.

Area/Volume. The use of square or cubic measurements is common for many materials. Estimators must be familiar with the units of measure used for various materials. This can be determined by checking the bid sheet, estimating software, or manufacturer literature.

Interior finish materials such as suspended ceilings, flooring, and gypsum drywall are taken off and priced based on square feet or square yards. See Figure 4-1. For example, room 255 has dimensions of 23′-6″ (23.5′) and 8′-10″ (8.8$\overline{3}$′). The area of the room is found by multiplying the two dimensions. The room has an area of 207.5 sq ft (23.5′ × 8.8$\overline{3}$′ = 207.51 sq ft), which is used for a suspended ceiling system estimate. The square foot value is divided by 9 to obtain square yards. The room has an area of 23.06 sq yd (207.51 sq ft ÷ 9 = 23.06 sq yd), which is used for a carpeting estimate.

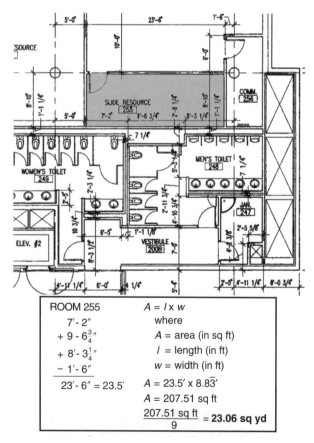

ROOM 255	$A = l \times w$
7′- 2″	where
+ 9 - 6$\frac{3}{4}$″	A = area (in sq ft)
+ 8′- 3$\frac{1}{4}$″	l = length (in ft)
− 1′- 6″	w = width (in ft)
23′- 6″ = 23.5′	$A = 23.5' \times 8.8\overline{3}'$
	$A = 207.51$ sq ft
	$\dfrac{207.51 \text{ sq ft}}{9} =$ **23.06 sq yd**

Figure 4-1. Areas for quantity takeoff are based on square feet or square yards depending on the material.

Earthwork is calculated in cubic yards of material to be moved or supplied. For example, takeoff and pricing for excavating a hole for placement of a retention pond is based on the number of cubic yards of earth to be moved. See Figure 4-2. Concrete for cast-in-place applications is also calculated in cubic yards. For example, a pier 2′ in diameter and 45′ deep has a volume of 141.37 cu ft (3.1416 × 1′ × 45′ = 141.37 cu ft). This value is divided by 27 to obtain cubic yards. The pier requires 5.24 cu yd of concrete (141.37 cu ft ÷ 27 = 5.24 cu yd).

Figure 4-2. Earthwork excavations are calculated in cubic yards.

Quantity. Items such as doors, windows, water closets (toilets), drinking fountains, air conditioners, boilers, and a variety of other fixtures are taken off by counting individual installations. See Figure 4-3. For example, eight doors are required for corridor 203. Thorough review of the specifications and accurate markup of plans are required to properly calculate quantities of individual items.

LABOR PRICING

Labor pricing is determined based on the number of labor hours required (labor quantity) and the cost of labor per hour (labor rate). The number of labor hours required for various operations can be related to the takeoff quantities. The costs of labor per hour are available from a variety of sources. Estimators should include all costs associated with labor hours including wages, fringe benefits, taxes, and all associated overhead costs.

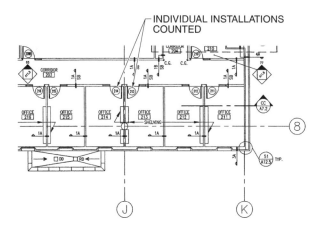

Figure 4-3. Items such as doors, windows, etc. are taken off by counting individual installations.

Labor Quantities

Labor quantities are the units of labor required for various construction operations. A common method of calculating labor quantities is to equate a level of labor cost with material quantities and develop unit pricing. For example, it may take a carpenter .75 hr to set a metal door jamb and install the metal door. The estimator calculates the number of metal doors to be installed in a particular job and uses this information to determine the number of carpenter labor hours. If 12 metal doors and jambs are to be set, 9 carpenter labor hours are included in the bid for this item (12 x .75 hr = 9 hr). Contractors often maintain their own labor quantity information based on data they have collected. This historical data provides a basis for labor costing for each material quantity.

For bidding using estimating software, the labor quantities may be entered into a database tied to the quantity takeoff. See Figure 4-4. As the estimator enters quantities of materials or number of units, the electronic database automatically calculates the labor quantity and enters this amount on the spreadsheet. Costs for labor are stored in another portion of the database. Labor quantities and labor costs are linked to the quantity takeoff in the estimating software system.

Standard Scheduling Information. Standardized labor unit (time) tables for various materials are available to determine labor quantities. See Figure 4-5. These tables provide information based on industry practices that equate various materials to the number of craft labor hours required for installation. The estimator may use these tables to determine labor quantities based on a variety of measures such as square

feet, cubic yards, or items of materials. Standardized labor unit tables are also available in electronic format to enter into databases.

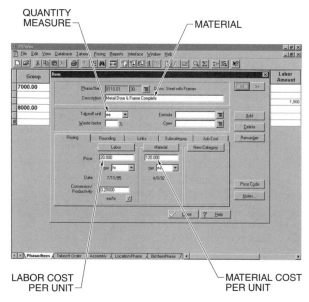

Figure 4-4. Estimating software allows labor quantities to be entered into a database tied to the quantity takeoff.

Labor Rates

Labor rates include all labor costs such as wages, taxes, fringe benefits, and other direct labor costs. Calculation of labor costs varies greatly depending on geographic area, market conditions, and craft. Labor rates are normally one of the high-risk items in calculating a construction bid. Labor rates may be higher or lower in various geographic locations depending on labor availability, production levels, quality, and skills. Sources for labor rate information include local trade associations, printed labor tables, and historical labor rate data.

Information Sources. Local construction trade associations provide information concerning labor rates, benefits, and tax rates in their locality. See Figure 4-6. Estimators may also rely on labor wage rate tables provided by government agencies. As with labor quantities, construction companies may also have their own historical data concerning labor rates.

> *Initial and periodic submittal requirements can represent a significant cost to the general requirements of a project. Always check the submittal specifications for any costs that should be included when estimating a project.*

Size	Copper					Aluminum	
	TW	**THW**	**THHN**	**XHHW**	**USE**	**THW**	**XHHW**
14	5.5	6	5.5	6.5	6.5	—	—
12	6.5	7	6.5	7.5	7.5	—	—
10	7.5	8	7.5	8.5	8.5	—	—
8	9.5	10	9.5	11	11	—	—
6	11.5	12	11	13	13	8	9
4	13.5	14	13	15	15	10	11
2	15.5	16	15	17	17	12	13
1	—	18	16	19	19	14	16
1/0	—	20	18	21	21	16	17
2/0	—	22	20	23	23	18	19
3/0	—	24	22	25	25	20	21
4/0	—	26	24	27	27	22	23
250	—	28	26	29	29	24	25
300	—	30	28	31	31	26	27
350	—	32	30	33	33	28	29
400	—	34	32	35	35	30	31
500	—	36	34	37	37	32	33
600	—	38	—	39	39	33	34
750	—	40	—	42	42	34	35
1000	—	45	—	48	48	35	36

WIRE INSTALLATION LABOR UNITS*

* per thousand feet for one conductor, solid or stranded, 600 V

Figure 4-5. Estimators may use standard industry information concerning labor units for installing various materials.

Labor productivity can be increased by proper supervision and coordination of trade work schedules.

Multiplication of the labor quantities and the labor rates produces a labor cost for each portion of the construction project. Labor costs are entered into the ledger sheet, spreadsheet, or estimating software along with the material costs to provide a total cost per item. The duration of the job may affect labor rates because allowances must be made for pay raises for labor during the construction process. On lengthy construction projects with many craftworkers, estimators should include pay rate increases in the overall labor costs. Equipment and overhead may also be included in these calculations.

> ▲ *When estimating equipment rental rates, reduce the risk of rental rate increases at the time of construction by obtaining guaranteed price quotations for major equipment from rental companies.*

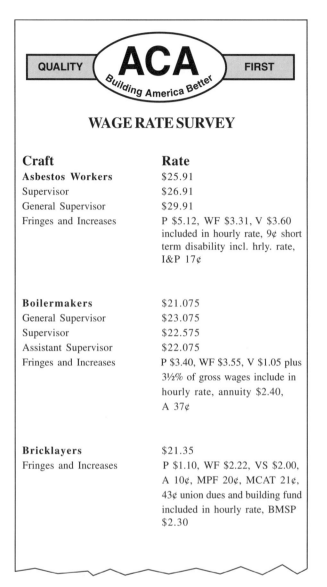

WAGE RATE SURVEY

Craft	Rate
Asbestos Workers	$25.91
Supervisor	$26.91
General Supervisor	$29.91
Fringes and Increases	P $5.12, WF $3.31, V $3.60 included in hourly rate, 9¢ short term disability incl. hrly. rate, I&P 17¢
Boilermakers	$21.075
General Supervisor	$23.075
Supervisor	$22.575
Assistant Supervisor	$22.075
Fringes and Increases	P $3.40, WF $3.55, V $1.05 plus 3½% of gross wages include in hourly rate, annuity $2.40, A 37¢
Bricklayers	$21.35
Fringes and Increases	P $1.10, WF $2.22, VS $2.00, A 10¢, MPF 20¢, MCAT 21¢, 43¢ union dues and building fund included in hourly rate, BMSP $2.30

Figure 4-6. Local construction trade associations provide information concerning labor rates, benefits, and tax rates in their locality.

EQUIPMENT

Equipment used on a construction project includes power equipment such as earth moving, hauling, and lifting equipment, temporary job site equipment, and communication equipment. See Figure 4-7. The quantities and costs of each piece of equipment are included in the overall job bid.

Power Equipment

Power equipment includes all machinery used in the construction process. The estimator determines the costs of owning and operating power equipment to determine direct costs and overhead costs for a project. A variety of factors are considered when calculating the hourly cost of owning and operating equipment. These costs are normally allocated to a standard quantity such as the cost per cubic yard for earth moving equipment or cost per ton for trucks.

Figure 4-7. Estimates include costs for equipment that is necessary to facilitate construction.

The total costs for hourly equipment operation include owning costs, operating costs, and wages. Owning costs are the fixed costs that are ongoing costs regardless of whether or not the equipment is in operation. Owning costs include depreciation, interest, insurance, and taxes. Operating costs are variable costs based on the amount of equipment use. Operating costs include fuel and lubricants, repairs including parts and mechanic costs, and tires.

Overhead costs for equipment are not included with owning and operating costs. Overhead costs for equipment include indirect costs such as supervision, transportation, storage, etc. Estimators may also choose to include items such as fuel, lubricants, parts, and mechanics in overhead costs rather than operating costs. Equipment overhead costs may be tied into operator labor costs and rated on a cost per hour of equipment operation.

Depreciation. *Depreciation* is the accounting practice of reducing the value of equipment by a standard amount each year. Each piece of equipment owned by a construction company has a depreciation cost per year. The cost of depreciation varies according to the original purchase price and the method of depreciation. Depreciation methods include straight-line, the sum-of-years' digits, and double-declining balance methods. Management and the company auditors determine the method used. Regardless of the depreciation method used, a cost for depreciation is included in equipment cost. The company auditor provides an annual depreciation cost for each piece of equipment. The depreciation cost can be prorated based on the amount of time each piece of equipment is used on a particular job.

Rental/Leasing. *Equipment rental* is the use of equipment for a fixed cost per a given unit of time. *Equipment leasing* is the use of equipment based on a contract that stipulates a set amount of time and a set cost for the equipment use. Renting or leasing equipment may be more economical for a job than purchasing the equipment. The costs of owning and maintaining equipment, including depreciation, are all included in the rental or lease price. When equipment is rented or leased, the estimator determines the amount of time the equipment is needed on the job site and the cost of renting or leasing. The time unit used may be by the day or month. Always check if the equipment rental or lease agreement includes an operator for the equipment. When renting equipment, the operator salary may be included in the hourly rate. For example, truck rental often includes the cost of the driver in the daily rental rate. Truck rental or usage rates may be priced on a unit price per ton.

Fuel. Equipment fuel includes gasoline, diesel fuel, or other energy sources used to operate a piece of equipment. Each piece of equipment should have some information concerning the amount and type of fuel consumed per hour of use. This information may be available from manufacturer literature or contractor historical data. In a manner similar to labor costs, the quantity of fuel is multiplied by the current market cost of the fuel to determine overall fuel costs per piece of equipment.

Cost Allocation. *Cost allocation* is the designation of costs to particular equipment. Cost allocation of equipment as an overhead cost to a project may be based on an internal rental rate based on usage factors for a specific job site. This internal rental rate may be based on company historical data or standard industry resources. Internal rental cost allocation rates may be based on hourly, daily, or weekly rates. The internal rental cost allocation rate may include factors such as the cost of purchasing the equipment minus any salvage value, the overall length of use of the equipment, the estimated idle and productive use of the equipment during its overall use time, and maintenance and insurance costs.

Temporary Job Site Equipment

The need for job site construction equipment and temporary construction materials is determined in conjunction with the building team. Temporary job site materials to aid in project construction include items such as field offices, access roads, signage, lumber for temporary fencing or sheds, heat and weather protection, and plumbing.

Field Offices. Large job sites require one or more trailers or offices to provide on-site recordkeeping and project management facilities. The field office contains office equipment, space for review of the plan drawings and specifications, telephone and other communication equipment, and possibly computer equipment. The cost for field office trailers is allocated to several jobs depending on company policies for length of use of each trailer. Job site trailers are rented, leased, or purchased and located where they do not interfere with the construction process.

Access Roads. An *access road* is a temporary road installed by the general contractor to facilitate transportation of material and craftworkers to and from a job site. An access road may be necessary depending on the job site location. Access road costs may include building and maintaining the road and removal of the road at the completion of construction. Access roads are commonly formed by digging a shallow path and filling it with stone.

Signage. Projects are often marked by a sign naming the general contractor, architect, and project. See Figure 4-8. Costs include sign design, erection, painting, maintenance throughout the life of the project, and removal at project completion. Signage costs are included in the temporary job site equipment portion of the estimate.

Figure 4-8. Costs such as signs naming the general contractor, architect, and project should be reviewed by the estimator and included in the overall estimate.

Fencing. Temporary fencing may be required at various locations on a job site for security, safety, etc. The estimator should include costs from a surveyor to get the proper fence lines. Information concerning temporary fencing is obtained in the general conditions of the specifications. For example, a specification may state that the entire area must be enclosed by a fence 7'-0" high above the respective grades as indicated on drawings attached to the specifications. The contractor also includes the cost of maintaining and providing fences for any existing trees. Existing trees are covered by additional drawings. Existing fence at the site may be used to enclose the job site. If the fence is built along the curb, free access from the street to hydrants, fire and police alarm boxes, and light standards must be maintained.

Fencing provisions are also made to allow for truck, worker, and equipment passage to and from the job site. Temporary fences are constructed and possibly painted to withstand the elements during construction.

Heat and Weather Protection. Temporary heat includes costs such as heating units, fuel, and possible ventilation during construction. See Figure 4-9. Heating costs vary depending on the construction project, location, and materials being installed. Projects such as commercial buildings with masonry, plaster, and other temperature-sensitive materials require additional heating equipment. Projects in cold climates also require additional heating equipment.

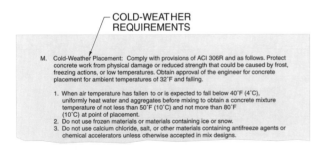

Figure 4-9. Contractors may be required to provide temporary heat in cold locations.

The contractor also provides weather protection. Weather elements include snow, sleet, rain, windstorms, tornados, hail, and floods. Weather elements are taken into consideration for the entire time the project is being constructed. Cost information is available from suppliers of heating units, weather protection materials, and contractor historical data. Weather protection material includes plastic sheets applied around areas to be heated.

The duration of the job is also considered for weather protection costs.

Plumbing. Local and state laws and union regulations must be considered in estimating a plumbing unit of cost. At least one toilet facility may be required for a given number of craftworkers to shelter them from view, the weather, and falling objects. The temporary facilities are commonly rented from a local supplier in a cost per month basis. The estimator must determine the length of the project and the number of workers to calculate a plumbing unit cost.

Communication Equipment

On large construction projects, production can be increased by using various communication equipment. Communication equipment allows field personnel at the job site to share current information concerning job site conditions with each other and the office staff. Communication equipment also helps all members of the construction team stay up-to-date on current job site conditions. Communication equipment includes voice communication and use of project-specific web sites.

Voice Communication. A variety of voice communication equipment is used on construction job sites. The most common voice communication equipment is the two-way radio. Cellular telephone use is also common. Two-way radio or cellular telephone equipment may be owned or rented depending on the equipment and the policy of the construction company. The estimator may be required to include voice communication equipment as a communication cost.

Project-Specific Web Sites. A project-specific web site uses Internet technology to create the ability for owners, architects, contractors, and suppliers to access job information through a web site established specifically for a construction project. See Figure 4-10. By using a password, those involved in the project can communicate concerning project schedules, change orders, and plan addenda, progress reports, and other important project information. Project-specific web sites are commonly priced based on the number of users and the length of the project. The cost for this web site should be included in communication costs.

> *Equipment should be operated near its maximum capacity to provide the largest possible benefit. The right equipment must be used for each job and all operators must be familiar with each piece of equipment.*

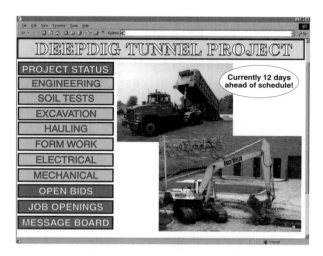

Figure 4-10. Project-specific web sites allow quick and accurate sharing of project information.

OVERHEAD

Overhead costs are expenses that are not attributable to any one specific operation. In addition to material quantities and labor costs, contractors incur a wide range of overhead costs. As a bid is generated, a certain percentage of the overhead costs are allocated to each construction project to ensure the company can continue operations. Overhead costs include temporary utilities, insurance, safety, bonds, permits, licenses, inspection fees, taxes, benefits, temporary job site support activities, environmental protection, progress reports, etc.

Temporary Utilities

Electric, natural gas, and other utility services are necessary on a construction site. Electrical power is provided for the field office, operation of power tools, lighting, and security. Natural gas may be necessary to heat a structure during construction. Costs for temporary utilities are included in the bid as overhead costs.

Light and Power. The contractor provides light and electrical power for the various trades and subcontractors under the jurisdiction of the contract. The light and power service must meet with approval of the county, township, or city in which the construction is taking place. This cost is determined by the price in relation to the size and duration of the project. The electrical contractor calculates costs for installation based on the amount of temporary power needed and the distance from the closest power source. Costs include the installation of the proper size meters, breaker panels, service boxes, and proper height electrical power poles.

Insurance and Safety

A contractor should not begin work until all required insurance has been obtained and the insurance policies have been approved by the architect and/or owner. The policies should be for the amount of coverage sufficient to cover any occurrences during construction. A contractor may be required to show receipts for complete payment of premiums to the architect and/or owner to indicate that the insurance policies are in force for the particular work of the contractor.

Many types, forms, and limits of insurance coverage are available. The estimator must take into account the costs to secure insurance policies that are required by the project contract or local, state, or federal law. The insurance coverage for various policies affects the final cost of construction. See Figure 4-11. The cost of these policies varies with the size, location, type, and hazards on the construction job. The unit cost applied to the estimate varies in percentage on each project. Contractor public liability insurance, fire and extended coverage insurance, and coinsurance are three types of insurance to consider during preparation of a bid. Other types of insurance included may be flood insurance, railroad protective insurance, or other specialized insurance coverage. Always consult with a reputable insurance agent for each project to determine insurance needs.

Figure 4-11. Insurance coverage for various policies affects the final cost of construction.

Contractor Public Liability Insurance. The contractor and subcontractor must carry public liability insurance covering bodily injury or death suffered as a result of any accident occurring from or by reason of, or in the course of operation of, the construction project. This includes accidents occurring by reason of omission or act of the contractor or any subcontractor or by people employed by the contractor or subcontractor. The amount of required coverage is usually determined by the size of the project and/or by the architect or the owner.

Fire and Extended Coverage Insurance. The contractor should carry insurance covering possible loss by fire, wind, tornado, and other natural disasters. This insurance covers a wide range of items and equipment and any damages done while work on the project is in progress. Special clauses are usually written into the policy by an insurance agent with a background in construction coverage liability. Fire protection insurance rates vary according to the rating organization having jurisdiction in a particular area.

Coinsurance. Contractors may choose to divide the insurance coverage among different companies in an attempt to obtain lower insurance rates. The estimator includes the actual coinsurance rates in the overhead expenses.

Safety Programs. Contractors may use a safety program for craftworkers and job site operations. Safety programs are an overhead cost that may be effective in reducing other overhead costs such as workers' compensation insurance. Contractors may hire their own safety training personnel or contract this service to a private firm or contractor association. Many contractors hold weekly meetings on the job site, referred to as tool box meetings, to meet requirements of the Occupational Safety and Health Act. Overhead costs for a safety program include common personnel employment costs such as wages and fringe benefits or contract costs based on safety training services provided.

Bonds

A *bond* is an insurance agreement for financial loss caused by the act or default of an individual or by some contingency over which the individual has no control. Bonding provides an assurance to the owner that the construction project will be completed as described in the specifications and print drawings. Bonding companies issue bonds for various amounts based on the

contractor qualifications and experience. Construction firms with good performance records may obtain bonds at a lower rate than inexperienced firms. Some large construction firms may be self-bonded, meaning that they do not purchase bonds from a third party, but guarantee the completion of the project with their own assets. Bonds include performance, completion, and street bonds.

Performance. A *performance bond* is a short-term insurance agreement that guarantees that a contractor will execute a construction job in the manner described in the contract documents. The owner and the architect of a project must be assured that the lowest competent bidder accepts the contract as bid. To ensure that the lowest competent bidder accepts the contract as bid, the architect and/or owner may require that the contractor or subcontractor post a performance bond. A performance bond can be for the entire cost of the job or a percentage of the job. See Figure 4-12. The amount of bonding required is noted in the specifications. Costs for bonds are obtained from a bonding agency.

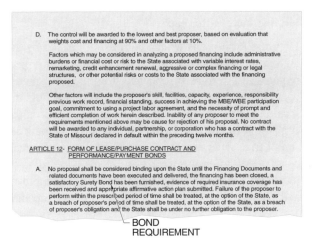

Figure 4-12. Bonding protects a project owner if a contractor or subcontractor is unable to complete a construction project.

Completion. A *completion bond* is a short-term insurance policy insuring the owner that a construction project will be fully completed and free from encumbrances and liens when turned over to the owner. Many contingencies such as weather, labor, and material availability may motivate the owner of a project to obtain additional protection ensuring that the contractor, once building has started, guarantees completion of the job. The additional protection can be provided by posting a completion bond. A completion bond gives the owner or bonding company the funds to hire another firm to complete the project if the contractor fails to complete construction as stated in the contract. The total amount of a completion bond is usually equal to the cost of the total estimate of a particular job. The amount of bonding required is noted in the specifications. Costs are obtained from a bonding agency.

Street. A *street bond* is a bond that guarantees the repair of streets adjacent to the construction project if they are damaged during construction. A street bond is purchased by the general contractor. The amount of bonding required is noted in the specifications. Costs are obtained from a bonding agency.

Permits, Licenses, and Inspection Fees

A variety of government permits, licenses, and inspection fees are required on construction projects. The permits, licenses, and inspection fees required vary depending on the project. An estimator should check the general conditions and be knowledgeable of local government codes and laws to determine which permits, licenses, and inspections are applicable and determine the cost for each.

Permits. A *permit* is written permission granted by the authority having jurisdiction. Every construction project requires that permits be obtained from the government agency having jurisdiction. Permits are obtained from the local building department. One necessary permit is the fire permit for the construction site required by the local fire district. After permits are obtained, inspectors track construction to ensure work is performed according to local building codes. Inspectors also issue final approval at the completion of the project prior to use or occupancy.

Licenses. A *license* is a privilege granted by a state or other government body to an individual or company to perform work or provide professional service. Contractors and subcontractors are often required to pay fees for a business or trade license. For example, electrical licenses or plumbing licenses are required by various localities. See Figure 4-13. Various localities have different requirements for licensing. The estimator should ensure that subcontractor bids are solicited only from firms meeting the necessary licensing requirements.

> ▲ *Material handling time can be reduced to a minimum by stacking material in the sequence of use with the first material needed on top.*

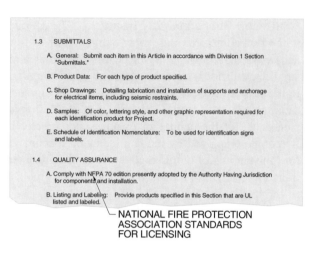

Figure 4-13. Various trades and contractors require licensing to perform work.

Inspection and Testing. A contractor may be required to furnish facilities and assistance for inspection, examination, and tests that inspectors may require. The contractor must also secure free access to factories and plants in which materials are being manufactured or prepared and to all parts of the construction project area for the inspectors. Contractors may also be required to give inspectors advance notice of preparation, manufacture, or shipment of any materials. Some overhead costs may be associated with these testing facilities and access issues. Concrete strength, soil compaction and composition, and torque on bolts are several items that may require testing.

Taxes and Benefits

In addition to labor costs, a variety of payroll taxes and benefits are calculated in the overhead costs for craftworkers and office staff. These overhead costs include Social Security, Medicare, workers' compensation, and health, welfare, and pension benefits. Additional state and local taxes may be taken into consideration by the estimator in calculating labor costs and overhead. Sales tax rates on material purchases are also included in overhead cost calculations.

Social Security. Contractors pay the assessments for Social Security insurance and unemployment benefits for their employees. The amount paid is determined by the wages that the employees of the contractor or subcontractor receive. The general contractor usually releases the owner from any responsibility for assessments that may be defaulted by the contractor or the subcontractor.

Workers' Compensation. The United States and the Canadian provinces require workers' compensation insurance protection. Federal laws cover federal employees and private employees in the District of Columbia. Workers' compensation provides employees with certain benefits in the event of injury or death and some diseases. A number of states have severe penalties for failure to carry workers' compensation insurance. Costs for workers' compensation insurance vary depending on the safety record of the construction company applying for the insurance. The estimator should consult an insurance agent to obtain costs for workers' compensation insurance coverage.

Health, Welfare, and Pension. Other contributions paid by contractors to craftworkers in addition to wages include health benefit coverage and pension plans. Costs for these coverages for craftworkers are based on local union agreements or company-based plans. Health and pension costs are commonly calculated on a cost per hour basis and are added to the labor cost and unit pricing. See Figure 4-14.

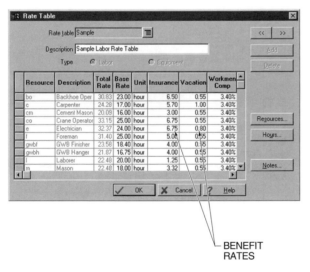

BENEFIT RATES

Figure 4-14. Benefit costs are added to the labor cost and unit pricing.

Sales Tax. States and municipalities have varying percentage rates of sales tax. Estimators should include sales tax as overhead expenses. Construction materials for nonprofit organizations may be exempt from sales tax. These exemptions vary according to state regulations. Estimators should consult the state taxing agencies for information concerning sales tax applicability on the construction project.

Temporary Job Site Support Activities

In addition to temporary job site facilities, temporary job site support activities are provided to ensure a successful construction project. Temporary job site support activities include job site office support staff, job site security for materials and workers, job site material handling, storage, and distribution, and provisions for worker protection for the duration of construction. Each of these are potential overhead costs to be considered by the estimator.

Office Support Staff and Facilities. The basic costs of the contractor and subcontractor office staff and office facilities are prorated across all jobs performed by the contractor or subcontractor. These costs include management, marketing, clerical services, and other costs associated with owning and operating an office, such as office supplies, utilities, insurance, custodial, and a variety of other costs. Management of the construction company adds a percentage onto the total bid of each job to pay for the cost of doing business.

Job Site Security. On many jobs, it is necessary for the contractor to provide adequate security for general day and night patrolling of the site, including Saturdays, Sundays, and holidays. Security helps prevent the theft of tools and materials. Security also prevents trespassing on the construction site that can lead to injuries for which the general contractor may be held responsible. Providing proper security may also help reduce some insurance rates.

Material Handling. Many materials on a construction site require handling and storage during the construction process prior to installation. These materials include gypsum drywall, hollow metal doors, finish hardware, millwork, and many other materials. Estimators may analyze the job site layout to include additional labor costs for unloading, storing, and distributing these materials.

Material handling on large jobs can amount to a substantial expense. The general contractor and subcontractors must agree on responsibilities for the unloading of various materials. This may be an important clarification that should be included in the scope of work letter and clearly understood by estimators prior to calculating labor and material handling costs.

Worker Protection. Worker protection includes job site security, safety equipment, and environmental protection. In hot weather, provisions must be made for supplying and distributing ice water or other hydrating liquids to workers on a job site. Costs include containers such as coolers and ice chests, cups, ice, and labor to distribute, collect, and clean the containers. In extremely cold weather, a heated space may be required to allow workers a place to warm up between work periods.

Environmental Protection

In estimating a project, the estimator must consider various environmental regulations surrounding the project area. The general conditions may require trash removal, general job site cleanup, erosion protection, and sedimentation protection. Estimators add these costs to overhead for the length of the job.

Trash Removal. Trash removal in most areas is governed by city and state codes. A careful study of codes, plans, and the construction site helps in determining the costs for trash removal. The general conditions may state that each contractor shall clean and gather all debris from the construction site daily and place the debris in containers or appropriate cans provided by the contractor at the entrance of the building unit or units, and have the containers removed by a trash removal contractor.

On some jobs, it may be necessary to construct a trash chute. The cost of erecting a trash chute for a multistory building increases as the height of the building increases. Costs associated with the removal of dirt and trash from a construction site are included in the general conditions of the estimate to ensure these costs are included in the overall bid.

Cleaning costs are incurred as part of the final preparation before turning a building over to the owner.

Cleaning. Cleaning includes final preparation of the building or structure for use and turning it over to the owner. This item is sometimes neglected by the general conditions but can be an important final cleanup cost. Windows are marked for safety reasons. When a building is complete, and before it is turned over to the owner, it must be presentable. The contractor may be required to clean all windows. An estimator should ensure this cost is in the bid along with the final cleanup cost.

Progress Reports

Some jobs may require a variety of progress reports for the owner and architect to track the work progress. See Figure 4-15. Overhead costs associated with generating progress reports include printing and possibly photography. Making a progress report requires the services of a competent person knowledgeable about the times required for each phase of the work and each trade.

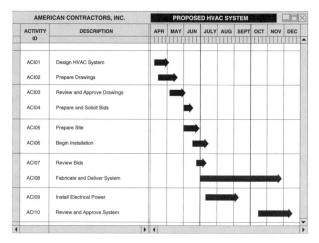

Figure 4-15. A contractor may be required to submit regular progress reports as a portion of the work, thereby creating additional overhead costs.

PROFIT

Profit is the amount of money in excess of project and company expenses. As with any business, the purpose of undertaking a construction project is to make a profit for the company after consideration of all project and overhead costs. Factors taken into consideration when estimating the amount of profit to be added to a bid include the overall work load within the company, the local construction market work load and outlook, the number of bidders on a particular project, and the risks inherent in a particular construction project. Companies may have a tendency to add a higher profit when the company has a large work load and does not necessarily need additional work. A low number of bidders on a project may allow a company to take a risk on a high profit margin. High-risk projects also require high profit margins because the chance of losing money is high. Three methods for calculating profit include percentage, cost-plus, and time and material basis. Companies may use other methods depending on company history and industry trends.

Percentage Basis

The *percentage basis* is a method of determining profit in which a percentage is added to the final estimated cost. The percentage basis method is a common method for adding profit to a job. The percentage is added after all material, labor, overhead, and equipment costs are calculated. This results in a fixed dollar amount being shown on the bid documents. For example, an estimator may add 3% onto the total project costs for profit. This number may be changed by management of the company based on a number of factors. The owner and architect know the exact cost of the project and the risk for profit or loss is with the contractor. Estimators may also add varying percentages of profit to individual bid items.

Risk Percentage. Contractors may measure the historical accuracy of their bids to assist in computing risks for various work and the proper profit percentages. A direct link exists between the level of profit in a construction project and the level of risk. A percentage of variance can be calculated by tracking estimated bids and comparing these figures to the actual job cost. A contractor using this system may break out the historical estimate accuracy on various items and assess risk and profit accordingly. For example, a contractor may determine by reviewing historical information that estimates for equipment have varied from actual costs by an average of 3.5%. This infers to the contractor that a minimum of 3.6% profit should be added to equipment pricing to balance profit and risk.

Cost Plus Basis

The *cost plus basis* is a method of determining profit in which costs are submitted for payment for labor, materials, equipment, and overhead, and an amount is added automatically for profit. Owners and architects may propose projects for contractors on a cost plus basis. This ensures a profit for the contractor, but does not ensure a fixed price for the owner.

Time and Material Basis

The *time and material basis* is a method of determining profit in which the owner pays a fixed rate for time spent on a project and pays for all materials. Small jobs may be performed on a time and material basis. Profit is built into the labor cost for time. This ensures a profit for the contractor but does not ensure a fixed price for the owner.

ESTIMATING SOFTWARE – DRAG AND DROP QUICK TAKEOFF

Quick takeoff is a Precision estimating software function that enables the estimator to enter quantities directly into the estimating software spreadsheet. Quick takeoff is the easiest way to place items into an estimate. The results are shown immediately because work is being done directly in the spreadsheet. When using the quick takeoff function, the estimating software performs the calculation using the cost information and formulas stored in the database. Quick takeoff is performed by applying a standard procedure. See Figure 4-16.

The general conditions contain information concerning earthmoving, hauling, and lifting equipment.

PRECISION ESTIMATING SOFTWARE – DRAG AND DROP QUICK TAKEOFF

1. Open the Quick Takeoff window by clicking the Quick Takeoff (⊞) button.

2. Select the items to be added to the spreadsheet by double-clicking through each level of the database (group, phase, and item). Double-click Group 1000.00 General Requirements, Phase 1101.00 Personnel: Supervision, and Item 10 Superintendent.

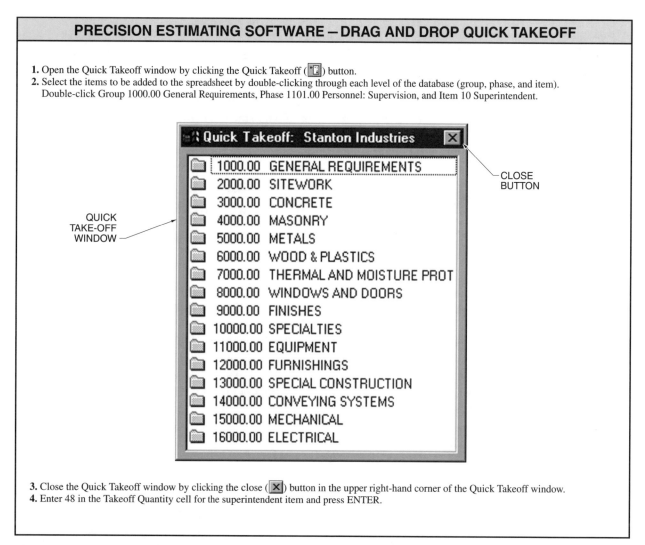

3. Close the Quick Takeoff window by clicking the close (☒) button in the upper right-hand corner of the Quick Takeoff window.

4. Enter 48 in the Takeoff Quantity cell for the superintendent item and press ENTER.

Figure 4-16. Quick takeoff enables the estimator to quickly add individual items and enter quantities directly into the estimating software spreadsheet.

Estimating

_____ **1.** _____ are the written agreements describing building components and various construction procedures included in the specifications for a construction project.

_____ **2.** _____ is the accounting practice of reducing the value of equipment by a standard amount each year.

_____ **3.** Equipment _____ is the use of equipment for a fixed cost per a given unit of time.
  A. rental   C. A and B
  B. leasing   D. neither A nor B

_____ **4.** Equipment _____ is the use of equipment based on a contract that stipulates a set amount of time and set cost for the equipment use.
  A. rental   C. A and B
  B. leasing   D. neither A nor B

_____ **5.** Cost _____ is the designation of costs to particular equipment.

_____ **6.** A(n) _____ road is a temporary road installed by the general contractor to facilitate transportation of material and craftworkers to and from a job site.

_____ **7.** _____ costs are expenses that are not attributable to any one specific operation.

_____ **8.** A(n) _____ is an insurance agreement for financial loss caused by the act or default of an individual or by some contingency over which the individual has no control.

_____ **9.** A(n) _____ is a privilege granted by a state or other government to an individual or company to perform work or provide professional service.

_____ **10.** A(n) _____ is written permission granted by the authority having jurisdiction.

_____ **11.** _____ is the amount of money in excess of project and company expenses.

_____ **12.** _____ is an estimating software function that enables the estimator to enter quantities directly into the estimating software spreadsheet.

T   F   **13.** Interior finish materials such as suspended ceilings, flooring, and gypsum drywall are taken off and priced based on square feet or square yards.

T   F   **14.** Earthwork is calculated in cubic yards of material to be moved or supplied.

_____ **15.** Items such as doors, windows, and fixtures are taken off by counting _____ installations.

_____ **16.** Labor pricing is determined based on the labor _____.
  A. quantity   C. A and B
  B. rate   D. neither A nor B

_____ **17.** A(n) _____ bond is a short-term insurance policy insuring the owner that a construction project will be fully completed and free from encumbrances and liens when turned over to the owner.

_____ **18.** The percentage _____ is a method of determining profit in which a percentage is added to the final estimated cost.

T   F   **19.** Labor rates include all labor costs such as wages, taxes, fringe benefits, and other direct labor costs.

T   F   **20.** The project owner provides light and electrical power for the various trades and subcontractors under the jurisdiction of the contract.

Wire Installation Labor Units

T   F   **1.** Aluminum conductors can be installed in less time that comparable copper conductors.

_____ **2.** No. 4 Cu USE requires _____ labor units for 500′ of conductors.

T   F   **3.** No. 14 through No. 8 Cu TW conductors require the same number of labor units as No. 14 through No. 8 Cu THHN conductors.

T   F   **4.** The chart may be used for determining labor units for installing solid or stranded conductors.

T   F   **5.** Wire installation labor units are given per 100 feet in the chart.

	WIRE INSTALLATION LABOR UNITS*						
Size	**Copper**					**Aluminum**	
	TW	**THW**	**THHN**	**XHHW**	**USE**	**THW**	**XHHW**
14	5.5	6	5.5	6.5	6.5	—	—
12	6.5	7	6.5	7.5	7.5	—	—
10	7.5	8	7.5	8.5	8.5	—	—
8	9.5	10	9.5	11	11	—	—
6	11.5	12	11	13	13	8	9
4	13.5	14	13	15	15	10	11
2	15.5	16	15	17	17	12	13
1	—	18	16	19	19	14	16
1/0	—	20	18	21	21	16	17
2/0	—	22	20	23	23	18	19
3/0	—	24	22	25	25	20	21
4/0	—	26	24	27	27	22	23
250	—	28	26	29	29	24	25
300	—	30	28	31	31	26	27
350	—	32	30	33	33	28	29
400	—	34	32	35	35	30	31
500	—	36	34	37	37	32	33
600	—	38	—	39	39	33	34
750	—	40	—	42	42	34	35
1000	—	45	—	48	48	35	36

* per thousand feet for one conductor, solid or stranded, 600 V

Activities
General Conditions

Name _____ Date _____

Activity 4-1 – Ledger Sheet Activity

Refer to Print 4-1 and Quantity Sheet No. 4-1. Take off the linear feet of steel tubular scaffolding required for the North and East walls of the commons building. The steel tubular scaffolding is 4′ wide.

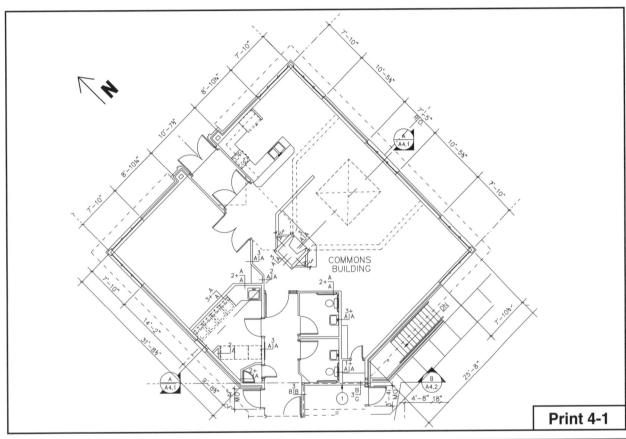

Print 4-1

QUANTITY SHEET

Sheet No. ____4-1____

Project: _____
Estimator: _____

Date: _____
Checked: _____

No.	Description	Dimensions				Unit		Unit		Unit		Unit
		L	W									
	Scaffolding											
	North wall											
	East wall											
	Total											

Activity 4-2 – Spreadsheet Activity

Refer to the cost data and Estimate Summary Sheet No. 4-2 on the CD-ROM. A new construction site requires a chain link fence, access road, and office trailer. The chain link fence must enclose property 180′ × 100′. The chain link fence is 6′ high. The access road is a temporary road 150′ long and 18′ wide consisting of 4″ of gravel. The office trailer is 20′ × 8′ and is needed for the project for a total of 2 years. Determine the total material, labor, and equipment cost for the chain link fence, access road, and office trailer.

COST DATA				
Material	**Unit**	**Material Unit Cost***	**Labor Unit Cost***	**Equipment Unit Cost***
Chain link fence	lf	3.71	5.04	—
Access road, gravel	sq yd	1.21	1.68	.34
Office trailer rental per month	ea	147	—	—

* in $

Activity 4-3 – Estimating Software Activity

Create a new estimate and name it Activity 4-3. Perform a quick takeoff of items **1101.10 Superintendent**, **1511.01 Temp Electricity**, **1705.10 Current Clean Up**, and **1730.10 Tools & Equipment**. These items are required for 4 months. Print a standard estimate report.

Sitework

Preparation of a construction site includes analysis of the existing site at grade level and below grade level, compatibility of the present site with the final site requirements, and the material and work necessary to make the change from existing to planned conditions. Soil engineers determine the bearing capacities and quality of the soil below grade. A structure that cannot be supported by surface soil requires a footing and foundation. Foundation material and methods are described in the specifications and shown on site plans, foundation plans, and detail drawings. After the majority of a building is constructed, the surrounding areas are finished to provide the appropriate access and landscape design.

SITE PREPARATION MATERIALS

Site preparation variables that affect the final bid and estimating procedures for a construction project include the type of soil, rock, and water at and below the surface of the property, the natural or existing contours of the property, existing structures, and the possible existence of hazardous materials. Factors involved in the preparation of a construction site include analysis of the existing site at grade level and below grade level, compatibility of the present site with the final site requirements, and the material and work necessary to make the change from existing to planned conditions. The estimator calculates the materials and amount of work necessary to make the planned changes. The materials to be studied include soil, rock, water, concrete, chemical, foundation, and plant materials, and paving systems.

Soil and Aggregates

Soil and aggregates include materials found at grade (surface) and below grade (subsurface) of a construction site. See Figure 5-1. Materials found at a construction site include rocks, boulders, gravel, sand, clay, silt, and soil. Solid rock normally provides a stable bearing foundation. Decayed rock may be compact and hard or fully decayed and soft. Loose rock has been detached from the rock layer in which it was originally formed.

Rock includes soft rock such as shale, medium-hard rock such as slate, and hard, sound rock such as schist or other metamorphic bedrock materials. Subsurface exploration information should be studied to determine if rock is present under a site because it may result in additional excavation costs if removal is necessary. For estimating purposes, rock is classified as soft rock, medium-hard rock, and hard, sound rock.

A *boulder* is a rock that has been transported by some geological action from its site of formation to its current location. *Gravel* is pieces of rock smaller than boulders and larger than sand. *Sand* is a loose granular material consisting of particles smaller than gravel but coarser that silt. Sand is classified as fine, medium, or coarse based on grain size. *Silt* is fine granular material produced from the disintegration of rock. *Clay* is a natural mineral material that is compact and brittle when dry but plastic when wet. Soil is measured for its depth and compaction at the construction site.

SOIL							
Division	**Symbols**			**Description**	**Value as Foundation Material**	**Frost Action**	**Drainage**
	Letter	**Graphic**	**Color**				
Gravel and gravelly soils	GW	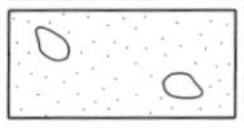	Red	Well-graded gravel or gravel-sand mixture. Little or no fines	Excellent	None	Excellent
	GP	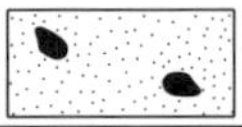	Red	Poorly-graded gravel or gravel-sand mixture. Little or no fines	Good	None	Excellent
	GM	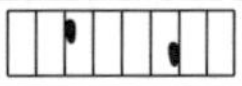	Yellow	Silty gravels, gravel-sand-silt mixture	Good	Slight	Poor
	GC	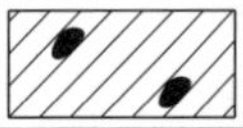	Yellow	Clayey-gravels, gravel-clay-sand mixtures	Good	Slight	Poor
Sand and sandy soils	SW	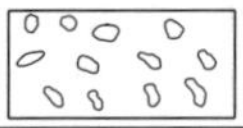	Red	Well-graded sands or gravelly sands. Little or no fines	Good	None	Excellent
	SP		Red	Poorly-graded sands or gravelly sands. Little or no fines	Fair	None	Excellent
	SM	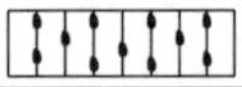	Yellow	Silty sands, sand-silt mixtures	Fair	Slight	Fair
	SC		Yellow	Clayey sands, sand-clay mixtures	Fair	Medium	Poor
Silts and clays	ML	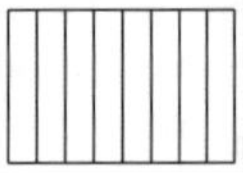	Green	Inorganic silts, rock flour, silty or clayey fine sands, or clayey silts with high plasticity	Fair	Very high	Poor
	CL	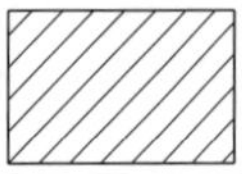	Green	Inorganic clays of low to medium plasticity, gravelly clays, silty clays, lean clays	Fair	Medium	Impervious
	OL		Green	Organic silt-clays of low plasticity	Poor	High	Impervious
Highly organic soils	Pt	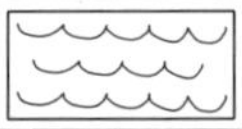	Orange	Peat and other highly organic soils	Not suitable	Slight	Poor

Figure 5-1. Various subsurface materials have different qualities for support of construction projects.

Soil is classified as light soil, medium soil, heavy soil, and hardpan. *Light soil* is soil that is readily shoveled by hand without the aid of machines. Light soil includes gravel and sand. *Medium soil* is soil that can be loosened by picks, shovels, and scrapers. Medium soil includes cohesive soil such as clay and adobe. When clay is encountered, an excavation is usually made several inches deeper than for sand, sandy soil, or gravel to allow for fill materials such as crushed gravel to provide a stable base. *Heavy soil* is soil that can be loosened by picks, but is hard to loosen with shovels. Machinery such as a backhoe is used to move heavy soil. Heavy soil includes compacted gravel, small stones, and boulders. *Hardpan* is a hard, compacted layer of soil, clay, or gravel. Hardpan includes boulders, clay, cemented mixtures of sand and gravel, and other substances difficult to loosen with a pick. Light blasting is used to loosen hardpan to make it easier for excavating equipment to pick up and load.

Angle of Repose. *Angle of repose* is the slope that a material maintains without sliding or caving in. Different soil varies greatly in its ability to hold in place at the edge of an excavation. *Excavation* is any construction cut, cavity, trench, or depression in the earth's surface formed by earth removal. Clay soil, unless subjected to the action of running water, stands almost vertically at the edge of an excavation. Soil with a high sand content slides in a manner that makes the bank of an excavation become a long slope. See Figure 5-2. Estimators must take the angle of repose into account when determining the need for temporary shoring during excavation and construction activities. The depth of the excavation, soil, angle of repose, and safety requirements are consulted to determine shoring needs. Various government regulations apply to shoring requirements for various depths of excavation.

Soil in its natural state is generally closely compacted. When soil is disturbed by excavation and loaded or piled, its volume increases due to swell. *Soil swell* is the volume growth in soil after it is excavated. Soil swell can increase the volume of soil by 50%. See Figure 5-3.

ANGLE OF REPOSE			
Soil	Degrees from Horizontal		
	Dry	Moist	Wet
Sand	20 – 35	30 – 45	20 – 40
Earth	20 – 45	25 – 45	25 – 30
Gravel	30 – 48	—	—
Gravel, sand and clay	20 – 37	—	—

Figure 5-2. The angle of repose for various soil affects excavation estimates.

SOIL SWELL*		
Soil	Before Excavation	After Excavation
Earth clay	100	125
Sand and gravel	100	150
Broken stones	100	150
Free stone	100	125
Rock	100	150

* in percent of initial volume

Figure 5-3. Soil swell adds to the cubic yard calculations for excavation.

Geotextiles. *Geotextile* is a sheet of material that stabilizes and retains soil or earth in position on slopes or other unstable conditions. Geotextile material is applied after rough grading is complete. Geotextile material may be covered with a thin layer of soil, rock, or other material to hold it in place. Geotextile material can be used to divert ground water and provide subsurface drainage and erosion control.

Hazardous Materials. A *hazardous material* is a material capable of posing a risk to health, safety, or property. Hazardous material handling includes the removal and proper disposal of any identifiable hazardous materials on a construction site. Soil analysis may indicate the existence of hazardous materials on the construction site. Hazardous materials include solid waste, toxic chemicals, and radiation. The Environmental Protection Agency (EPA) has identified specific hazardous

waste materials and classified them according to corrosivity, ignitability, reactivity, and toxicity. Special remediation steps must be taken to contain or safely remove hazardous material when found. *Remediation* is the act or process of correcting. The location(s) of contaminated soil areas and location(s) for installation of geotextiles and groundwater monitoring wells are indicated on the site plan and specifications. See Figure 5-4.

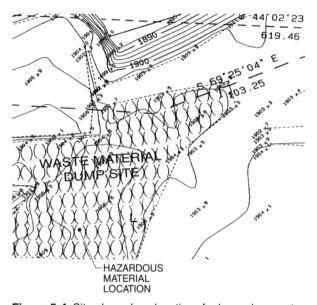

Figure 5-4. Site plans show locations for hazardous materials to be monitored or removed.

Soil Amendments. A *soil amendment* is a material added to existing job site soil to increase its compaction and ability to maintain the desired slope at the completion of construction. In some cases, existing soil characteristics do not allow for achievement of specified soil stabilization requirements. Soil amendments such as clay, slurry, and other locally available materials may be mixed into existing soil to achieve the required results. *Slurry* is a mixture of water and fine particles.

SITE PREPARATION METHODS

Soil engineers determine the bearing capacities and quality of the soil below grade. Removal of the appropriate trees and shrubs and protection of the trees and shrubs that are to remain are planned. Rock removal may require heavy blasting. Operating engineers grade, remove, add, and compact the soil on the site in preparation for new construction. Below-grade and at-grade drainage systems are laid out and installed.

Subsurface Investigation

Subsurface investigation is the analysis of all materials below the surface of a construction site to a predetermined depth. Soil engineers analyze samples taken from the construction site to determine their composition. A report from the soil engineer to the owner and architect describes the various materials below the surface at specific points on the construction site. Determination of the composition of the subsurface materials provides valuable information for structural engineers when selecting foundation systems. Structural engineers calculate the live and dead loads that the structure places on the bearing soil at the construction site. A *live load* is the predetermined load a structure is capable of supporting. A *dead load* is a permanent, stationary load composed of all building materials, fixtures, and equipment permanently attached to a structure. Loading information is compared to the soil engineer samples to determine the steps necessary to ensure that the footings and foundation provide adequate support to bear all imposed loads. Subsurface information is used by estimators when estimating drilling or excavation work. Information considered in the estimating process includes the number, depth, and diameter of holes to be drilled, and rock and soil type.

Core Drilling. *Core drilling* is the process of making holes in the ground to retrieve soil samples for analysis. Soil engineers determine a layout for drilling at the construction site. A core drill is used to drill holes into the earth at predetermined points. The layers of soil, rock, and other subsurface materials are measured for depth and analyzed for type. The holes are often made to determine the depth at which solid bearing is reached.

Groundwater. *Groundwater* is water present in subsurface material. A *well point* is a pipe with a perforated point that is driven into the ground to allow groundwater to drain. Well point pumps are used to dewater an excavation by pumping the water up vertical (riser) pipes. Horizontal pipes are connected to the vertical pipes to discharge the water at a location away from the work area. This maintains a low water level at the excavation, which allows construction work to continue. Piping may be permanent or temporary depending on the type of construction.

Groundwater can create problems unless the water flow is controlled. Improper handling of groundwater can cause sand excavations to be much more costly than rock excavations. Removing groundwater is a cost included in a bid. Dewatering contractors bid on dewatering systems by calculating items such as cost for driving well points, installing and maintaining vertical and horizontal pipes, maintaining and fueling pumps, and labor involved in operating the system. Removal of a temporary dewatering system may also be included in the dewatering cost.

Demolition

A construction site may have structures to be demolished and removed. *Demolition* is the organized destruction of a structure. The cost of demolition is included in the bid along with the cost of hauling debris from the site. Demolition costs include heavy equipment operation, equipment ownership or leasing, labor, and overhead. Hauling costs include loading of trucks, truck rental or leasing, labor, dumping fees, and overhead. Demolition contractors may need to analyze the structure being removed to determine if hazardous materials exist in the structure.

Clearing. *Clear and grub* is the removal of vegetation and other obstacles from a construction site in preparation for new construction. Trees and large vegetation are removed from areas indicated on the site plan. A site visit may be required to observe the types and quantities of vegetation to be removed.

Shoring

Shoring is wood or metal members used to temporarily support formwork or other structural components. Shoring provides temporary support for various loads during the construction process. See Figure 5-5. Shifting, unstable soil conditions, construction sites closely set to existing structures, or other factors may create the need for temporary or permanent shoring at a construction site. Shoring may be accomplished with removable boxes or jacks, various types of piling, grouting and other soil stabilization techniques, or soil and rock anchors. Estimators must ensure that the shoring system meets industry and government agency standards for the application selected. Shoring must be sufficient enough to provide complete protection for workers and materials based on the depth of the excavation and the materials being supported.

> ▲ *Trees and shrubs that are to remain during construction should be marked prior to clearing operations and verified with the site plan. Trees and shrubs that are to remain should be protected from accidental damage.*

Figure 5-5. Shoring is wood or metal members used to temporarily support formwork or other structural components.

Sheet Pile. A *pile* is a structural member installed in the ground to provide vertical and/or horizontal support. *Sheet pile* is interlocking vertical support driven into the ground with a pile driver to form a continuous structure. Sheet piling is designed to retain the horizontal thrust of earth that could cause excavation walls to collapse. Vertical retaining walls of wood or steel may be used when excavating adjacent to an existing structure or where protection from cave-in is necessary. The most common forms of sheet piling include wood planking (lagging) placed horizontally between vertical I-beams driven into the ground or interlocking steel sheets. Precast concrete sheet piling may be driven when sheet piling is to become a permanent part of a structure.

A pile driving rig is used to drive piles to the point of refusal where the earth is soft enough to permit driving. See Figure 5-6. This prevents sheet piling from being moved by horizontal thrusts. Sheet piling may also be installed in a previously dug hole. Sheet piling installed in a previously dug hole is braced, shored, anchored to the bank, or backfilled.

Temporary Soil Stabilization. *Temporary soil stabilization* is the process of holding existing surface and subsurface materials in place during the construction process. Various methods are used to hold soil adjoining an excavation in place during the construction process. Cementious grout or slurry may be shot or pumped into adjoining areas to stabilize the soil. Grout may be shot onto banks following excavation. Holes may be drilled prior to excavation and pumped with grout or slurry. After the grout or slurry sets, excavation may take place with the bank stabilized. Refrigeration systems may be used to freeze the soil to retain a stable bank where water and soil conditions permit.

Figure 5-6. Sheet piling provides support around the perimeter of an excavation.

Permanent Soil Stabilization. *Permanent soil stabilization* is the process of providing long-term stabilization of the ground at a construction site through mechanical compaction or the use of infill materials. *Infill material* is material including sand, gravel, and crushed rock installed to increase the stability of the soil. Mechanical methods of permanent soil stabilization include dynamic compaction and vibratory consolidation. Infill materials are introduced into the soil or used to replace existing soil where existing soil conditions do not permit the proper compaction level to be reached. The gradation, quality, and final compaction of infill material affects the final stability. Standard infill materials include sand, gravel, and crushed rock with a maximum particle size of 2.5″, and concrete and soil-cement mixes.

Cofferdams. A *cofferdam* is a watertight enclosure used to allow construction or repairs to be performed below the surface of water. Cofferdams are comprised of a series of sheet piles that are driven to form a cell or interlocking cells. Cofferdams provide protection for projects built far below grade or below the surface of water such as bridge piers. The material inside the cell, such as soil, rock, or water, is removed to allow worker access during construction. Cofferdams are commonly used in marine construction to allow access to river and lake bottoms in the construction of bridges and dams. See Figure 5-7.

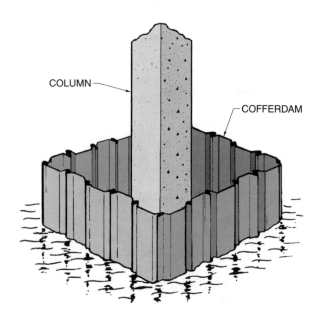

Figure 5-7. Cofferdams allow foundation construction in rivers, lakes, and other marine applications.

Site Utilities

Site utilities for a project include items such as storm sewers, sanitary sewers, water service, fire protection service, natural gas service, electrical service, and telephone service. Each site utility is identified on the site plans. See Figure 5-8. Site preparation for utility installations must take place prior to placement of foundations and streets to ensure full accessibility to the construction site.

Storm Sewers. A *storm sewer* is a piping system designed to collect storm water and channel it to a retention pond or other means of removal. Storm sewers are commonly made of precast concrete pipe. The sections of precast concrete pipe are set in place according to plans and specifications that provide information concerning elevation, slope, and connection. Excavations are made for placement of the pipe. Gravel or rock may be placed in the trench and shoring may be required. Cranes and other excavation equipment lift pipe sections and set them in place.

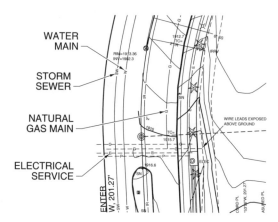

Figure 5-8. Site plans indicate the location of utilities such as electrical service, storm sewers, and natural gas and water mains.

Sanitary Sewers. A *sanitary sewer* is a piping system designed to collect liquid waste from a building and channel it to the appropriate removal or treatment system. Sanitary sewer systems may be formed of precast concrete or plastic pipe. All sanitary sewer connections must be secure at the building, throughout the piping system, and at the point of connection to the treatment system or removal piping.

Water, Fire, and Gas Mains. Piping and connection information for water supply piping, fire hydrants, and natural gas piping is shown on site plans. Water supply piping from the main service line to the structure may be copper, cast iron, or plastic. Fire hydrant piping is most commonly made of cast iron. Natural gas piping may be copper, cast iron, or plastic (polyethylene). Site plans and specifications provide information concerning water, fire hydrant, and natural gas piping. Local building and fire codes should be consulted to ensure that the specified piping is correct according to the applicable code.

Electric and Telephone Service. Electric and telephone service may be underground or overhead. Connection points and possible transformer needs are shown on the site plan and possibly the electrical plans and specifications. For underground installations, trenches are placed prior to other surface construction such as pavement and sidewalks. Trenchless technology may be used to install electric and telephone cables by means of drilling through the soil without the need for further excavation.

A site inspection is used to determine that the site drawings are complete. All existing utilities are checked off on the site plan as they are located. The site plan is marked up for further investigation if discrepancies are found.

SITE PREPARATION QUANTITY TAKEOFF

After the existing conditions, materials, and location of new construction is determined, the estimator takes new construction information from specifications and site plans. New construction information includes placement of new structures, utilities, surface grading, paving, curbs, walks, and landscaping.

Site plans are referred to as site survey, site map, site drawings, and civil drawings. Site plans include topographical, paving, landscape, and detail drawings for drainage and piping. Site plans are commonly noted with the prefix C (civil). For example, drawing sheet C1.1 is the first page of the site plans (civil drawings).

Earthwork Estimation

Earthwork estimating calculations include the amount of excavation, cut and fill, final grading, and trenching required on a construction site. Estimators should review the specifications, including the geotechnical report and site plans prior to beginning quantity takeoff. The estimator studies the site, topographic maps, plans, and specifications, which may reveal opportunities for increased efficiency in earthwork operations.

The quantity takeoff for excavation is most commonly calculated in cubic yards. The quantity takeoff may be calculated in cubic feet when the excavation requires the removal of a few inches of soil or work performed by hand labor. The volume of material to be removed or added must be calculated accurately for the best determination of all costs involved. Additional earthwork items taken into account include the costs involved in backfilling, utility line excavation, and cradling. *Cradling* is the temporary supporting of existing utility lines in or around an excavated area to properly protect the lines. Cradling is accomplished by placing timbers or shoring under utility lines and jacking up the timbers or shoring as the excavation goes deeper to keep the lines at their original level. Utility companies may also be contacted to install protective devices on overhead electrical lines to prevent arcing. See Figure 5-9.

Gridding. *Gridding* is the division of a topographical site map for large areas to be excavated or graded into small squares or grids. Grids may be drawn on an overlay sheet of tracing paper, directly on the site plan drawing, or charted on a computer screen when CAD drawings are available. The size of the grid is determined by

the nature of the terrain. Grids may represent squares of 100′ on each side if the terrain is gradually sloped. Grids may represent squares of 25′ on each side if the terrain is irregular. The small grid size gives a more accurate picture of the cut and fill needed for a particular site. See Figure 5-10.

Figure 5-9. Utility protection costs such as electrical arc protection are included in sitework estimates.

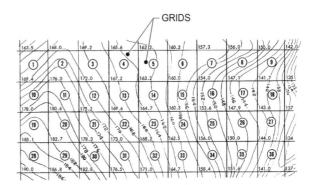

Figure 5-10. Gridding is the division of a topographical site map into small squares or grids to assist in cut and fill calculations.

The approximate elevation at each corner of each grid is established by using the nearest contour line. A *contour line* is a dashed (existing grade) or solid (finished grade) line on a plot plan used to show elevations of the surface. Interpolation between adjoining contour lines may be required when the corner of a grid does not fall directly in alignment with a contour line. Cut and fill averages are calculated for each grid. The difference between the totals for the cut and fill indicate the amount of fill or cut that must be done on the site. Cut and fill numbers are entered in a ledger sheet, spreadsheet, or estimating software cell. See Figure 5-11. The total cut or fill value is calculated and multiplied by the excavation cost. Soil and rock types may also be taken into consideration when determining a final cost for excavation.

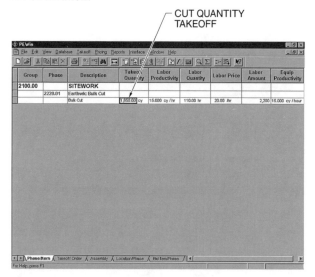

Figure 5-11. Gridding calculations determine the amount of cubic yards for cut and fill operations.

Excavation. Excavation is any construction cut, cavity, trench, or depression in the earth's surface formed by earth removal. Engineering improvements include footings, foundations, basements, tunnels, sub-stories, ramps, exterior stairways, areaways, pipe trenches, and pits. General excavation includes all excavation, other than rock and water removal, that can be done by mechanical equipment or any general piece of equipment such as a bulldozer, clam shell crane, backhoe, scraper, power shovel, or loader. The use of trucks to remove the excavated material is also included in this portion of the estimate calculations.

Calculating the volume of excavation material may be accomplished using the cross-section method or the average end area method. The cross-section method is used when the shape of the excavated area is roughly square or rectangular. The average end area method is used when the sides of the excavation are irregularly-shaped and not parallel. The estimator determines the method used based on the shape of the excavation.

When using the cross-section method, the volume excavated (in cubic yards) is calculated by determining the average depth of excavation from various known points and multiplying by the total surface area. The

excavation volume for a grid square is calculated by determining the difference between the existing elevation and the proposed elevation at each corner of the grid. See Figure 5-12. For example, a 75′ × 75′ square grid with corners at elevations of 90.85′, 95.02′, 87.28′, and 85.13′ requires excavation to 80.5′. The planned elevation of 80.5′ is subtracted from each corner elevation to determine the difference between the existing elevation and planned elevation at each corner. The four differences (depths) are added and the sum is divided by four to obtain the average excavation depth of 9.07′ at each corner. The average excavation depth (9.07′) is multiplied by the surface area of the square (75′ × 75′ = 5625 sq ft) to determine the cubic feet of excavation (5625 sq ft × 9.07′ = 51,018.75 cu ft). The total cubic feet of excavation is divided by 27 to obtain 1889.58 cu yd of excavation (51,018.75 cu ft ÷ 27 = 1889.58 cu yd).

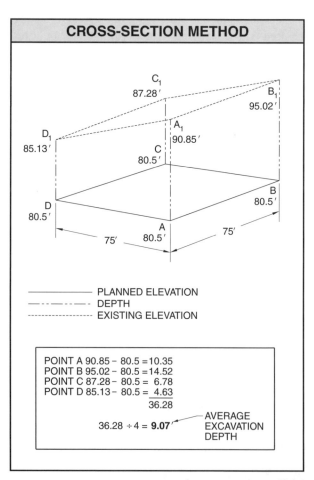

Figure 5-12. Cross-section volume calculations use average excavation differences in the corners of an area multiplied by the area of excavation to determine the volume excavated.

When using the average end area method, the average area of each end of an excavation is calculated. The difference between the existing and planned elevation is determined in each corner of the excavation area. Imaginary triangular planes are drawn representing the two ends of the excavation. The area of each triangle is calculated in square feet. The two areas are added and divided by two to determine the average end area. The average end area is multiplied by the length of the excavation to determine the total cubic feet of excavation. The total cubic feet is divided by 27 to obtain the number of cubic yards. For example, a 50′ long excavation area has a width of 12′ at one end and 6′ at the other. Both edges of the excavation area require 10′ of cut. See Figure 5-13. The area of the triangular planes created by the ends of the excavation area equal 60 sq ft (½ × [12′ × 10′] = 60 sq ft) and 30 sq ft (½ × [6′ × 10] = 30 sq ft) respectively. These two values are added and divided by two to determine the average end area ([60 sq ft + 30 sq ft] ÷ 2 = 45 sq ft). The average end area is multiplied by the length of the excavation (50′) to determine the cubic feet of excavation (50′ × 45 sq ft = 2250 cu ft). The total cubic feet is divided by 27 to obtain the number of cubic yards (2250 cu ft ÷ 27 = 83.33 cu yd).

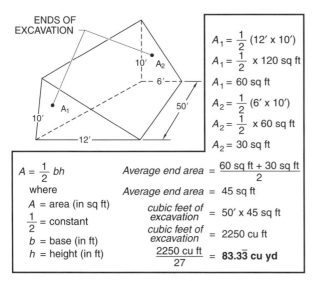

Figure 5-13. Average end area calculations are used for irregularly-shaped excavations.

When a subsurface structure is to be built, excavation is often larger than the actual structure to allow space for craftworkers to get to the work area for activities such as setting and removing concrete forms. *Overdig* is the amount of excavation required beyond the dimensions of a building to provide an area for construction activities. This additional work area must be filled with earth at the completion of construction. *Backfilling* is the replacing of soil around the outside foundation walls after the walls have been completed. When backfilling, the estimator must ensure that the ground does not contain large boulders or debris that could damage the new construction. Backfilled areas must also be compacted at the conclusion of the project. The required compaction levels are stated in the specifications.

Excavation done by hand, special machines, blasting, or by the use of special methods is considered special excavation. An example of special excavation is a pipe trench required for a water line. A pipe trench is normally dug by a trencher. A *trencher* is a machine with a conveyor to remove the soil as it digs a narrow, long hole into the earth. The hole is normally 8″ to 12″ wide and several feet deep, depending on soil conditions and the depth of trench required. Another example of special excavation is the installation of telephone, electrical, and utility poles. Pole excavation may require a vertical boring rig.

A *payline* is a measurement used by excavation contractors to compute the amount of necessary excavation as stated in the specifications. The payline establishes the exact amount of excavation for which the contractor is responsible. The amount of excavation may be described in the specifications as: the payline shall be 2′ from the top edge of the footing out, and then up on a slope as required by OSHA excavation standards for the proper soil type.

Slope cut is an inclined or sloped wall excavation. A slope cut is used for deep excavations in soft, shifting soil. The angle of a slope cut is determined by the angle of repose for soil in the area according to standard safety guidelines.

Excavated earth is stockpiled on the site and represents a cost. The cost of moving earth around a construction site is added to the other excavation costs. The unit cost and total cost depend on the type, size, and efficiency of the equipment used.

When excavations are required in rock, a series of regularly spaced holes are drilled into the rock layers. Explosive charges are carefully placed into the holes. The spacing, number of holes, and amount of explosives in each blasting charge hole is determined by an analysis of the type and strength of the rock to be removed. After blasting, the loose rock is removed by the appropriate excavating equipment. See Figure 5-14. Blasting costs include drilling equipment, blasting caps and explosive charges, and removal of the blast material.

Figure 5-14. Blasting with blasting caps and explosive charges loosens rock for excavating.

Excavation costs are calculated based on the cubic yard. Unit cost per cubic yard of excavation, trenching, or other site preparation activity is based on expenses including the cost of machine ownership or rental, fuel, labor to operate the machine, possible cost of transporting the machinery during the course of construction, and other equipment overhead costs. The total expense (machine, labor, transportation, overhead) is divided by the number of cubic yards of excavation to determine the unit cost of excavation per cubic yard. The unit cost of excavation per foot may also be determined from industry resource materials or company historical records. Excavation estimate values are entered in a ledger sheet, spreadsheet, or estimating software cell.

Grading. *Grading* is the process of lowering high spots and filling in low spots of earth at a construction site. The estimator reviews site drawings and uses gridding to determine earthwork calculations for cut and fill. The cross-section method of volume calculation is commonly used for cut and fill calculations. Negative numbers are used to indicate fill in areas where the planned elevation is above the existing elevation. A *zero line* is a line connecting points on a topographical map where existing and planned elevations are equal. Depending on the grid layout, some grids may require all cut or all fill, and some grids may require cut and fill. The estimator may calculate these as separate portions of the takeoff to simplify overall cut and fill calculations.

The contractor may be required to buy fill (borrow) to bring the level to the specified grade if the amount of fill is greater than the amount of cut. This represents an additional cost. Hauling and removal of additional cut materials is also a cost if required. These costs are based on cost per cubic yard calculations derived from industry resource materials or company historical data. *Final grading* is the process of smoothing and sloping a site for paving and landscaping. Final grading quantities are taken off in square feet or square yards of surface to be graded.

Depending on the soil and surface condition, steps may be required to control the creation of dust and to minimize soil erosion during the grading process. Costs associated with this include water spray on dusty areas and temporary erosion protection of plastic fencing or straw bales. Grading estimate values are entered in a ledger sheet, spreadsheet, or estimating software cell.

Tunneling. Tunneling costs include the boring of the tunnel, bracing and jacking of the tunnel walls, tunnel wall lining, and removal of the earthen debris. Tunnel boring costs vary depending on the material through which the tunnel is bored (rock, soil, or other material). Tunnel walls may require steel or concrete lining and are shown on architectural drawings or specifications. Other tunnel considerations include blasting cost, dewatering, and safety provisions including air supply and worker protective equipment. Recent developments in tunneling use systems in which a large conduit is put in place and followed by a carrier pipe. Other trenchless technology systems allow for placement of piping underground without the need for tunneling. Tunneling estimate values are entered in a ledger sheet, spreadsheet, or estimating software cell.

Demolition Estimation

Quantities calculated for demolition estimation include the sizes and types of structures to be removed, access to the demolition site, protection of surrounding structures, and structural analysis of the structure to be demolished to determine the demolition sequence and possible need for temporary shoring. In addition, analysis may be required for the possible presence of hazardous materials, removal of existing hazardous materials, calculation of the amount (cubic yards) of demolition material to be removed, the hauling distance from the site to the nearest disposal site, dumping fees, and overhead costs. Each of these items is calculated based on a study of the structure to be demolished through personal observation and study of any existing architectural drawings. Pavement removal is taken off as the number of square yards to be removed.

Selective demolition of individual components of a structure is taken off in a variety of methods. Quantities may be taken off in linear, square, or cubic measure, or components may be taken off as individual units. Selective demolition takeoff is a specialized area that requires experience and expertise in selective demolition work.

Implosion is used for large structures when economically feasible. Structural engineers analyze the existing structure and determine locations for the placement of explosive charges. Certain structural elements may be weakened or strengthened prior to activating the explosive charges to control the direction of debris during blasting. Implosion contractors are specialized in performing this analysis and estimating. Costs include the analysis, explosives, labor for structural modifications prior to blasting, debris removal, and overhead. Due to the specialized nature of this work, implosion contractors use ledger sheets, spreadsheets, and estimating software that is custom-designed for this type of work.

Clearing and Grubbing. Clearing and grubbing costs are determined by the acre (hectare for metric calculations). Local regulations concerning burning, burying, or hauling of the removed vegetation are key elements in determining costs. Costs include removal equipment and disposal costs. Some provisions may also be required for temporary erosion control fencing or straw bales to prevent loose soil from being washed into the groundwater supply. Clearing costs are entered in a ledger sheet, spreadsheet, or estimating software cell.

Hazardous Material Handling. Subcontractor bids may be required when hazardous materials must be removed or remediated. Removal and remediation must be performed by a company that meets local, state, and federal regulatory requirements for licensing, worker training, and recordkeeping. Costs included in hazardous material handling include meeting licensing and training requirements, specialized machinery for removal and cleanup, specialized worker protective equipment, additional fees associated with transporting hazardous materials, and fees required for safe disposal of removed material. All of these costs vary depending on the severity, contamination, and location of the construction site. These considerations are required when working in areas of past fuel handling or storage such as airports or gasoline service stations. Hazardous material

handling estimate values are entered in a ledger sheet, spreadsheet, or estimating software cell.

Shoring Estimation

Items to be considered for shoring include the linear feet, type, and required strength of the shoring, and the depth of the excavation. Small excavations may be shored by the installation of trench boxes. See Figure 5-15. A *trench box* is a reinforced wood or metal assembly used to shore the sides of a trench. Trench boxes are set into an excavation by a crane to prevent worker injury and equipment damage in case of a cave-in. Costs for trench boxes include the ownership or rental costs and installation costs. Trench boxes are normally set in place and removed with a crane. Costs vary for piling installations and for soil stabilization.

Figure 5-15. Small excavations may be shored with trench boxes on a temporary basis.

Piling. Items considered during piling takeoff include the subsurface analysis to determine piling depth, the linear feet of piling to be installed, the piling (I-beam and lagging, steel sheet, or precast concrete), and site access. The linear feet of piling and the depth determine the piling quantities. *Lagging* is vertical planks used to support earth on the side of an excavation. For I-beam and lagging, the number and type of I-beams and the length and square feet of lagging materials are determined. For steel sheet piling, the number of pieces, lengths, and designs are determined. A variety of steel sheet piling designs are available depending on the strength requirements. See Figure 5-16. For precast concrete piling, the design and length of the piling is calculated based on specifications and linear feet of piling required.

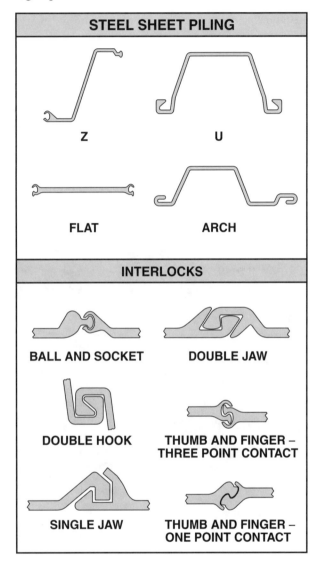

Figure 5-16. Various steel sheet piling designs are used for different shoring applications.

Piling removal costs are included in the estimate for pulling the piling out of the ground and hauling the piling from the site. For all sheet piling installations, costs in addition to material include the pile driving equipment and labor. For marine applications such as cofferdams, barges, support boats, and other access and anchoring equipment is required and included in the estimate. Piling estimate values are entered in a ledger sheet, spreadsheet, or estimating software cell.

Soil Stabilization. Temporary soil stabilization may be sufficient to hold an excavated bank in place where the angle of repose permits. Estimators calculate the linear feet of area to be stabilized and the depth of the excavation. When chemicals or slurry are used for temporary or permanent soil stabilization, costs include chemical material and application methods, drilling of slurry holes, and material costs for concrete slurry. When excavated banks are shot with cementious materials, costs include the ownership or rental of the shotcrete equipment and concrete material. This material is left in place and seldom removed after construction. Costs of mechanical stabilization include equipment costs for rental, leasing, owning, maintaining, and labor costs for operation of the equipment.

Refrigerant systems are complex and specialized. Estimating calculations for refrigerant systems include installation and removal of piping, refrigerant material, and maintenance and operation of pumping equipment. Soil stabilization estimate values are entered in a ledger sheet, spreadsheet, or estimating software cell.

Site Utility Estimation

Estimating considerations for site utilities include the linear feet of excavation, depth of excavation, existing job site conditions, connection requirements, and inspection. For piped utilities such as storm and sanitary sewers, water, fire, and natural gas, the pipe, pipe diameter, and required fittings and connections are taken off for the estimate. These calculations are commonly made in linear feet of each pipe and the individual fitting requirements.

Utility Excavation. Excavation takeoff for site utilities is based on the linear feet of trenching required to install the desired utility. The existing condition of the construction site at the time of the utility installation can affect the amount of trenching required and accessibility. Site utilities are best installed after preliminary clearing and grubbing and site grading and compaction. Pipe placed in areas that have not reached proper compaction limits may shift and fail. A site visit and coordination with other excavation contractors is essential to properly plan site utility excavations.

Connection. Estimators base utility pipe and connection information on the linear feet of pipe shown on plan drawings, connection fittings required, and local building codes. Inspectors check the connections prior to completion of the work. In cases such as connections to water service, telephone service, and electrical service, crews from the utility company make the final connection. Connection fees for these services are included in the estimate.

FOUNDATION MATERIALS AND METHODS

A structure that cannot be supported by surface soil requires a footing and foundation. A *footing* is the section of a foundation that supports and distributes structural loads directly to the soil. A *foundation* is the primary support for a structure through which the imposed load is transmitted to the footing or earth. Footings and foundations hold the structure in place and provide stability. Foundation systems and materials included in the foundation portion of the CSI MasterFormat™ include deep foundation systems consisting of piling and poured concrete footings and walls.

Foundation materials and methods are described in the specifications and shown on site plans, foundation plans, and detail drawings. Specifications indicate the engineering requirements of the foundation support system and the materials to be installed. See Figure 5-17. Site plans show foundation dimensions, locations, and quantities. Detail drawings show items such as the diameter of reinforced concrete pile, steel reinforcing, foundation footing, and wall width.

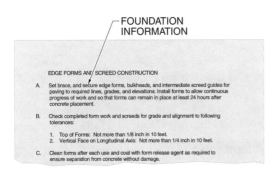

Figure 5-17. Estimators locate information concerning foundations in Division 2 of the specifications.

Piling

A *caisson* is a poured-in-place concrete piling of large diameter created by boring a hole and filling it with cast-in-place concrete. Piles or caissons may be required by the structural engineer and architect for deep foundations where soil bearing is poor. Piling includes driven pile and bored pile.

Driven Pile. A *driven pile* is a steel or wood member that is driven into the ground with a pile-driving hammer. Driven pile includes pipe pile, H-pile, timber pile, and precast concrete pile. *Pipe pile* is pile made of steel pipe that is driven into the ground. A cap is placed on the tip of the pipe to prevent it from filling with soil during driving. The pipe is filled with concrete after the proper depth is reached. An *H-pile* is a steel H-beam driven into the ground. H-piles commonly rest on a subsurface support layer at a lower level. *Timber pile* is pile made of tapered, treated timber poles driven into the ground to a predetermined point of refusal. *Precast concrete pile* is pile cast with reinforcing steel and designed to withstand the impact of the pile-driving hammer during installation. Precast concrete piles are commonly square in cross-section. All driven piles are designed to be cut off at a given elevation after the desired amount of foundation support is obtained at the designed level.

Bored Pile. *Bored pile* is a foundation support formed by drilling into the ground to a predetermined depth and diameter and filling the hole with concrete. Bored piles are usually stepped to various diameters specified for a certain depth. Cast-in-place concrete pile sides are formed by the surrounding soil, a casing that is removed as concrete is placed, or steel pipe. For concrete placed into the surrounding soil, a hole is drilled to a specified depth, reinforcing steel is set in place, and the hole is filled with concrete. The bottom of the pile may be belled out to create a large bearing surface if soil bearing is poor. A metal sleeve is set into the hole as drilling proceeds in areas where the sides of the drilled hole are not stable enough to remain in place until concrete is placed. Reinforcing steel is hoisted into the drilled hole and held in place. The metal sleeve is removed immediately after the concrete is placed in the hole.

Pile Caps. A *pile cap* is a large unit of concrete placed on top of a pile or group of piles. A pile cap distributes the load of the structure to the pile. Dimensions for pile caps on structural drawings include the width, length, depth, and reinforcing requirements. See Figure 5-18. Pile caps may be very large and require many cubic yards of concrete.

 A pile is considered defective if it is driven out of position, bent, or bowed along its length. A defective pile may be withdrawn and replaced by another pile or left in place with another pile driven adjacent to it.

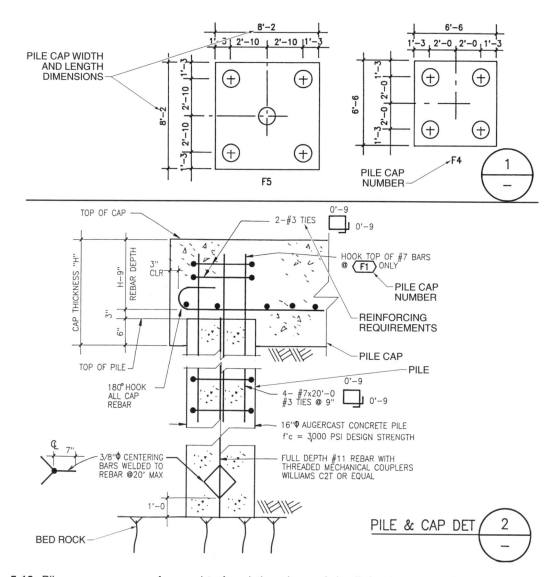

Figure 5-18. Pile caps are cross-referenced to foundation plan and detail drawings to obtain information concerning concrete and reinforcing requirements.

Concrete footings are commonly used to support foundation walls.

Footings and Foundation Walls

A footing is a section of a foundation that supports and distributes structural loads directly to the soil. Materials used for structural footings and foundation walls include wood, masonry, and concrete. Concrete is the most commonly used material for commercial construction footings and foundation walls. Treated wood and concrete masonry units are more commonly used in residential or light commercial applications.

Concrete. Wood or steel forms are set to hold poured-in-place concrete while it is in a plastic state. The forms are set according to elevations and dimensions given on architectural foundation plans. Reinforcing steel is set in the footings and foundation wall forms. Concrete footings are placed prior to concrete walls. A keyway is formed in the top of the footing to provide an interlock between the footing and the foundation wall. After the concrete has reached an initial set, the forms are removed to allow construction to proceed.

Footing and foundation wall forms are commonly reused many times. Fasteners and pins are removed and the faces of the forms are cleaned and oiled to allow release from the footing or wall surface. A *gang form* is a large form constructed by joining small panels. Gang forms help reduce the labor cost of assembling many small forms repeatedly. Interlocking foam block forms are used in light construction as permanent forms that stay in place and provide additional insulation to the concrete foundation wall.

FOUNDATION QUANTITY TAKEOFF

The information required to perform foundation quantity takeoff is contained in CSI MasterFormat™ Division 2 – Site Construction, site plans, foundation plans, and related detail drawings. Estimators should consult these sources and other appropriate construction documents to obtain information about foundation types, sizes, and materials. This helps to determine the required amount of labor and materials needed to construct the foundation system. See Figure 5-19.

Figure 5-19. Elements of a foundation system include formwork, reinforcing material, shoring, and foundation material.

Driven Pile Estimation

A grid of letters and numbers is provided at regular intervals in a plan view. The letter and number grid intersection points provide references for pile placement. Piles are indicated on foundation plans with information including the elevation to the top of the pile cap, the elevation of the top of the pile, and the depth of the lower tip of the pile. See Figure 5-20.

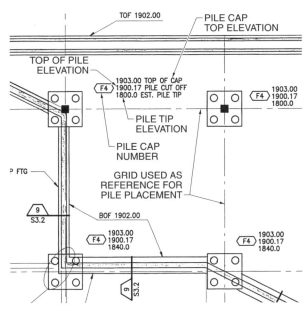

Figure 5-20. Foundation plans provide information concerning pile depths and elevations for estimating calculations.

Precast Concrete. Specifications provide information about concrete strength, placement requirements, and reinforcing steel. Takeoff quantities are calculated by counting the number of cast-in-place or precast concrete piles on site plans or foundation plans. Depths may be noted on the site plans or provided in a schedule. The cubic yards of concrete required are calculated based on pile diameter or cross-section and the depth of each pile. Labor cost tables indicate labor costs based on pile type and depth. Costs should also take into account site accessibility for pile placement.

When using a spreadsheet, the number and type of piles are entered into the proper cell. Formulas may be used in the spreadsheet to calculate concrete quantities based on the depth and diameter. When using estimating software, concrete piles can be entered as an item into the item listing in a manner that allows the estimator to enter the diameter and depth, use the database to calculate concrete and reinforcing material quantities and labor costs, and enter them into the proper cell.

Steel. Pipe pile and H-pile are taken off by counting the number of each pile required. Depths are determined from site plans or provided in a schedule. Estimators note the number of each length of steel pile material needed. As with concrete, labor costs are calculated based on pile type, depth, and construction site accessibility. Steel pile estimate values are entered in a ledger sheet, spreadsheet, or estimating software cell.

Timber. Timber pile is sized by diameter and length. The estimator totals the number of each diameter and length to develop takeoff quantities and costs. Labor costs are based on the number of piles and the depth of penetration required. Metal shoes may be used to protect the points of timber piles when the piles must be driven into hard strata. The metal shoes should be used only when permitted by the specifications. Takeoff methods are similar for ledger sheet, spreadsheet, and estimating software.

Footings and Foundation Wall Estimation

Foundation plans, elevation drawings, and detail drawings provide the quantity takeoff information for footings and foundation walls. The length, width, and height of the footings and foundation walls is found in these sources.

Concrete. The amount of concrete required for footings and foundation walls is determined from the length, width, and height of the footings and walls. The volume of concrete required is calculated by multiplying footing and wall length by width by height and dividing by 27. See Figure 5-21. Dimensions expressed in feet are divided by 27 to convert to cubic yards. *Note:* 1 cu yd = 27 cu ft. For example, the amount of concrete required for a wall 32'-0" long, 8" wide, and 7'-6" high is determined by multiplying $32' \times .6\overline{6}' \times 7.5'$. The result is 158.4 cu ft ($32' \times .6\overline{6}' \times 7.5' = 158.4$ cu ft). The cubic foot value is divided by 27 to obtain $5.8\overline{6}$ cu yd required (158.4 cu ft ÷ 27 = $5.8\overline{6}$ cu yd). The $5.8\overline{6}$ cu yd value is rounded up to 6 cubic yards.

Formwork for concrete footings and foundation walls is a major portion of the calculations for concrete footings and foundation walls. Formwork materials include snap ties, plywood, studs and bracing for job-built systems, and prefabricated form sizes and types for patented systems. Calculations for formwork materials are based on the number of square feet of contact area for footings and foundation wall forms, and foundation wall thickness.

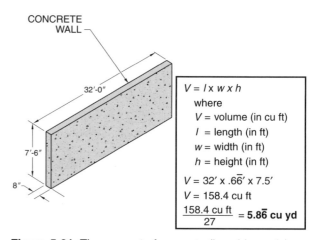

Figure 5-21. The amount of concrete (in cubic yards) required for foundation footings and walls is determined from the length, width, and height dimensions.

PAVEMENT AND PLANTING MATERIALS AND METHODS

After the majority of a building is constructed, the surrounding areas are finished to provide the appropriate access and landscape design. Many local building codes require a certain amount of landscaping and green space to be provided on new construction projects. The paving portion of the site plans provides the dimensions and locations for planters and open areas.

Pavement

Paving material for walks, roads, and parking areas includes concrete, asphalt, brick, stone, and rubberized material. The design of the paving material depends on the planned use. Heavy-use drives are designed to withstand great loads. Details in the site plans indicate the surface and subsurface paving materials required. Suitable paving performance requires proper compaction of the subgrade, installation and compaction of structural fill material, and application of surface material. Other construction items required in addition to the actual paving surface include curbs, pavement marking and sealing, fencing, gates, and signage.

Concrete. Poured-in-place concrete is used for all types of paving. Poured-in-place concrete is comprised of cement, sand, water, and aggregate. *Aggregate* is granular material such as gravel, sand, vermiculite, or perlite that is added to cement to form concrete, mortar, or plaster. Different aggregate may be used and various surface finishes applied to give concrete paving a variety of surface finishes. Coloring may also be added, along with surface stamping to give concrete paving the appearance of brick or stone.

Shallow excavations are made and formwork is set in place according to site plans and specifications for the construction of pavement. Concrete is placed in the formwork after surface preparation is complete. Reinforcing may be placed before or during concrete placement, depending on the reinforcing. Reinforcing bars are commonly set in place prior to concrete placement. Welded wire mesh is commonly set into concrete during placement. The concrete surface is finished according to the site plans and specifications.

Asphalt. *Asphalt* is dark-colored pitch primarily composed of crushed stone and bituminous materials. A layer of asphalt is commonly placed on a subsurface of compacted limestone or other material that provides a solid base. After initial placement, asphalt is compacted to achieve a finished surface. After the bituminous materials that bind the crushed stone in the asphalt have set, striping and other markings are painted onto the paving surface as shown on site plans.

Planting

Planting is the landscaping process of excavating, treating and preparing soil, and installing plant material in specified locations. Landscaping plans give information about plant material, methods for planting, and final surface treatments. Landscaping plans indicate the plant material to remain in place and be protected during construction, plants to be removed, and new plant material. Locations for exterior signage and irrigation systems may also be shown on detail drawings or on the topographical portion of the site plans.

Lawns. Unpaved and unlandscaped areas are sodded, seeded with grass seed, or left in their natural state. When sod is laid, the surface is prepared by grading and rough raking. Additional surface treatment of the soil is required where seeding is used. Surface protection may be required to allow germination time for the seed. Specifications may require a certain type of grass planting. Treatment of these areas is also indicated on landscape plans.

Trees and Shrubs. Various trees, shrubs, and ground cover are shown on landscape plans, site plans, or in a schedule in the specifications. Existing plantings at a construction site are noted as to whether they are to remain or be removed. The plantings to remain may require protection to avoid damage during construction.

Irrigation Systems. Site plans indicate sprinkler piping, size, and head location. See Figure 5-22. Site plans and detail drawings show piping connections, valve boxes, and sprinkler head details. Irrigation systems are commonly made of high-grade plastic pipe that is placed slightly below the surface of the ground. Installation occurs after all other construction work is completed at the site due to the light-gauge nature of the piping and the care needed to protect sprinkler heads.

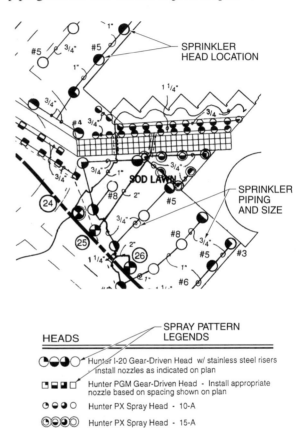

Figure 5-22. Sprinkler piping, size, and head location are determined by multiplying.

Site Amenities

Division 2 of the CSI MasterFormat™ also contains information concerning a variety of other site finish items including fountains, fencing, gates, retaining walls, concrete median barriers, guide rails, seating, tables, shelters, athletic and recreational screening, and various traffic control devices and signs.

Fencing. Fencing includes wood, chain link, ornamental metal, and plastic fence. Each type is indicated on site plans by an architectural line symbol. This portion of the CSI MasterFormat™ refers to permanent fencing designed to remain after completion of construction and does not include temporary security fencing placed during construction. Fencing placed on unpaved ground is fastened to posts spaced at regular intervals. Posts are commonly set in concrete to a depth below the frost line in locations where there is a possibility of freezing and thawing cycles moving the posts. Posts are set at a distance sufficient to support horizontal rails and at locations necessary to support gates. Chain link fencing is fastened to horizontal rails by stretching into place and fastening with wire. Wood fencing is fastened to horizontal rails with nails or screws. Some ornamental metal fencing may require prefabricated sections or welding in place.

PAVEMENT AND PLANTING QUANTITY TAKEOFF

Streets, parking lots, walks, and curb takeoff information is included in the specifications, detail drawings, and site plans. Area is calculated to determine quantities for the various subsurface and paving materials and labor. Landscaping locations and materials are given in the specifications and site plans.

Pavement Estimation

The primary information necessary for pavement takeoff is the exterior dimensions of the paved area and the dimensions of any voids. These dimensions are used to determine the number of square feet of pavement to be put in place, subgrade preparation, pavement thickness and possible reinforcement, surface finish, and possible surface protection.

Subgrade preparation includes a granular subbase of crushed gravel or sand. In some cases, stabilization fabric may also be required. When using a ledger sheet, calculations of the number of square feet of area to be paved with each material are determined by multiplying length by width. Deductions are made for any internal void areas such as planters, etc. A spreadsheet may include a calculator function that allows for perimeter dimension entry and automatic area calculations. Estimating software allows for length and width dimension entry and automatic item calculation of materials, reinforcing, and labor costs. A digitizer can be used when taking off large and irregularly-shaped paving areas containing curved dimensions that may require additional geometric calculations.

Concrete Pavement. Estimating for concrete pavement is made by determining the number of square feet of pavement required. The primary items considered include the thickness of the pavement, the surface finish, aggregate, and protection to be provided after placement of the concrete. See Figure 5-23. For example, to calculate the volume of concrete required in a driveway 140.9′ long, 20′ wide, and 6″ thick, multiply the length by the width by the thickness. The driveway contains 1409 cu ft of concrete (140.9′ × 20′ × .5′ = 1409 cu ft). This value is divided by 27 to determine that 52.18 cu yd of concrete is required (1409 cu ft ÷ 27 = 52.18 cu yd). The cubic yards of concrete determines the material cost. The number of square feet of surface area determines the labor cost. Specifications provide information concerning concrete mix and the required reinforcing and compressive strength.

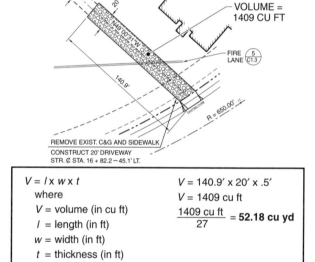

Figure 5-23. The volume of paving material required is calculated by multiplying the length by the width by the thickness.

When reinforcing steel is required, the number of each type of reinforcing steel bar is determined by calculating the spacing in each direction and the number of bars required in a given number of square feet. For example, when bars are spaced 16″ OC (1.33′) in both directions, a 16′ square section of pavement (256 sq ft) requires 24 bars ([16 ÷ 1.33] × 2 = 24 bars). Additions are made for the amount of bar overlap required in the specifications or detail drawings. Welded wire mesh is calculated by the number of square feet of coverage required.

Asphalt. Many different types and designs of asphalt mixes are available. The design of the asphalt pavement application is based on the traffic loads imposed and climatic conditions. Selection of the proper asphalt mixture for traffic loads is made by the architect and engineer and is given in the specifications or noted on the site plans. The number of square feet of asphalt coverage and depth determines the quantity of asphalt required. The number of cubic yards of material required is calculated by multiplying the square feet of coverage by the depth. Asphalt material is priced by the ton of material delivered to the job site. Conversion amounts vary based on the type of mixture specified. Items taken into consideration when estimating costs and quantities for hot mix asphalt (HMA) include mix design and type, surface preparations required, mix placement, screeding methods, joint construction, compaction, and spreading and compacting equipment used. The number of square feet determines the labor cost. Labor costs are based on industry standard information or company historical data.

Curbs and Gutters. Various curb designs are used depending on the need to match existing curbs or to protect against damage in heavy use conditions. Estimators calculate the number of linear feet of each type of curb from the site plans. Standard costs per linear foot of curb include materials such as concrete (including possible reinforcing), asphalt, and labor. The amount of formwork required for curbs and gutters depends on radiuses, curves, aprons, and close tolerances required horizontally and vertically.

Planting Estimation

Landscaping costs include finish grading, sodding, seeding, excavations for trees and shrubs, plant materials, and plant protection. Specifications, schedules, and detail drawings show required planting methods and materials, including the depth of planting, soil amendments, and mulch. Lawn, tree and shrub, and irrigation system costs are entered in a ledger sheet, spreadsheet, or estimating software cell.

Lawns. Specifications indicate if the lawn areas are to be sodded or seeded. Costs are calculated based on the number of square feet of lawn area. Possible costs include finish grading and final surface preparation, laying of sod or spreading of seed, fertilizer, watering, and straw to protect the surface during germination.

Trees and Shrubs. The scientific name, common name, and size of each plant is given on a plant schedule along with the quantity required. See Figure 5-24. Plant nurseries commonly provide a cost per plant that includes plant material, planting, and staking. Estimators should check the specifications for protective barriers placed around trees and shrubs during and following construction.

Irrigation Systems. Takeoff for irrigation systems includes calculation of the linear feet of each pipe and the number of each valve and spray head in the system. Control valves, timers, drains, and other specialty items are also included in the quantity takeoff. Labor costs are based primarily on company historical data for installation times and labor rates.

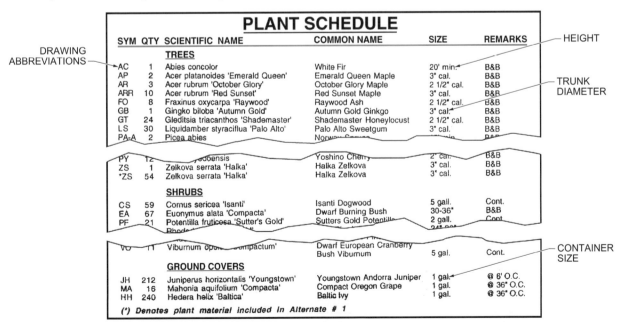

Figure 5-24. Estimates for landscaping material can be calculated from plant schedules.

Site Amenity Estimation

Costs for site amenity items such as shelters, seating, or athletic screening are obtained from suppliers on an as-needed basis. Costs for site amenities such as traffic control devices and signage are calculated on an item basis. Estimators determine the number of each sign, traffic device, or other item required. The number of each site amenity item is entered in a ledger sheet, spreadsheet, or estimating software cell.

Fencing. The linear feet of fencing is found on the site plans. Estimators determine the linear feet of each type of fencing required. Variables for the estimator include the fence material (wood or metal), height, finish, and items such as swinging or sliding gates. Material costs per linear foot include fence posts, railings, and fencing. Material and labor rates for fencing and gates are commonly based on company historical data or industry standards.

ESTIMATING SOFTWARE – QUICK TAKEOFF ENTER DIMENSIONS FUNCTION

In Precision estimating software quick takeoff, the enter dimensions function acts as a switch that controls whether the enter dimensions window opens when selected items are dragged to the spreadsheet. A check mark next to the function causes the enter dimensions window to open whenever an item is taken off. Leaving the function unmarked loads the selected items directly into the spreadsheet. Quantities can then be entered in the takeoff quantity column of the spreadsheet. The quick takeoff enter dimensions function is used by applying a standard procedure. See Figure 5-25.

> ▲ *Seeded lawn landscaping costs includes fertilizer, seed, water, and straw or special cloth covering to retain moisture and reduce erosion.*

PRECISION ESTIMATING SOFTWARE – QUICK TAKEOFF ENTER DIMENSIONS FUNCTION

1. Open the Quick Takeoff window by clicking the Quick Takeoff button.
2. Right-click in the Quick Takeoff window.
3. A check mark is displayed next to the function name, indicating that the Enter Dimensions function is turned ON.

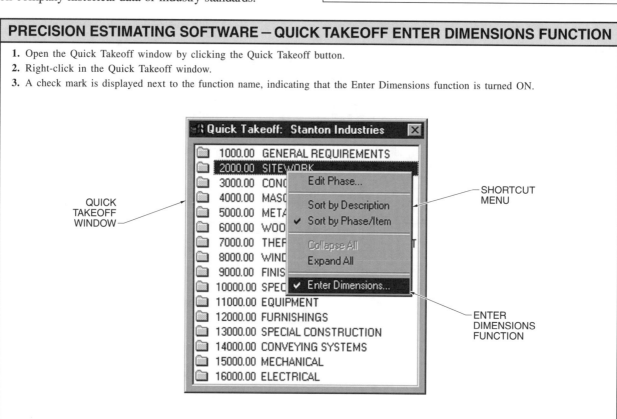

4. Double-click on 2000.00 Sitework and 2220.25 Earthwork: Footings (Mach). Choose item 10, excavate footing by machine, by double-clicking on the item or dragging and dropping the item to the spreadsheet.
5. Item 10 has an associated formula. The Enter Dimensions function prompts the user with the appropriate variables based on this formula. In the Enter Dimensions window, enter 1 for Quantity, 50 for Length, 4 for Width, and 4 for Depth. Click OK to send the item to the spreadsheet.
6. Next, disable the Enter Dimensions function by right-clicking in the Quick Takeoff window and choosing Enter Dimensions from the shortcut menu. The check mark no longer displays next to the function name.

Figure 5-25. The enter dimensions function acts as a switch that controls whether the Enter Dimensions window opens when selected items are dragged to the spreadsheet.

Review Questions
Sitework

Estimating

_____ 1. A(n) _____ is a rock that has been transported by some geological action from its site of formation to its current location.

_____ 2. _____ is a loose granular material consisting of particles smaller than gravel but coarser than silt.

_____ 3. _____ is a natural mineral material that is compact and brittle when dry but plastic when wet.

T   F   4. Hardpan is a hard, compacted layer of soil, clay, or gravel.

T   F   5. Geotextile material is applied after rough grading is complete.

_____ 6. A soil _____ is a material added to existing job site soil to increase its compaction and ability to maintain the desired slope at the completion of construction.

_____ 7. _____ is water present in subsurface material.

_____ 8. _____ is the organized destruction of a structure.

_____ 9. A(n) _____ load is the predetermined load a structure is capable of supporting.

_____ 10. A(n) _____ load is a permanent, stationary load composed of all building materials, fixtures, and equipment permanently attached to a structure.

_____ 11. A(n) _____ is a structural member installed in the ground to provide vertical and/or horizontal support.

T   F   12. Shoring is wood or metal members used to permanently support formwork or other structural components.

T   F   13. Clear and grub is the removal of vegetation and other obstacles from a construction site in preparation for new construction.

_____ 14. A(n) _____ is a watertight enclosure used to allow construction or repairs to be performed below the surface of the water.

_____ 15. A _____ sewer is a piping system designed to collect storm water and channel it to a retention pond or other means of removal.

A. sanitary   C. A and B
B. storm   D. neither A nor B

_____ 16. _____ is the temporary supporting of existing utility lines in or around an excavated area to properly protect the lines.

_____ 17. A(n) _____ line is a dashed (existing grade) or solid (finished grade) line on a plot plan used to show elevations of the surface.

_____ 18. _____ is the process of lowering high spots and filling in low spots of earth at a construction site.

_____ **19.** _____ is vertical planks used to support earth on the side of an excavation.

_____ **20.** A(n) _____ is a poured-in-place concrete piling of large diameter created by boring a hole and filling it with cast-in-place concrete.

_____ **21.** A(n) _____ form is a large form constructed by joining small panels.

_____ **22.** _____ is granular material such as gravel, sand, vermiculite, or perlite that is added to form concrete, mortar, or plaster.

T   F **23.** Planting is the landscaping process of excavating, treating and preparing soil, and installing plant material in specified locations.

T   F **24.** Angle of reprise is the slope that a material maintains without sliding or caving in.

_____ **25.** Soil _____ is the volume growth in soil after it is excavated.

_____ **26.** A(n) _____ material is a material capable of posing a risk to health, safety, or property.

_____ **27.** _____ is a mixture of water and fine particles.

_____ **28.** _____ material is material including sand, gravel, and crushed rock installed to increase the stability of the soil.

_____ **29.** _____ is the division of a topographical site map for large areas to be excavated or graded into small squares or grids.

_____ **30.** _____ grading is the process of smoothing and sloping a site for paving and landscaping.

_____ **31.** A(n) _____ pile is a steel or wood member that is driven into the ground with a pile-driving hammer.

_____ **32.** _____ pile is a foundation support formed by drilling into the ground to a predetermined depth and diameter and filling the hole with concrete.

_____ **33.** _____ is the act or process of correcting.

_____ **34.** _____ is a dark-colored pitch primarily composed of crushed stone and bituminous material.

_____ **35.** Temporary soil _____ is the process of holding existing surface and subsurface materials in place during the construction.

Soils

_____ **1.** Light soil

_____ **2.** Medium soil

_____ **3.** Heavy soil

A. A granular substance that is readily shoveled by hand without the aid of machines

B. Soil that can be loosened by picks, shovels, and scrapers

C. Soil that can be loosened by picks, but is hard to loosen with shovels

Cross-Section Volume Calculation

_____   **1.** The depth to be excavated at A is _____'.

_____   **2.** The depth to be excavated at C is _____'.

_____   **3.** The greatest depth is excavated at _____.

_____   **4.** The average excavation depth is _____'.

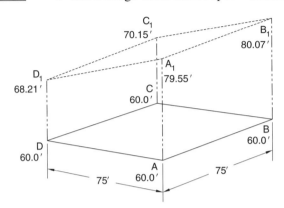

———————— PLANNED ELEVATION
—·—·—·— DEPTH
----------------- EXISTING ELEVATION

Volume of Concrete

_____   **1.** The wall is _____" thick.

_____   **2.** The wall is _____' long.

_____   **3.** The wall is _____' high.

_____   **4.** The wall contains _____ cu ft of concrete.

_____   **5.** The wall contains _____ cu yd of concrete.

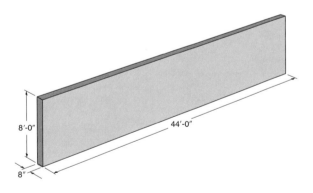

Steel Sheet Piling

_____   **1.** Z

_____   **2.** U

_____   **3.** Flat

_____   **4.** Arch

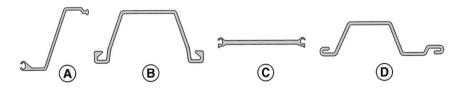

Excavation

_____	**1.** Overdig	
_____	**2.** Backfilling	
_____	**3.** Trencher	
_____	**4.** Payline	
_____	**5.** Slope cut	

A. An inclined or sloped wall excavation

B. A measurement used by excavation contractors to compute the amount of necessary excavation as stated in the specifications

C. A machine with a conveyor to remove the soil as it digs a narrow, long hole into the earth

D. The amount of excavation required beyond the dimensions of a building to provide an area for construction activities

E. The replacing of soil around the outside foundation walls after the walls have been completed

Utilities

_____ **1.** Electrical service

_____ **2.** Natural gas main

_____ **3.** Water main

_____ **4.** Storm sewer

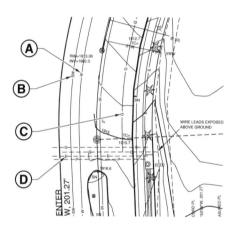

Activities
Sitework

Name _____ Date _____

Activity 5-1 – Ledger Sheet Activity

Refer to the cost data, Print 5-1, and Estimate Summary Sheet No. 5-1. Take off the acres of clear and grub, cubic yards of stripping topsoil, and acres of lawn seeding for the recreation grove. Determine the total labor and equipment cost for clear and grub, stripping topsoil, and lawn seeding.

COST DATA			
Material	Unit	Labor Unit Cost*	Equipment Unit Cost*
Clear and grub cut and chip light trees to 6″ dia	acre	1210	1292
Stripping topsoil	cu yd	.15	.40
Lawn seeding, 215 lb/acre	acre	159	147

* in $

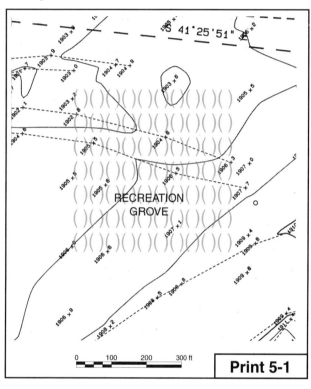

Print 5-1

ESTIMATE SUMMARY SHEET

Sheet No. _____5-1_____

Project: _____

Estimator: _____

Date: _____

Checked: _____

No.	Description	Dimensions			Quantity		Labor			Equipment			Total	
						Unit	Unit Cost	Total	Unit Cost	Total	Unit Cost	Total		
	Clear and grub													
	Stripping topsoil													
	Lawn seeding													
	Total													

Activity 5-2 – Spreadsheet Activity

Refer to the cost data and Estimate Summary Sheet No. 5-2 on the CD-ROM. A construction site requires the demolition of a 20′ × 1′ × 6′ masonry block wall with brick veneer and the excavating and backfilling of a trench 60′ × 3′ × 3′. Take off the cubic feet of masonry wall demolition, cubic yards of excavating, and cubic yards of backfilling. Determine the total labor and equipment cost for the demolition, excavating, and backfilling.

COST DATA			
Material	**Unit**	**Labor Unit Cost***	**Equipment Unit Cost***
Masonry wall demolition	cu ft	.81	.64
Excavating	cu yd	2.93	1.61
Backfilling	cu yd	.38	1.02

* in $

Activity 5-3 – Estimating Software Activity

Create a new estimate and name it Activity 5-3. Perform a quick takeoff of item **2220.50 10 Backfill Footings** for a 275′ long, 3′ wide, and 4′ deep footing excavation. Take off item **2511.00 20 H Asphalt Material** for a 100′ long, 20′ wide, and 6″ deep (thick) road. Print a standard estimate report. *Note:* The Enter Dimensions function must be turned ON.

Concrete

Concrete consists of portland cement, aggregate, and water, which solidifies into a hard, strong mass. Architectural drawings and specifications contain notes and information concerning specific concrete performance requirements. Concrete may be cast-in-place or precast concrete. Concrete quantity includes the type and volume of concrete required, reinforcing, transportation, surface finish, and possible climatic protection. Formwork is the system of support for freshly-placed concrete. Formwork may be job-built or patented. Concrete formwork design takes into account all forces to be placed on the form. The estimator must calculate the amount of formwork materials required when formwork drawings are not provided.

FORMWORK MATERIALS AND METHODS

Concrete is a material consisting of portland cement, aggregate, and water, which solidifies through chemical reaction into a hard, strong mass. *Reinforced concrete* is concrete to which reinforcement has been added in the form of steel rods, bars, or mesh to increase its strength and resistance to cracking. Reinforced concrete members may be cast-in-place, precast on the job site, or precast off site. Concrete may be used with other construction methods such as masonry and structural steel. Architectural and structural drawings contain information for reinforced concrete construction. The majority of reinforced concrete information is shown on foundation, floor, and structural plans, and in elevation and detail drawings.

Formwork is the system of support for freshly-placed concrete. Formwork includes forms, hardware, and bracing. Many methods are used to create the formwork that allows for the placement of cast-in-place concrete. The formwork used depends on the overall size of the concrete structure. A *form* is a temporary structure or mold used to retain and support concrete while it sets and hardens. Forms may be set on or in the ground for pilings, footings, and on-grade slabs. Forms may be set on concrete slabs to support walls, columns, and above-grade slab systems. Elevated forms may be set for above-grade slabs. Each cast-in-place concrete application requires specific print information about dimensions of the finished concrete, reinforcing requirements, inserts, and concrete properties.

Walls

Cast-in-place concrete walls are placed on footings or slabs. A *footing* is the section of a foundation that supports and distributes structural loads directly to the soil. A *slab* is a horizontal or nearly horizontal layer of concrete. Wall forms may be made of wood or metal faces reinforced with wood or steel frames. See Figure 6-1. Wall forms are held in alignment with walers, strongbacks, and braces. A *waler* is a horizontal member used to align and brace concrete forms or piles. A *strongback* is a vertical support attached to concrete formwork behind the horizontal walers to provide additional strength against deflection during concrete placement. A *brace* is a structural piece, either permanent or temporary, designed to resist weights or pressures of loads. Reinforcing steel and blockouts for doors and windows are set before the opposing side of the wall forms are completed. A *blockout* is a frame set in a concrete form to create a void in the finished concrete structure.

Figure 6-1. Wall forms may be made of wood or metal faces reinforced with wood or steel frames.

A *form tie* is a metal bar, strap, or wire used to hold concrete forms together and resist the pressure of wet concrete. Different form ties are used to hold two wall forms at the proper distance and distribute concrete loads during placement. The form ties used depend on the width and design of the wall. Form ties include snap ties, she bolts, and coil ties. A *snap tie* is a concrete form tie that is snapped off when the concrete is set. A *she bolt* is a special bolt that threads into the ends of a tie rod. A *tie rod* is a steel rod that runs between forms to hold them together after concrete is placed. The she bolt is removed when the forms are removed. A *coil tie* is a concrete form tie used in heavy construction in which a bolt screws into the tie and is removed during form removal. For large commercial jobs with complicated form requirements, concrete form manufacturers provide detailed drawings of forms, ties, and bracing placement to ensure proper form construction.

In some applications, one side of a wall is formed and the opposite side is supported by soil or rock. This may be done for an abutment, slurry wall, or retaining wall. An *abutment* is the supporting structure at the end of a bridge, arch, or vault. A *slurry wall* is a wall built around a foundation to hold back ground water. A *retaining wall* is a wall constructed to hold back earth. Reinforcing is set in place and the forms are set and braced. Additional bracing of one-sided wall forms is necessary to support the hydrostatic pressure (weight) that is not offset by the other side of the wall as with a conventional form tie system.

Blockouts are placed in wall forms to create openings for doors and windows. Blockout placement is determined from elevation drawings. Large or numerous blockouts in concrete forms affect form costs and concrete volume. Concrete wall forms include job-built and patented wall forms.

Job-Built Wall Forms. Job-built wall forms are composed of framing made of 2″ × 4″ lumber studs and facing made of ½″ or ¾″ plywood. See Figure 6-2. Job-built wall forms are commonly 4′ × 8′ in size to make maximum use of plywood materials. Studs are spaced 12″ or 16″ on center (OC). Holes are drilled at a regular spacing to allow for placement of snap ties.

Figure 6-2. Plywood, framing lumber, and snap ties are used for job-built wall forms.

Job-built wall forms may be built for reuse or built for a single application. In addition, job-built wall forms may be built as a large unit and fastened together to form a gang form. A *gang form* is a large form constructed by joining small panels. Gang forms are set in place by cranes and reused in several applications to reduce the labor costs of rebuilding large forms several times.

Patented Wall Forms. Many manufacturers provide patented wall form systems having a variety of panel designs, tie systems, and form fastening systems. Panels are manufactured with metal frames faced with plywood or metal cladding. Panels are available in standard sizes to allow these systems to form walls of various lengths, heights, and widths. See Figure 6-3. Patented form tie systems vary according to the manufacturer and the wall being formed.

Figure 6-3. Manufacturers of patented wall form systems provide a variety of form sizes, designs, and shapes.

Job-built and patented form bracing is commonly nominal 2″ × 4″ or 2″ × 6″ lumber with a variety of anchoring and support methods. Bracing varies from the simple system of a stake driven into the ground and a 2″ × 4″ nailed to the back of the form and stake to a complex system with metal channels bolted to the back of forms with additional bracing of screw jacks fastened to metal or wood braces. Design of wall form bracing systems can be a complex portion of estimating concrete form material and labor costs.

Slabs

Concrete slabs may be placed on the surface of the ground to be used as a foundation system in light-bearing applications. Slabs-on-grade are also used for paving and basement and ground-level floors. Paving such as highways and roads is covered in Division 2 of the CSI MasterFormat™ for site construction and is normally performed by contractors and subcontractors who specialize in road construction. Prior to placement of concrete, information obtained from plans and specifications includes treatment of the area below the slab, items placed in the slab, blockouts, elevations, and finishes.

Slab-On-Grade. A *slab-on-grade* is a concrete slab that is placed directly on the ground. A trench may be dug around the perimeter of a floor slab to provide support and protection against movement during freezing and thawing cycles. This trench is shown on architectural drawings and allows the slab to support light- to medium-weight loads. See Figure 6-4. Spread footings or piles are necessary where heavy load support is required. A *spread footing* is a foundation footing with a wide base designed to add support and spread the load over a wide area. A *pile* is a vertical structural member installed in the ground to provide vertical and/or horizontal support.

Welded wire fabric may be used to reinforce lightweight concrete slabs.

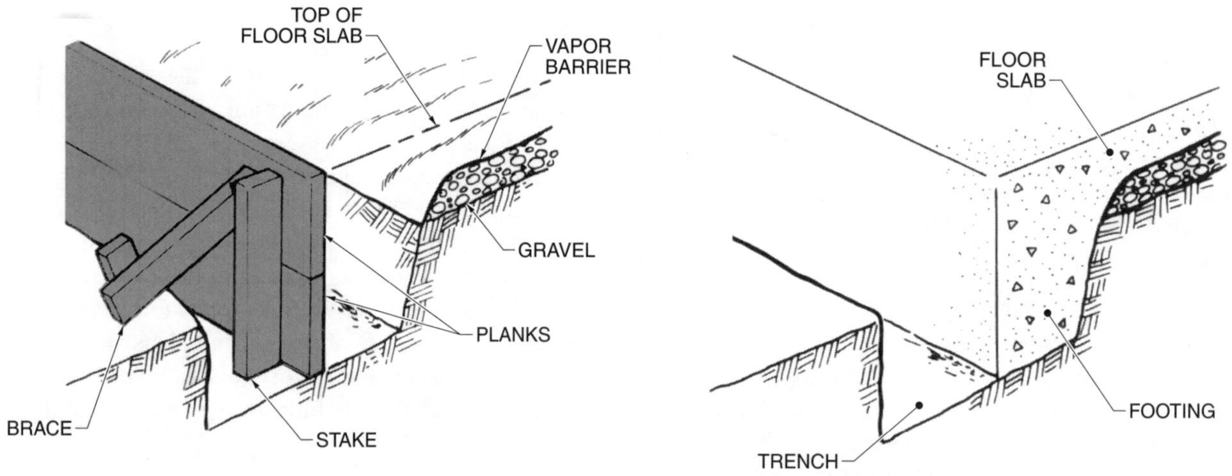

Figure 6-4. Footings provide additional foundation support at the perimeter of a slab-on-grade.

Forms may be built for perimeter footings and slabs where they project above the surface of the soil. Other sections of the slab may be thickened to support other loads. Locations of the thickened sections are commonly indicated with dashed lines on plan views and with additional section views.

Forms are not required where a slab-on-grade is placed between existing walls or other structural members. Isolation joints may be installed at these points to help prevent movement of structural supporting members from creating cracks in the slab. An *isolation joint* is a separation between adjoining sections of concrete used to allow movement of the sections. The location and type of isolation joints are indicated on detail drawings. See Figure 6-5. The edge design of a keyway or dowel is shown if the slab must join to other members. A *contraction (control) joint* is a groove in a vertical or horizontal concrete surface to create a weakened plane and control the location of cracking due to imposed loads. Control joints placed in slabs are also shown on plan and detail drawings. Control joint spacing is critical to properly control random slab cracking. Control joint spacing varies depending on the thickness of the concrete slab. Locations for cutting the slab are given for paving.

Elevated Slabs. Concrete slabs placed above grade may be built in several different designs. These include a monolithic design with beams and the above-grade slab placed at the same time, supporting the above-grade slab with precast concrete or structural steel beams, or supporting the slab with steel beams and corrugated decking.

Steel or precast concrete beams may be designed to be placed independently of the above-grade concrete slab. Detail drawings indicate the thickness of the slab and reinforcing placement. Forms are hung from beams after columns, walls, and beams are in place. Hangers are placed on the beams and allow for temporary joists and bracing brackets to be supported and set to the proper slab height. Threaded rods are lubricated and used to support brackets for joists. Deck forms are placed on the joists. The slab is tied to the beam by projecting reinforcing steel from a precast concrete beam or welded studs on top of a structural steel beam. An edge form is built around the perimeter of the deck. Electrical piping, mechanical blockouts, and reinforcing are set in place and the concrete is placed and finished. The temporary joists, bracing brackets, and deck forms are removed from below after the concrete has reached its design strength. Threaded lubricated rods are removed and the remainder of the hanging hardware is placed into the concrete deck and remains in place.

A concrete slab-on-grade normally requires little bracing. Formwork for a slab-on-grade is normally only several inches deep, and small stakes and wood bracing are sufficient.

> ▲ *Concrete wall formwork must be constructed following safe and established construction procedures. Information regarding safe construction procedures can be obtained from the American Concrete Institute (ACI) and the Occupational Safety and Health Administration (OSHA).*

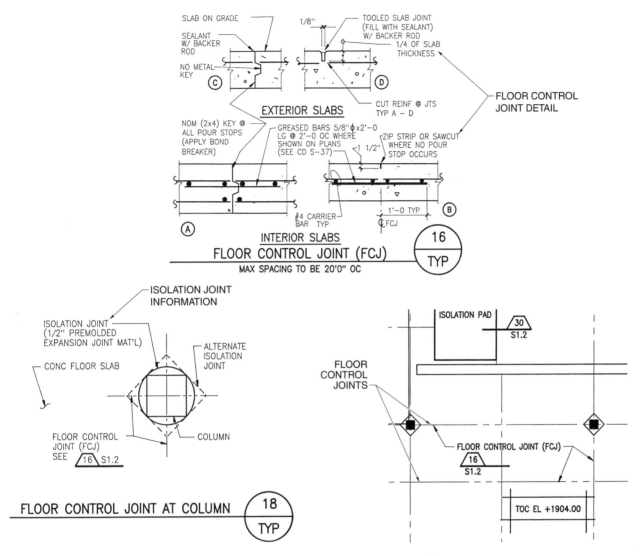

Figure 6-5. Isolation joints in a slab-on-grade require additional labor and material costs.

Elevated slabs use shoring and decking to form a flat surface at the elevation of the bottom of the cast-in-place concrete beams. *Shoring* is wood or metal members used to temporarily support formwork or other structural components. *Decking* is light-gauge metal sheets used to construct a floor or deck form or used as floor or deck members. Shoring may be made of wood posts, steel pipe, or panelized shoring systems. Beams and joists are placed on these shores to support deck forms. Domes are used to form voids in the concrete slab that create the beams in either one or two directions. See Figure 6-6. A *spandrel beam* is a beam in the perimeter of a structure that spans columns and commonly supports a floor or a roof. Spandrel beams are formed around the outside edge of the deck to pro-

vide structural support. Perimeter edge forms are built to hold the concrete in place and keep the top of the slab at the proper elevation.

Formwork is designed by specialists to ensure that the proper amount of shoring is in place and the proper amount of camber has been designed into the form. *Camber* is the slight upward curve in a structural member designed to compensate for deflection of the member under load. Shoring must support the weight of the form itself, the wet concrete, all reinforcing, electrical and mechanical materials, and the workers placing and finishing the concrete. See Figure 6-7. Shoring is set on the slab below and set to the proper elevation. Beams, joists, decking, edge forms, and other blockouts are set in place on the shoring.

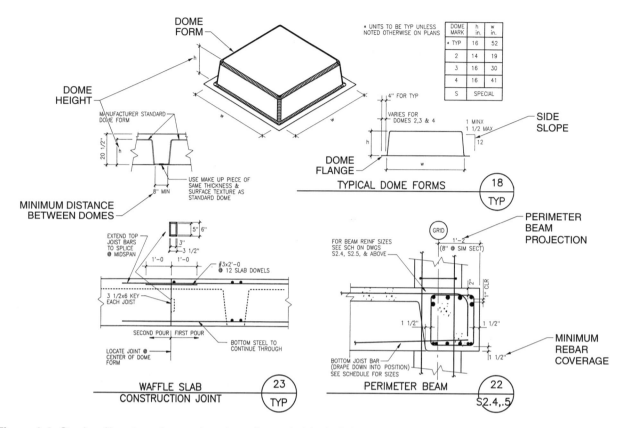

Figure 6-6. Steel or fiberglass domes placed on elevated slab deck forms create beams for cast-in-place slabs.

Screw jacks or some form of adjustment mechanism is normally installed in the shoring system to allow for final adjustment of height prior to concrete placement. Large construction jobs may use a system in which the shoring is designed as large units that can be moved in a single piece. This allows shoring, decking, and bracing to be set up and removed as a single unit, saving labor costs. Shoring and bracing costs are significant for elevated slab construction estimates.

FORMWORK QUANTITY TAKEOFF

Information concerning the type and design of formwork is not given on structural drawings. For large projects, a separate set of formwork drawings is commonly provided by the concrete form supplier. See Figure 6-8. Formwork drawings are developed by formwork design specialists. Formwork design is based on concrete dimensions given in the architectural drawings. Formwork drawings indicate manufacturer form identification numbers and type, placement, form fastening systems, form ties, and shoring and bracing information. Concrete formwork design takes into account all forces to be placed on the form.

The estimator must calculate the amount of formwork materials required when formwork drawings are not provided. This includes forms for footings and walls, slabs-on-grade and above grade, and all components including columns, beams, and shoring.

Walls

Cast-in-place concrete wall information used in estimating formwork materials includes elevations to the top and bottom of the wall, the thickness of the wall, placement of wall surface features, all embedded items, and reinforcing steel. Length and width dimensions are given on foundation plan views, elevation drawings, and detail section views. Concrete formwork is commonly taken off in linear feet or square feet of contact area. Estimators should check elevation drawings for blockouts that could require additional form materials and labor.

> ▲ *Deflection of wall forms occurs when panels of inadequate thickness are used or when studs or walers are spaced too far apart. A wall form must be constructed to hold the concrete to a straight and true surface.*

Figure 6-7. Shoring must support the weight of the form itself, the wet concrete, all reinforcing, electrical and mechanical materials, and the workers placing and finishing the concrete.

Snap ties are calculated as a specified number per square foot of wall area. Snap tie spacing is determined from the specifications, elevation drawings, patented form manufacturer recommendations, or the snap tie loading information available from the manufacturer. For example, this information may state that snap ties are spaced 24″ OC for an 8″ thick wall. Each 4′ × 8′ section of wall (32 sq ft) requires 8 ties with allowance for adjoining forms. See Figure 6-9. This results in one snap tie for each 4 sq ft of wall area. The total square feet of foundation wall area is divided by 4 to determine the number of snap ties needed for the foundation walls.

Labor costs for the building of wall forms are normally based on the cost per square foot of contact area. These costs are available from industry resource materials or company historical data. Variables for these costs include the formwork used such as job-built, patented, or gang forms, lifting equipment for large formwork

such as cranes, the height of the wall that may require scaffolding or fall protection equipment, job site accessibility for erecting and removing formwork, scheduling of the work in relation to other construction activities, special architectural details, and climatic conditions.

Job-Built Walls. The amount of form material needed for a concrete foundation is calculated by multiplying wall length by wall height and then multiplying this total by 2. See Figure 6-10. Dimensions must be expressed in feet. This gives the total square footage of forms needed for both sides of the foundation wall. For example, the square footage of forms required for a wall 22′-0″ long, 8″ wide, and 8′ high is determined by multiplying 22′ × 8′ × 2. The square footage of forms required equals 352 sq ft (22′ × 8′ × 2 = 352 sq ft). The number of 4′ × 8′ panels required for form material is determined by dividing the square footage of forms required by 32. The number of 4′ × 8′ panels required equals 11 (352 ÷ 32 = 11 panels). Any fractional number of sheets is rounded up to the next full sheet.

Snap ties are noted to be spaced 12″ OC horizontally and 2′ OC vertically. Each 4′ long panel requires 8 ties. The number of panels needed for one side of the wall is multiplied by the number of ties per panel to obtain 44 ties required for the wall (5.5 panels × 8 ties/panel = 44 ties).

Increte Systems

Form liners may be used with traditional concrete wall formwork to create the look of natural rock in a poured concrete wall.

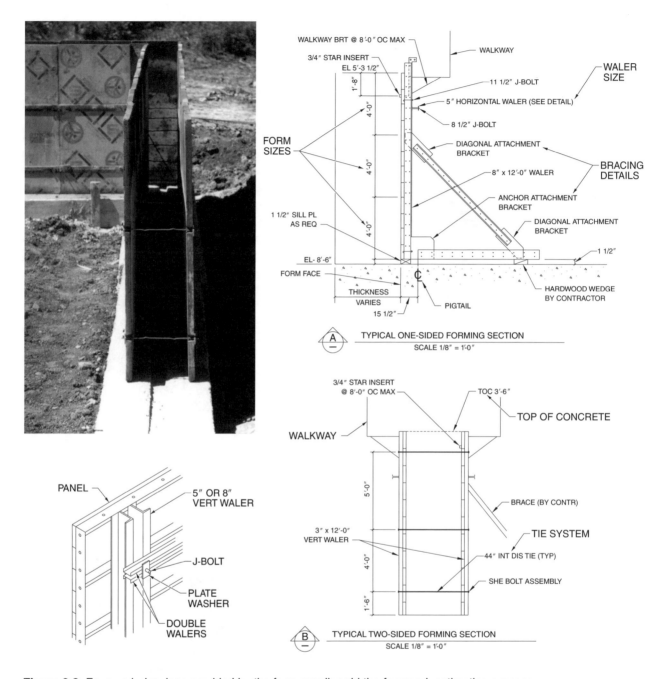

WALKWAY BRT @ 8'-0" OC MAX
WALKWAY
3/4" STAR INSERT
EL 5'-3 1/2"
1'-8"
11 1/2" J-BOLT
WALER SIZE
4'-0"
5" HORIZONTAL WALER (SEE DETAIL)
8 1/2" J-BOLT
FORM SIZES
DIAGONAL ATTACHMENT BRACKET
4'-0"
8" x 12'-0" WALER
BRACING DETAILS
ANCHOR ATTACHMENT BRACKET
1 1/2" SILL PL AS REQ
DIAGONAL ATTACHMENT BRACKET
4'-0"
1 1/2"
EL- 8'-6"
FORM FACE
HARDWOOD WEDGE BY CONTRACTOR
THICKNESS VARIES
15 1/2"
PIGTAIL

A — TYPICAL ONE-SIDED FORMING SECTION
SCALE 1/8" = 1'-0"

3/4" STAR INSERT @ 8'-0" OC MAX
TOC 3'-6"
TOP OF CONCRETE
WALKWAY
5'-0"
BRACE (BY CONTR)
3" x 12'-0" VERT WALER
TIE SYSTEM
4'-0"
44" INT DIS TIE (TYP)
SHE BOLT ASSEMBLY
1'-6"

PANEL
5" OR 8" VERT WALER
J-BOLT
PLATE WASHER
DOUBLE WALERS

B — TYPICAL TWO-SIDED FORMING SECTION
SCALE 1/8" = 1'-0"

Figure 6-8. Formwork drawings provided by the form supplier aid the formwork estimating process.

Framing lumber is calculated separately depending on the spacing of the studs. Top and bottom plates may also be part of the form panel design, similar to a wood wall frame. The number of vertical bracing studs for wall forms is calculated as the length of the wall divided by the OC distance of the studs. The number of studs is doubled if there are two studs at each position. This number is doubled to determine the total number of studs for both sides of the wall. For example,

the number of studs needed for the vertical braces on a wall 22'-0" long and 8'-0" high having two studs 16" OC is determined by multiplying 22' × 2 × .75. A total of 33 8' studs (22' × 2 × .75 = 33 studs) are required for each side of the wall. A total of 66 studs (33 studs × 2 = 66 studs) are required for both sides of the wall. Additional studs are added for corners.

The linear feet of plates is determined by multiplying the linear feet of wall by 2 for a single top and bottom plate. This value is multiplied by 2 for both sides of the wall. The linear feet of wall is multiplied by 4 for forms requiring double top and bottom plates. For example, a wall 22'-0" long requiring single top and bottom plates contains 44 lf (22' × = 2 = 44 lf). The wall requires a total of 88 lf (44 lf × 2) of lumber for plates for both sides of the wall.

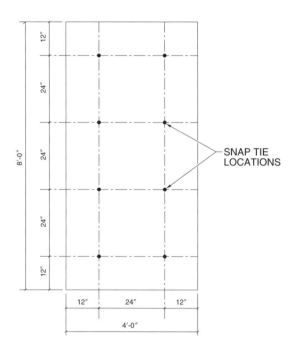

Figure 6-9. Snap ties are calculated as a specified number per square foot of wall area.

Wall walers are spaced at an OC distance determined by the wall width, rate of concrete placement, and other structural considerations in formwork design. The lower rows of walers may be doubled because of the greater hydrostatic pressure at the bottom of the form. The greater the height of the form, the greater the number of double rows of walers. The number of walers needed is calculated by taking the overall height of the form wall and dividing by the OC spacing to determine the number of rows of walers, then multiplying by the number of walers in each row (single or double). For example, a wall 22'-0" long and 8'-0" high has double walers placed at a 2' vertical spacing. The wall requires 4 rows of double walers (8' ÷ 2' = 4). With double walers, the wall requires 8 (4 rows × 2 walers = 8 walers) continuous 2" × 4" rows the entire length of the wall on both sides of the wall. The linear feet of walers is found by multiplying the length of the wall by the number of walers times 2 for both sides of the

wall. A total of 352 lf of 2" x 4" lumber is required for the walers (22' × 8 × 2 = 352 lf). An allowance may be made for overlap at corners.

Patented Walls. Calculating the number of patented wall forms normally requires development of a wall elevation drawing and a schedule of the forms needed for each wall. The formwork schedule notes the size and type of forms and the quantity of each required for a specific job. The estimator must also calculate the amount of fastening hardware required based on the fastening method specifications. Snap ties and walers are estimated in the same manner as job-built forms.

Slabs

A slab-on-grade commonly requires a minimal amount of formwork materials. Formwork may be more complex in special applications such as ice rinks, stepped seating, or freezer slabs. The estimator must take into consideration the proper subsurface preparation and reinforcing. Elevated slabs require more planning and information to perform the takeoff because of the equipment and materials required for formwork.

Formwork for slabs depends on the slab to be placed and whether the sides of the slabs are to be open or framed. The slab formwork is an extension of the exterior wall formwork and calculated as such if the slab is to be a continuous member and placed as part of the walls.

Concrete slab formwork includes an edge form set to the required height of the slab that is braced to provide support while the concrete sets.

WALL FORM CALCULATIONS

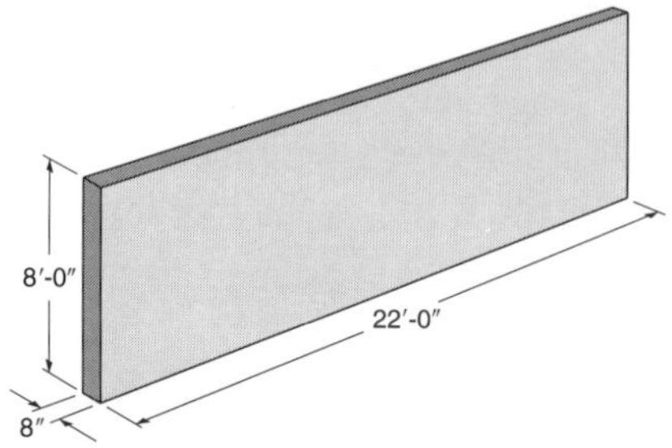

2″ x 4″ STUDS FOR JOB-BUILT FORM 16″ OC

2 STUDS AT EACH POSITION

22 x 2 x .75 = 33 STUDS

33 x 2 = **66 STUDS**

NOTE: ADD ADDITIONAL STUDS FOR CORNERS

FRAMING LUMBER

1 TOP PLATE

1 BOTTOM PLATE

22′ x 2 = 44 lf

44′ x 2 = **88 lf**

PLATES

22′-0″ x 8′-0″ x 2 = 352 sq ft

NUMBER OF 4′ x 8′ PANELS REQUIRED

$\frac{352'}{32'}$ = **11 PANELS**

SURFACE AREA OF EACH PANEL

WALL FORM MATERIAL

2′ OC WALER SPACING

8′ ÷ 2′ = 4 ROWS — DOUBLE WALERS

4 ROWS x 2 WALERS = 8 WALERS

22′ x 8 x 2 = **352 lf**

NOTE: ADD ALLOWANCE FOR OVERLAP AT CORNERS

DOUBLE WALERS

8 SNAP TIES PER PANEL

NUMBER OF SNAP TIES PER PANEL

5.5 x 8 = **44 TIES**

NUMBER OF PANELS FOR ONE SIDE OF WALL

SNAP TIES

Figure 6-10. Wall length and height are necessary for wall form calculations.

Slab-On-Grade. Formwork for a slab-on-grade consists of a stop for the concrete placement at the perimeter of the slab and bracing of wood or metal stakes. Calculating the linear feet around the perimeter of the slab area to be placed results in the linear feet of forms needed. See Figure 6-11. The height of the slab determines the type of lumber needed for the forms. Wood or metal bracing consisting of wood or metal stakes is used to hold the forms in alignment.

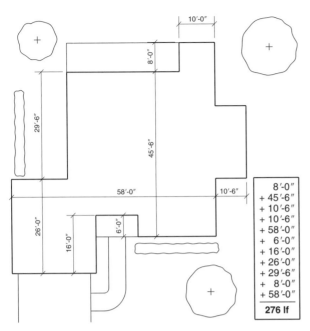

| 8'-0" |
| + 45'-6" |
| + 10'-6" |
| + 10'-6" |
| + 58'-0" |
| + 6'-0" |
| + 16'-0" |
| + 26'-0" |
| + 29'-6" |
| + 8'-0" |
| + 58'-0" |
| **276 lf** |

Figure 6-11. Plan views provide perimeter information for slab formwork takeoff.

Elevated Slabs. Formwork components for elevated slabs include beams, columns, decking, domes, and shoring. The formwork for elevated slabs may be calculated as the square feet of contact area (SFCA) of the underside of the slab.

Formwork for beams is calculated as the SFCA of open surface, beam sides, and beam bottom. A spandrel beam has formwork on three sides. The outside beam side includes a slab edge form. An intermediate beam has formwork on three sides (beam sides and bottom). For example, to calculate the amount of formwork needed for a spandrel beam, the widths of the three open sides are added and the total width is multiplied by the total length of the beams of that size.

For columns, the estimator must note that column dimensions vary. The first step in determining formwork requirements for columns is to develop a schedule of column dimensions and heights and determine the total number of each type of column to be filled at each placement. See Figure 6-12. Column formwork is commonly removed and reused.

For columns, formwork is calculated in the same manner as wall formwork. The face area is calculated in square feet or the number of patented forms required for each. For square or rectangular columns, multiply the column perimeter by the column height to determine the square feet of formwork needed. Circular columns are commonly formed using fiberboard or

metal forms. The height and diameter of each column should be calculated and recorded. For decking, the square feet of slab area is determined by multiplying slab length by slab width.

Where domes are used for a one-way or two-way beam system, the estimator should develop a schedule in a manner similar to a column schedule. Using the architectural drawings, the estimator determines the number of each dome necessary and enters the totals in a ledger sheet, spreadsheet, or estimating software cell.

Shoring requirements are specified by an engineer based on the live and static loads placed on formwork. Estimators should consult shoring manufacturers and formwork designers for detailed costs concerning loading capabilities of various shoring systems. Costs per square foot for shoring depend on the height between floors for an elevated slab, the thickness of the slab, and the concrete placement method. These costs are based on manufacturer and supplier information or company historical data.

CAST-IN-PLACE CONCRETE MATERIALS

Architectural drawings and specifications contain notes and information concerning specific concrete performance requirements. This information includes 28-day strength, aggregate size, slump, water/cement ratio, air entrainment, and admixtures. See Figure 6-13. Overall concrete strength is stated as the pounds per square inch (psi) of pressure that the concrete can withstand after 28 days.

Concrete

Concrete is composed of portland cement, aggregate (sand and gravel), admixtures, and water. By varying the quantity and type of each component, concrete properties are changed in regard to workability, strength, and setting time. Chemical admixtures may also be introduced into the mix to change the properties of the basic mix of ingredients. An *admixture* is a material other than water, aggregate, fiber reinforcement, and portland cement used as an ingredient in a batch of concrete.

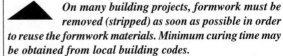
On many building projects, formwork must be removed (stripped) as soon as possible in order to reuse the formwork materials. Minimum curing time may be obtained from local building codes.

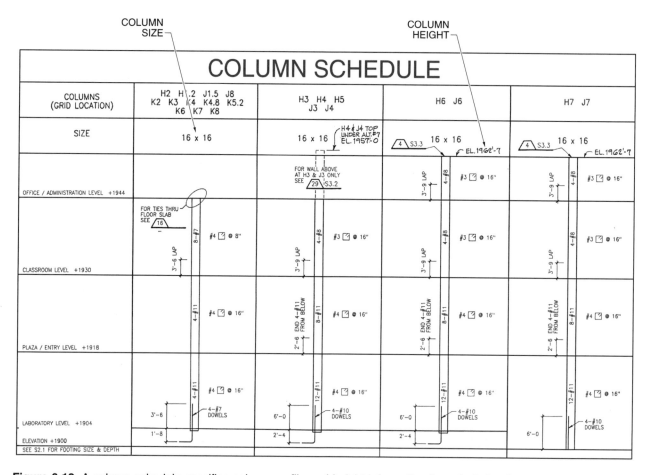

Figure 6-12. A column schedule specifies column profile and height information for calculating formwork.

Slump is the amount of sagging that occurs in a 12″ high conical fresh concrete sample. The slump of concrete measures the consistency of the fresh mix and can be used as a general measure of water content. The amount of water in the concrete mix is important in determining final strength. The proper amount of water is required to fully hydrate the cement, while excessive water weakens the mix. The water/cement ratio is the weight of the water divided by the weight of the cement contained in one cubic yard of concrete.

Portland Cement. The five different types of portland cement are Type I, Type II, Type III, Type IV, and Type V. Type I is a general-purpose cement used where no special design properties are required. Type II has a slightly lower temperature production during hydration and provides some protection in areas where the concrete may be exposed to sulfate attack. Type III is used where forming needs to be removed rapidly, cold temperatures are prevalent during placement, or high strengths are needed at an early stage. Type IV has a

low hydration heat generation and is suitable for large placements. Type V has a high resistance to sulfates.

Aggregates. *Aggregate* is granular material such as gravel, sand, vermiculite, or perlite that is added to cement to form concrete, mortar, or plaster. Aggregates are divided into fine and coarse categories. Particles less than $3/8″$ are graded fine, and those equal to or greater than $3/8″$ are graded coarse. Rock materials used for aggregate include sand, crushed stone or gravel, shale, slate, clay, or slag. Aggregates, including sand and gravel, represent approximately 70% of the overall volume of concrete. Aggregate information in the specifications affects final concrete strength and finishing properties. Specifications may describe the type and maximum size of aggregate used.

> ▲ *In a concrete mixture, components such as rock, sand, and water can separate if improperly handled. Separation of these compounds can severely reduce the strength of the concrete.*

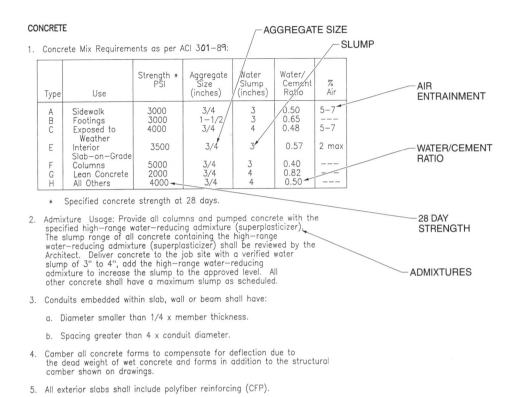

Figure 6-13. Concrete design specifications have an effect on concrete costs and placement requirements.

Admixtures. Admixtures may be added to concrete to affect air content, setting time, resistance to freezing, and flow. The addition of admixtures for air entrainment creates minute air pockets in concrete that increase workability and resistance to freezing and thawing and decrease separation of the concrete components. See Figure 6-14. Information concerning use and types of admixtures is included in the specifications.

Reinforcement. The most common method of reinforcing piles, roadways, footings, columns, walls, and beams is with steel bars having deformed surfaces. These bars (rebar) are tied together in various configurations to reinforce large concrete structures. Ridges are formed into the surface of the bars to improve their adhesion to the surrounding concrete. Number 3 through #8 reinforcing steel bars are sized in increments of $1/8''$ diameter. For example, a #4 bar is $4/8''$ ($1/2''$) in diameter. See Figure 6-15.

Steel grades used for reinforcing bars include grade 40, grade 50, grade 60, and grade 75. Grade 40 has a minimum yield strength of 40,000 psi. Grade 50 has a minimum yield strength of 50,000 psi. Grade 60 has a minimum yield strength of 60,000 psi.

Grade 75 has a minimum yield strength of 75,000 psi. Reinforcing steel bars may be coated with epoxy for exterior applications. Epoxy-coated reinforcing bars are used in road and bridge construction to minimize the effects of corrosion on the steel.

RECOMMENDED AIR CONTENT*		
Nominal Maximum Size of Coarse Aggregate**	**Exposure**	
	Mild	Extreme
$3/8$	4.5	7.5
$1/2$	4.0	6.0
$3/4$	3.5	6.0
1	3.0	6.0
$1 1/2$	2.5	5.5
2	2.0	5.0
3	1.5	4.5

* in %
** in in.

Figure 6-14. Aggregate and exposure variables affect the proper amount of air content in concrete for maximum effectiveness.

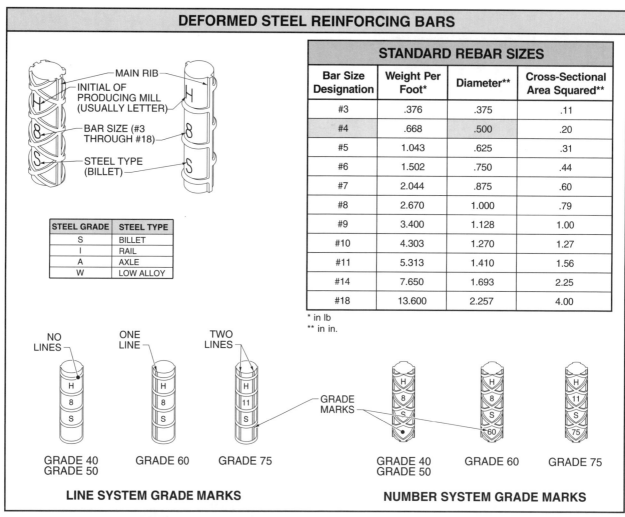

DEFORMED STEEL REINFORCING BARS

STANDARD REBAR SIZES

Bar Size Designation	Weight Per Foot*	Diameter**	Cross-Sectional Area Squared**
#3	.376	.375	.11
#4	.668	.500	.20
#5	1.043	.625	.31
#6	1.502	.750	.44
#7	2.044	.875	.60
#8	2.670	1.000	.79
#9	3.400	1.128	1.00
#10	4.303	1.270	1.27
#11	5.313	1.410	1.56
#14	7.650	1.693	2.25
#18	13.600	2.257	4.00

* in lb
** in in.

STEEL GRADE	STEEL TYPE
S	BILLET
I	RAIL
A	AXLE
W	LOW ALLOY

LINE SYSTEM GRADE MARKS

NUMBER SYSTEM GRADE MARKS

REINFORCING STEEL STRENGTH AND GRADE

ASTM Specification	Minimum Yield Strength*	Ultimate Strength*
Billet Steel ASTM A-615		
Grade 40	40,000	70,000
Grade 50	50,000	90,000
Grade 75	75,000	100,000
Rail Steel ASTM A-616		
Grade 50	50,000	80,000
Grade 60	60,000	90,000
Axle Steel ASTM A-617		
Grade 40	40,000	70,000
Grade 60	50,000	90,000
Deformed Wire ASTM A-496		
Welded Fabric	70,000	80,000
Cold Drawn Wire ASTM A-82		
Welded Fabric < W1.2	56,000	70,000
Size ≥ W1.2	65,000	75,000

* in psi

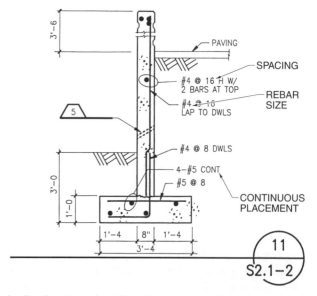

Figure 6-15. Reinforcing bars (rebar) are identified with a standardized system denoting size and strength characteristics.

Reinforcing bars are shown on section views with solid circles for end views and solid lines for edge views. Detail drawings for cast-in-place concrete give specific reinforcing installation information including rebar size, spacing, and placement within the finished concrete.

Welded wire fabric is also used to reinforce concrete slabs. Rolls of welded wire fabric are stretched out and set in place to reinforce lightweight slabs such as sidewalks. Sizing of welded wire fabric is designated by W number (spacing and cross-sectional area) or the wire gauge. See Figure 6-16. Wires may be smooth or deformed. Welded wire fabric has yield strengths between 60,000 psi and 70,000 psi. Welded wire fabric is shown on detail drawings as a dashed line.

Each portion of the architectural, structural, and detail drawings for concrete structures provides reinforcing steel information. Cages of upright bars with intermediate stirrup ties are formed for columns. The size of the upright and stirrup tie bars and the spacing of the bars in both directions are indicated on the column schedule. See Figure 6-17. A variety of different bar sizes and shapes are required for beams. A schedule for sizes, shapes, and lengths is provided in the beam schedule. Schedules for the placement of reinforcing bars in cast-in-place concrete joists and slabs are also included in the structural drawings.

> ▲ *In slab construction, concrete should be placed so each batch is deposited against previously placed concrete. Concrete should not be placed in separate piles and worked together.*

CAST-IN-PLACE CONCRETE METHODS

Proper transportation, handling, and placement of concrete is required for the production of the final concrete structure. Estimators must ensure that equipment and provisions are included in the bid for accessibility of the concrete source to the final placement site, equipment required for moving the concrete from the supply source to the forms, proper placement in the forms, vibration after placement, and possible protection from the elements if additional heating, cooling, or hydration water is required.

Placement

Concrete is commonly placed in walls and slabs from a ready mix truck. Concrete is mixed or batched and transported to the construction site. It may be more economical to rent or construct a batch plant at the construction site for large projects due to the size of the job or accessibility from existing batch plants. Estimators should determine the most economical method for delivery and placement of concrete into forms for walls and slabs.

Walls. For most walls, concrete can be directly discharged from the ready mix truck into the wall forms. Where walls are far below grade and cannot be accessed by a ready mix truck, a tremie or chute is used to place concrete and prevent long drops that could result in separation of the cement paste and aggregate. A *tremie* is a pipe or tube through which concrete is placed into vertical formwork or under water. For slabs or walls above grade, a concrete pump or bucket attached to a crane is used for lifting and placing concrete. See Figure 6-18.

COMMON STOCK STYLES OF WELDED WIRE FABRIC				
New Designation (W-Number)	Old Designation (Wire Gauge)	Steel Area*		Weight**
		Longitudinal	Transverse	
6 × 6 – W1.4 × W1.4	6 × 6 – 10 × 10	.028	.028	21
6 × 6 – W2.0 × W2.0	6 × 6 – 8 × 8	.040	.040	29
6 × 6 – W2.9 × W2.9	6 × 6 – 6 × 6	.058	.058	42
6 × 6 – W4.0 × W4.0	6 × 6 – 4 × 4	.080	.080	58
4 × 4 – W1.4 × W1.4	4 × 4 – 10 × 10	.042	.042	31
4 × 4 – W2.0 × W2.0	4 × 4 – 8 × 8	.060	.060	43
4 × 4 – W2.9 × W2.9	4 × 4 – 6 × 6	.087	.087	62
4 × 4 – W4.0 × W4.0	4 × 4 – 4 × 4	.120	.120	85

* in sq in./ft
** in lb per 100 sq ft

Figure 6-16. Floor slabs that support light loads are often reinforced using welded wire fabric.

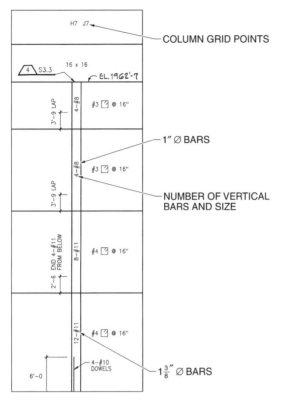

Figure 6-17. A column schedule provides information concerning reinforcing steel bars for each concrete column.

Figure 6-18. Concrete pumps deliver concrete to hard to reach areas that are inaccessible by concrete ready mix trucks.

Slabs. Concrete for a slab-on-grade is commonly discharged into wheelbarrows or motorized buggies for placement. For slabs-on-grade and elevated slabs, concrete pumps or buckets attached to cranes are also used where height or accessibility due to reinforcement or other impediments do not permit access by other means.

Finishes

Estimators should consult the elevation drawings and specifications to determine the finishes for exposed concrete surfaces of walls, slabs, columns, beams, and decks. See Figure 6-19. The surface finish required can affect the form material provided, form removal times, and cost per square foot for surface finish.

Figure 6-19. Concrete slab finish requirements, which affect labor costs, are described in the specifications.

Walls. Exposed concrete wall surfaces may have a variety of finishes including smooth, textured, and various architectural designs. Smooth concrete is produced when the surface of the concrete is rubbed and stoned smooth after form removal. Any voids are patched and filled to produce a smooth surface. Textured concrete is commonly produced by inclusion of decorative aggregate and sandblasting of the concrete surface. This produces a rough texture. Architectural designed concrete can be produced by lining the inside of the concrete forms with wood, metal, or plastic form liners that produce various patterns, such as vertical or horizontal grooves or a simulated brick pattern. See Figure 6-20. Finishes increase the cost per square foot for formwork and finishing of wall surfaces.

Slabs. Slab surface finishes are smooth, exposed aggregate, patterned, or superflat. Broom finish slabs are used for sidewalks, roadways, and other common applications. Concrete is placed, screeded to grade, and finished with trowels or brooms as noted in the specifications to produce the required finish. Exposed aggregate surfaces are produced by including decorative aggregate in the concrete and washing and brushing the cement paste off of the slab surface prior to reaching a final set. Patterned concrete slabs are created by adding concrete coloring and using metal patterns to create brick, stone, or other decorative finishes. Superflat floors are used in large industrial applications where automated equipment or machinery requires a very flat surface to ensure proper operation after installation. Laser screeds are used to create an extremely flat surface during finishing of the concrete. The surface finish of the slab creates cost variables which the estimator must include in the slab cost portion of the bid.

Figure 6-20. Exposed concrete wall surfaces may have a variety of finish designs.

Elevated slab components also include columns, beams, and the underside of the slab. In a manner similar to concrete walls, the specifications are checked to ensure provisions are included in the bid for surface finishing of these components.

CAST-IN-PLACE CONCRETE QUANTITY TAKEOFF

Items to consider for concrete quantity takeoff include the type and volume of concrete required, reinforcing, transportation, surface finish, and possible climatic protection. Each of these elements is commonly unit priced according to the standard concrete volume measurement of a cubic yard. Some hidden costs for concrete takeoff include waiting time during concrete delivery and discharge into forms and items in the concrete mix such as special sand or cement. All of these items should be included in the final concrete bid.

Volume Calculations

Concrete volume is specified in cubic yard measurements. Knowledge of basic geometric volume formulas is essential when calculating these volumes. See Appendix. In estimating software, entry of the basic dimensional information results in automatic volume calculations. The volume calculations must include divisions for the various concrete materials used and variables in the mixes where changes in aggregates, cement, admixtures, or reinforcing may affect costs. Each type of concrete placed should be priced separately according to the cost per cubic yard from the supplier. Concrete volumes may be reduced for applications containing large or numerous blockouts. A 1% to 2% waste allowance should be added to each concrete calculation to allow for spillage during placement.

Columns. The volume of concrete for columns is computed by multiplying the height by the cross-sectional area taken from floor to ceiling. See Figure 6-21. For example, a 2'-0" square column 8'-6" high supports a 6" concrete slab. The volume of the slab is ignored when calculating the volume of the column. To calculate the volume of the column, multiply the height by the cross-sectional area of the column to obtain 34 cu ft ($8.5' \times 2' \times 2' = 34$ cu ft). Adding a 2% waste factor results in a total of 34.68 cu ft (34 cu ft $\times$ 1.02 = 34.68 cu ft). The total cubic feet is divided by 27 to obtain 1.28 cu yd (34.68 $\div$ 27 = 1.28 cu yd) of concrete per column. Estimators enter this total for this type of column and calculate the total number of each type of column to enter on a ledger sheet or spreadsheet. When using estimating software, entry of the column dimensions, height, and the number of columns results in a final cubic yard calculation. Estimators must ensure that column heights and cross-sectional dimensions are checked for each column because these dimensions vary.

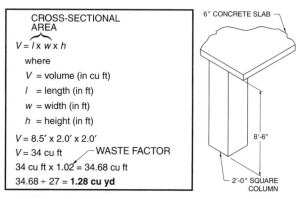

Figure 6-21. Column concrete volume is based on the column height and cross-sectional area.

Schedules in the prints and notes on architectural details provide information concerning reinforcing steel for various columns. The quantities and sizes of bars for each column are determined and entered in a ledger sheet, spreadsheet, or estimating software cell. The total number of each type of bar is calculated based on the individual totals.

Walls. Concrete foundation walls and sizes are often grouped by the estimator. Concrete wall volumes are based on wall length, width, and height dimensions. For example, the volume of concrete for a wall 30′ long, 8″ wide, and 6′ high is determined by multiplying 30′ × .66̄′ × 6′ for a total of 120 cu ft (30′ × .66̄′ × 6′ = 120 cu ft), which equals 4.44̄ cu yd (120 cu ft ÷ 27 = 4.44̄ cu yd). When using estimating software, the length, width, and height dimensions are entered into the program, which automatically calculates the concrete volume and adds it to the overall total.

Wall reinforcing steel placement, spacing, and sizes are determined from detail drawings. The linear feet of each bar is determined by adding the linear feet of wall and the reinforcing per linear foot of each wall.

Slabs. The volume of concrete for slabs is found by multiplying the surface area in square feet by the slab depth. For example, a 10′ by 20′ slab that has a depth of 6″ is taken off by multiplying 10′ × 20′ × .5′ to obtain 100 cu ft (10′ × 20′ × .5′ = 100 cu ft), which equals 3.7 cu yd (100 cu ft ÷ 27 = 3.7 cu yd). When using estimating software, entry of the slab width, length, and depth results in an automatic calculation of the concrete volume. See Figure 6-22.

SLAB DIMENSION
INFORMATION

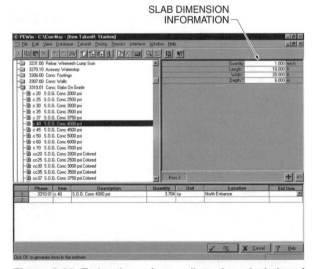

Figure 6-22. Estimating software allows the calculation of a number of slab components by entering slab dimension information.

The number of reinforcing steel bars of each type is determined by calculating the spacing in each direction and calculating the number of bars required in a given number of square feet. Additions are made for the amount of bar overlap required in the specifications or detail drawings. Reinforcing steel bars are commonly priced as the cost per ton of steel, with costs varying with the diameter of the bar and steel required. Welded wire mesh is calculated by the number of square feet of coverage required.

Stairs. Concrete volume for stairs is calculated by multiplying the number of treads by the width of the stairway. Always consult the detail drawings for concrete stair volume calculations. A rough estimate can be made by using 1 cu ft of concrete per linear foot of tread. For example, to find the amount of concrete needed for a concrete stairway with 10 treads each 3′-6″ wide, multiply the number of treads by the width of the stairway. This results in 35 cu ft of concrete (10 × 3.5′ = 35 cu ft) which equals 1.3 cu yd (35 cu ft ÷ 27 = 1.3 cu yd).

PRECAST CONCRETE MATERIALS

Developments in concrete design and lifting equipment have led to the widespread use of precast concrete members. Precast concrete members are used for beams, walls, slabs, piling, and a variety of other construction members. Precast concrete members may be precast at the job site and lifted into place or formed and placed at a casting yard and transported to the job site by truck. All precast members require special reinforcing to ensure they meet the stresses of transportation and installation in addition to final structural loads. See Figure 6-23. Concrete members that are precast and transported to the job site include beams, piling, exterior wall finish panels, paving materials, and various piping and masonry units.

Beams

Precast concrete beams are used for building and bridge construction. Precast concrete beams include rectangular, L-shaped, and inverted tee beams. See Figure 6-24. The beam used is indicated on detail drawings. Structural engineers design the placement of reinforcing steel based on loading to be imposed on the beam. Beams are commonly reinforced by prestressing or post-tensioning. A *prestressed beam* is a beam containing steel reinforcing cables that are put in tension before the concrete is placed. After the concrete is set, the tension on the cables is transferred to the concrete, putting the beam in tension. A *post-tensioned beam* is a beam containing steel reinforcing cables that are tensioned after the concrete has set.

Figure 6-23. Precast concrete members minimize the need for extensive job site formwork for walls and other members.

Rectangular beams are the most commonly used beams due to their ease of forming and casting. L-shaped beams are incorporated as perimeter beams for spandrels and around openings. Inverted tee beams are used for spanning between column caps, beams, or girders in bridge construction. Inverted tee beams may also be formed as a double tee with two vertical sections. Precast tees may be joined side by side to create an integrated unit for bridge beams and road surfaces.

Precast slabs are available in a variety of thicknesses and designs. Precast slabs are used for above-grade floor slab systems and for roof panels. Precast slabs may have a hollow core or solid core. See Figure 6-25. The type and size of precast slab panels are noted on plan and elevation drawings. The thickness of the precast slab member is noted in inches preceded by HC for hollow core or FS for flat slab. For example, a slab designated as FS6 is a solid flat slab 6″ thick.

Identification may also be accomplished with a manufacturer code number or name. Methods for fastening the slabs to each other and to supporting members are shown on detail section drawings. Fastening may be accomplished by bolting or welding steel inserts.

Walls

The three primary concrete precast walls are structural, finish, and curtain wall panels. Structural precast wall panels are most commonly cast on the job site and tilted up into place. Finish precast wall panels are cast off site with various concrete colors, aggregate sizes and types, and surface finishes. A *curtain wall* is a non-bearing prefabricated panel suspended on or fastened to structural members.

Structural precast wall panels are cast with door and window openings in place and provisions for lifting and supporting other structural members cast into the wall. Many surface textures for precast concrete panels are used for structural, finish, and curtain walls. For curtain walls, steel angles or plates are cast into the panels to allow them to be welded to other members such as structural steel columns and beams or reinforced concrete members with adjoining steel inserts. Precast wall panels are lifted into place with a crane, braced, and fastened in place.

Precast concrete wall panels may be cast with a layer of insulation to increase thermal resistance.

PRECAST CONCRETE BEAM SAFE IMPOSED LOADS*													
Beam	Designation	Strand No.	H**	H1/H2**	Span†								
					18	22	26	30	34	38	42	46	50
RECTANGULAR (B = 12" OR 16")	12RB24	10	24	—	6726	4413	3083	2248	1684	1288	1000	—	—
	12RB32	13	32	—	—	7858	5524	4059	3080	2394	1894	1519	1230
	16RB24	13	24	—	8847	5803	4052	2954	2220	1705	1330	—	—
	16RB32	18	32	—	—	—	7434	5464	4147	3224	2549	2036	1642
	16RB40	22	40	—	—	—	—	8647	6599	5163	4117	3332	2728
L-SHAPED	18LB20	9	20	12/8	5068	3303	2288	1650	1218	—	—	—	—
	18LB28	12	28	16/12	—	6578	4600	3360	2531	1949	1524	1200	—
	18LB36	16	36	24/12	—	—	7903	5807	4405	3422	2706	2168	1755
	18LB44	19	44	28/16	—	—	—	8729	6666	5219	4166	3370	2754
	18LB52	23	52	36/16	—	—	—	—	9538	7486	5992	4871	4007
	18LB60	27	60	44/16	—	—	—	—	—	—	8116	6630	5481
INVERTED TEE	24IT20	9	20	12/8	5376	3494	2412	1726	1266	—	—	—	—
	24IT28	13	28	16/12	—	6951	4848	3529	2648	2030	—	—	—
	24IT36	16	36	24/12	—	—	8337	6127	4644	3598	2836	2265	1825
	24IT44	20	44	28/16	—	—	—	9300	7075	5514	4378	3525	2868
	24IT52	24	52	36/16	—	—	—	—	—	7916	6326	5132	4213
	24IT60	28	60	44/16	—	—	—	—	—	—	8616	7025	5800

* safe loads shown indicate 50% dead load and 50% live load and 800 psi top tension requiring additional top reinforcement
** in in.
† in ft

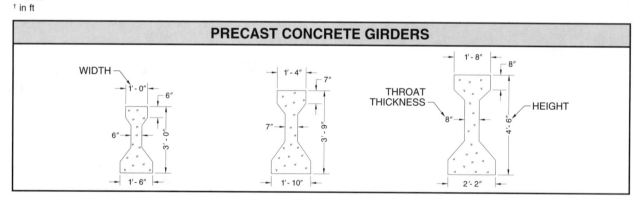

PRECAST CONCRETE GIRDERS

Figure 6-24. Different precast beam designs are used depending on building loading requirements.

PRECAST CONCRETE CONSTRUCTION METHODS

Precast concrete construction methods include plant precast and site precast. Plant precast concrete materials cast at an off-site plant are designed according to industry standards or job-specific specifications. These materials are transported to the job site and set in place. Site precast components are normally too large to be transported. These components are cast at a site that minimizes formwork and concrete movement and are lifted or jacked into place after reaching a set with sufficient strength to allow moving.

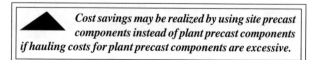

Cost savings may be realized by using site precast components instead of plant precast components if hauling costs for plant precast components are excessive.

PRECAST SLABS				
Hollow Core		**Solid Core**		
Thickness*	**Designation**	**Thickness***	**Designation**	
6	4HC6	4	FS4	
8	4HC8	6	FS6	
10	4HC10			
12	4HC12	8	FS8	

* in in.

Figure 6-25. Precast slabs are available in a variety of thicknesses and may have a hollow core or solid core.

Plant Precast

Concrete piping, curtain walls, beams, and pavers may be plant precast and transported to the job site. Estimators determine from the architectural drawings and specifications if the architect has used standard-size components or if special casting arrangements are required.

Site Precast

Site precast components include floor slabs and walls. Floor slabs are set with perimeter forms and jacked into place. Walls are precast and tilted up into place with cranes. Site precast construction methods include lift slab and tilt-up construction.

Lift Slab. *Lift slab construction* is a method of concrete construction in which horizontal slabs are cast on top of each other, jacked into position, and secured to columns at the desired elevation. A *casting bed* is the forms and support used for forming precast concrete members. A chemical bond breaker is applied to the surface of a slab to allow for the upper slab to be lifted when necessary. A series of slabs are cast with the appropriate blockouts for columns and mechanical equipment runs. As the slabs are cast, they are jacked to the proper elevation and supporting columns and beams are placed. This method minimizes shoring costs associated with conventional elevated slab construction. Details for reinforcing steel and lifting anchors are included in the detail drawings.

Tilt-Up. *Tilt-up construction* is a method of concrete construction in which concrete members are cast horizontally at a location close to their final position and tilted in place after removal of the forms. A *tilt-up panel* is a panel that is precast and lifted into place at the job site. See Figure 6-26. This method of precast construction is used for construction of low-rise concrete structures of one to three stories. Print information used by estimators for tilt-up panels includes wall and opening dimensions, placement and types of lift anchors and reinforcement, and interior and exterior wall finish.

Precast concrete wall panels are placed in perimeter edge footings and braced until floors and roofs are constructed.

Low forms are set on the slab after placement of the first floor slab. The height of the forms is equal to the thickness of the wall to be precast. The outside perimeter of the forms is set to the dimensions of the wall. Provisions are made around the perimeter of the form for the projection of reinforcing steel. A bond breaker is applied to the floor slab to ensure concrete placed for the wall does not adhere to the first floor slab. Blockouts are set in place for door and window openings and beam pockets. Any surface treatment to be provided on the downward facing side of the wall is set in place, such as welding plates for connecting to adjoining panels or form liners to create decorative finishes.

Figure 6-26. Precast tilt-up wall panels allow large sections of concrete walls to be precast and lifted into place at the job site.

Detail drawings for reinforcing steel for precast tilt-up wall panels are similar to those given for cast-in-place concrete. Lift anchors are set in the concrete wall for the attachment of lifting hardware and crane cables. Brace inserts are also set in the concrete wall to allow for attachment of braces. Special symbols on detail drawings and manufacturer manuals describe the lift anchors and brace inserts placed in the concrete.

A crane with a spreader bar to facilitate lifting is used to set the finished tilt-up wall panels in place after the concrete has reached the proper hardness. Cables are fastened to the lift anchors cast in the wall. The wall is lifted and pivoted into final position. Braces are then bolted to the brace inserts cast into the wall. Each wall section is braced for final positioning. Braces for tilt-up wall sections remain in place until panels are fastened together and floors and roofs are in place. Floors and roofs tie the tilt-up wall sections together and provide stability to the entire structure.

Tilt-up wall panels are fastened together using flush-poured pilasters, poured columns, precast columns, steel columns, or flush steel plates. See Figure 6-27. Detail drawings indicate the panel joining method used.

PRECAST CONCRETE QUANTITY TAKEOFF

The estimator begins precast concrete quantity takeoff by consulting a schedule for the various precast materials. Each precast component is entered on a separate row with columns for cost information, including material, labor, transportation, and lifting costs.

Plant Precast

Estimators consult with the casting plant to obtain costs for plant precast components. Costs include cost per unit and transportation costs. Placement costs at the job site are not included in the plant precast costs and must be added in pricing these components.

Site Precast

Material quantities for site precast components are determined in a manner similar to cast-in-place concrete items. Concrete volume, reinforcing, and surface finish are calculated the same as for cast-in-place slabs and walls. Differences in takeoff include forming costs and lifting costs.

 Concrete takeoff requires careful review of plans and specifications. Information required by the estimator can appear on drawings other than structural drawings.

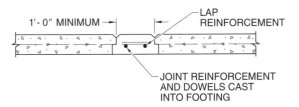

FLUSH-POURED PILASTER

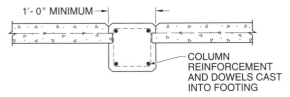

POURED COLUMN

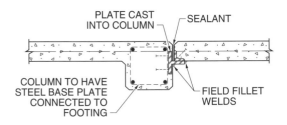

PRECAST COLUMN

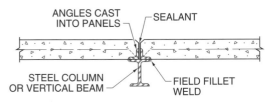

STEEL COLUMN

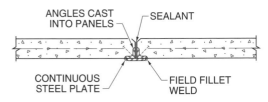

FLUSH STEEL PLATE

Figure 6-27. Flush-poured pilasters, poured columns, precast columns, steel columns, and flush steel plates are methods used to join tilt-up wall panels after placement.

Lift Slab. Concrete volume, reinforcing, and finishing for lift slabs is calculated the same as for a slab-on-grade. Additional expenses are determined for the additional reinforcing required for lifting of the slab, jack-ing and lifting equipment operations, and shoring during placing and setting of support columns.

Tilt-Up. Tilt-up walls are placed similarly to slabs so finishing labor costs are similar to slab finishing labor costs. Additional tilt-up costs include crane costs for lifting and temporary bracing of the wall required until support columns are set or placed between the tilt-up wall panels.

Several tilt-up wall panels may be similar in design for large projects. Estimators may create a schedule of tilt-up wall panels to minimize duplication of efforts in takeoff for identical wall panels.

ESTIMATING SOFTWARE – MULTIPLE ITEM QUICK TAKEOFF

The Precision estimating software quick takeoff function can be used to take off multiple items and enter dimensions for all of the items at one time. Multiple item quick takeoff is accomplished by applying a standard procedure. See Figure 6-28.

The concrete volume of column and floor slabs is calculated separately.

PRECISION ESTIMATING SOFTWARE — MULTIPLE ITEM QUICK TAKEOFF

1. Open the Quick Takeoff window.

2. Take off items 3111.00 10 Footing Forms, 3111.00 50 Keyway in Footing, and 3306.00 c 30 Footing Conc 3000 psi. Hold down the CTRL key and click on item 10 Footing Forms, item 50 Keyway in Footing, and item c 30 Footing Conc 3000 psi.

QUICK TAKEOFF WINDOW →

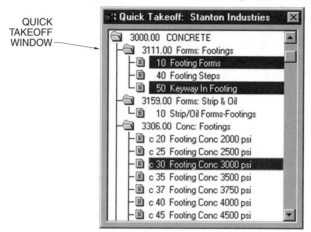

Note: In addition to taking off items by double-clicking each item or dragging and dropping the items to the spreadsheet, time is saved by holding down the CTRL key and clicking on multiple items to be taken off. Once all items have been tagged, drag and drop them to the spreadsheet.

3. Once all items have been added to the spreadsheet, the dimensions can be entered for all items at once. Position the cursor in the takeoff quantity cell for the Footing Forms and while keeping the mouse button depressed, drag the cursor down to the last item taken off (Footing Concrete).

Description	Takeoff Quantity
GENERAL REQUIREMENTS	
Personnel: Supervision	**wk**
Superintendent	48.00 wk
SITEWORK	
Earthwk: Footings (Mach)	**cy**
Excavate Footing By Machine	29.63 cy
CONCRETE	
Forms: Footings	**sf**
Footing Forms	sf
Keyway In Footing	lf
Conc: Footings	**cy**
Footing Conc 3000 psi	cy

Once these cells have been highlighted, right mouse click anywhere in the block of cells and select Enter Dimensions from the shortcut menu. Enter 50 for Length, 3 for Width, and 4 for Depth of concrete.

4. Click the Add to Quantity button. *Note:* Each item chosen has an associated formula. The Enter Dimensions function prompts the estimator with the appropriate variables needed to calculate the formula.

Figure 6-28. The estimating software quick takeoff function can be used to take off multiple items and enter dimensions for all of the items at one time.

Estimating

_____ 1. _____ is a material consisting of portland cement, aggregate, and water, which solidifies through chemical reaction into a hard, strong mass.

_____ 2. A(n) _____ is the section of a foundation that supports and distributes structural loads directly to the soil.

_____ 3. A(n) _____ is a temporary structure or mold used to retain and support concrete while it sets and hardens.

_____ 4. A(n) _____ bolt is a special bolt that threads into the ends of a tie rod.

_____ 5. A(n) _____ tie is a concrete form tie that is snapped off when the concrete is set.

_____ 6. A(n) _____ tie is a concrete form tie used in heavy construction in which a bolt screws into the tie and is removed during form removal.

T   F   7. A blockout is a structural piece, either permanent or temporary, designed to resist weights or pressures of loads.

T   F   8. A brace is a frame set in a concrete form to create a void in the finished concrete structure.

_____ 9. _____ is the amount of sagging that occurs in a 12″ high conical fresh concrete sample.

_____ 10. _____ is granular material that is added to cement to form concrete, mortar, or plaster.

_____ 11. A(n) _____ beam is a beam containing steel reinforcing cables that are put in tension before the concrete is placed.

_____ 12. A(n) _____-tensioned beam is a beam containing steel reinforcing cables that are tensioned after the concrete has set.

T   F   13. A curtain wall is a nonbearing prefabricated panel suspended on or fastened to structural members.

T   F   14. A casting bed is the forms and support used for forming precast concrete members.

T   F   15. Formwork is the system of support for freshly-placed concrete.

_____ 16. A(n) _____ is a horizontal or nearly horizontal layer of concrete.

_____ 17. A(n) _____ is a horizontal member used to align and brace concrete forms or piles.

_____ 18. A form _____ is a metal bar, strap, or wire used to hold concrete forms together and resist the pressure of wet concrete.

_____ 19. A(n) _____ wall is a wall constructed to hold back earth.

_____ 20. A(n) _____ wall is a wall built to hold back ground water.

_____   **21.** A(n) _____ is the supporting structure at the end of a bridge, arch, or vault.

_____   **22.** A(n) _____ form is a large form constructed by joining small panels.

T   F   **23.** A spread footing is a foundation footing with a wide base designed to add support and spread the load over a wide area.

T   F   **24.** A pile is a horizontal structural member installed in the ground to provide vertical and/or horizontal support.

T   F   **25.** An isolation joint is a separation between adjoining sections of concrete used to allow movement of the sections.

T   F   **26.** Shoring is wood or metal members used to temporarily support formwork or other structural components.

_____   **27.** _____ is the slight upward curve in a structural member designed to compensate for deflection of the member under load.

_____   **28.** A(n) _____ beam is a beam in the perimeter of a structure that spans columns and commonly supports a floor or a roof.

_____   **29.** _____ is light-gauge metal sheets used to construct a floor or deck form or used as floor or deck members.

_____   **30.** _____ construction is a method of concrete construction in which concrete members are cast horizontally at a location close to their final position and tilted in place after removal of the forms.

Wall Form Calculations

_____   **1.** A total of _____ panels is required.

_____   **2.** At eight snap ties per panel, _____ snap ties are required.

_____   **3.** At two studs per position, _____ studs are required.

_____   **4.** _____′ of top and bottom plates is required.

_____   **5.** At 2′ OC waler spacing, _____ walers are required.

Precast Concrete Girders

_____   **1.** Width

_____   **2.** Throat thickness

_____   **3.** Height

Panel Joining

_____   **1.** Flush steel plate

_____   **2.** Steel column

_____   **3.** Flush-poured pilaster

_____   **4.** Precast column

_____   **5.** Poured column

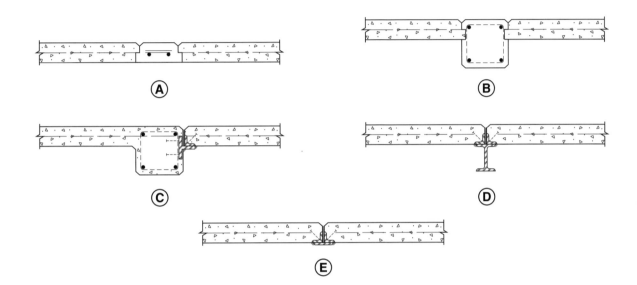

Ⓐ   Ⓑ   Ⓒ   Ⓓ   Ⓔ

Column Concrete Volume

_____   **1.** A total of _____ cu yd of concrete is required for Column A.

_____   **2.** A total of _____ cu yd of concrete is required for Column B.

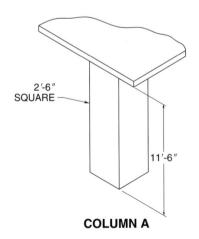

2'-6" SQUARE

11'-6"

COLUMN A

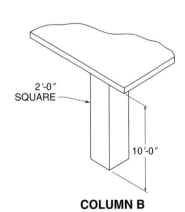

2'-0" SQUARE

10'-0"

COLUMN B

Deformed Steel Reinforcing Bars

_____ **1.** Initial of producing mill

_____ **2.** Bar size

_____ **3.** Steel type

_____ **4.** Main rib

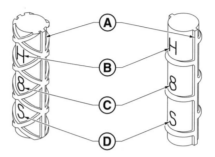

Activities
Concrete

Name _____ Date _____

6
Chapter

Activity 6-1 – Ledger Sheet Activity

Refer to Print 6-1 and Quantity Sheet No. 6-1. Take off the cubic yards of concrete and linear feet of reinforcing for the eight footings. The reinforcing consists of four #5 rebars 3′-0″ long equally spaced along the bottom of each footing.

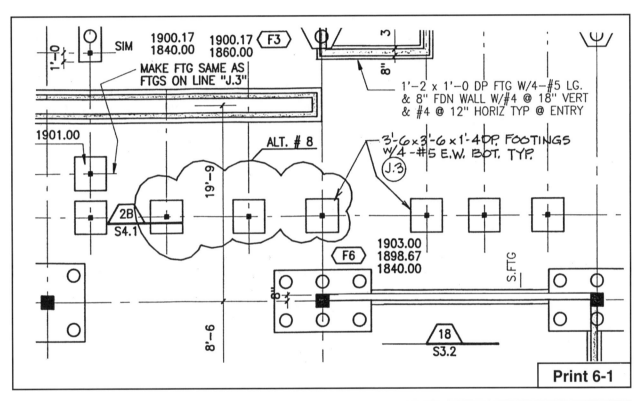

Print 6-1

QUANTITY SHEET

Sheet No. _6-1_

Project: _____
Estimator: _____

Date: _____
Checked: _____

No.	Description	Dimensions				Unit		Unit		Unit		Unit
		L	W	D								
	Footings (8)											
	Concrete											
	Reinforcing (4 × 8)											

137

Activity 6-2 – Spreadsheet Activity

Refer to the cost data, Print 6-2, and Estimate Summary Sheet No. 6-2 on the CD-ROM. Take off the square feet of wall forms and cubic yards of ready mix concrete for the walls of the elevator pit. Determine the total material and labor cost for the wall forms and ready mix concrete.

COST DATA			
Material	**Unit**	**Material Unit Cost***	**Labor Unit Cost***
Wall forms, job built	sq ft contact area	1.28	4.25
Concrete, ready mix, 3500 psi	cu yd	67	—

* in $

Activity 6-3 – Estimating Software Activity

Create a new estimate and name it Activity 6-3. Perform a quick takeoff of items **3111.00 10 Footing Forms, 3111.00 50 Keyway In Footing, 3159.00 10 Strip/Oil Forms – Footings,** and **3306.00 c40 Footing Conc 4000 psi** for a 190'-6" long, 30" wide, and 8" deep footing. Print a standard estimate report.

Masonry

Masonry units include brick, structural clay tile, and block. Brick and block masonry units are made from clay, concrete, and glass. Masonry units are commonly used with other construction methods such as wood framing, structural steel, or structural concrete. Mortar is a mixture of fine aggregate and cement paste used to fill voids between aggregate or masonry units and reinforce a structure. Stone is used as a veneer material, in structural applications, and decorative placements. Stone may be placed dry and fitted without mortar or set with mortar similar to brick and concrete masonry units. When estimating masonry material, an estimator commonly performs the takeoff from the bottom of the structure up, and from the outside of the structure in.

MASONRY UNIT MATERIALS

Brick and block masonry units are available in many sizes and shapes. Brick and block masonry units are made from materials such as clay, concrete, and glass. Masonry units are joined with mortar that may be reinforced with steel members. Structures of large size and weight-bearing capacity can be built with masonry unit materials. See Figure 7-1. Masonry unit materials are fire-resistant and are often used as fire breaks between adjoining areas of a structure. Considerations must be made to provide openings for piping and ductwork for mechanical and electrical systems during masonry installation.

A *masonry control joint* is a continuous vertical joint in a masonry wall without mortar in which rubber, plastic, or caulking is installed. Locations and types of masonry control joints are shown on elevation and detail drawings. Wall section views indicate methods for attachment of masonry veneer to structural members and decorative effects. Architects may require that a sample section of masonry wall be laid prior to application to obtain a view of the finished product.

Figure 7-1. A variety of brick and block masonry units are used in masonry construction.

Bricks

Brick is a masonry unit used for building or paving purposes. Brick materials include face and building brick, ceramic glazed clay masonry units, clay tile, and terra cotta materials. These materials are manufactured in a variety of shapes and sizes and are designed for load-bearing and non-load bearing applications. Brick is set in a variety of designs to create various patterns for different building applications. Specifications and exterior elevation drawings provide information concerning brick designs and types.

Common bricks include standard modular, Norman, SCR, engineer, and economy brick. See Figure 7-2. Each brick has different dimensions and architectural applications. Standard modular bricks are designed to work into standard 4″ units when mortar joints are included. Norman bricks have a greater length than most other bricks. SCR bricks are thicker than other bricks and allow walls to be laid in a single course. Engineer bricks have a greater height than other designs, but retain standard thickness and length dimensions. Economy bricks are available in full 4″ modules for rough work applications.

Face brick is brick commonly used for an exposed surface. Face brick is manufactured in a variety of colors and shapes. Manufacturing of face brick creates uniform hardness, size, and strength to create a quality finished surface. *Firebrick* is brick manufactured from clay containing properties that resist crumbling and cracking at high temperatures. Firebrick is used in areas subjected to high temperatures.

Glass block is hollow opaque or transparent block made of glass. Glass block is available in many sizes and colors. Glass block is not used in load-bearing applications. Most brick materials are packaged and delivered on pallets, with each pallet containing 250 bricks.

Structural Clay Tile

Structural clay tile is a hollow or solid clay building unit. Glazed and unglazed structural clay tiles are used for structural and decorative applications. A broad range of tile designs and shapes are available for stretchers, sills, caps, miters, corners, jambs, and cove bases. See Figure 7-3. Structural clay tiles are used as a structural back-up material to face brick in older buildings and for interior construction applications in high wear areas such as public kitchens and bathrooms.

Concrete Masonry Units

A *concrete masonry unit (CMU)* is a precast hollow or solid masonry unit made of portland cement and fine aggregate, with or without admixtures or pigments. Concrete masonry units are used for many applications, including foundations and exterior and interior walls. Actual dimensions of concrete masonry units are $^3/_8″$ to $^1/_2″$ less than the nominal size. Concrete masonry unit materials are high in fire-resistant qualities and are commonly used for fire break walls.

BRICK							
Designation	Nominal Dimensions*			Joint Thickness*	Actual Dimensions*		
	t	h	l		t	h	l
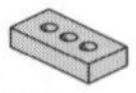 STANDARD MODULAR	4	$2^2/_3$	8	$^3/_8$	$3^5/_8$	$2^1/_4$	$7^5/_8$
				$^1/_2$	$3^1/_2$	$2^3/_{16}$	$7^1/_2$
 NORMAN	4	$2^2/_3$	12	$^3/_8$	$3^5/_8$	$2^1/_4$	$11^5/_8$
				$^1/_2$	$3^1/_2$	$2^3/_{16}$	$11^1/_2$
 SCR	6	$2^2/_3$	12	$^3/_8$	$5^5/_8$	$2^1/_4$	$11^5/_8$
				$^1/_2$	$5^1/_2$	$2^1/_4$	$11^1/_2$
 ENGINEER	4	$3^1/_5$	8	$^3/_8$	$3^5/_8$	$2^{13}/_{16}$	$7^5/_8$
				$^1/_2$	$3^1/_2$	$2^{11}/_{16}$	$7^1/_2$
 ECONOMY	4	4	8	$^3/_8$	$3^5/_8$	$3^5/_8$	$7^5/_8$
				$^1/_2$	$3^1/_2$	$3^1/_2$	$7^1/_2$

* in in.

Figure 7-2. Common bricks include standard modular, Norman, SCR, engineer, and economy, which have different thickness, height, and length dimensions.

Masonry **141**

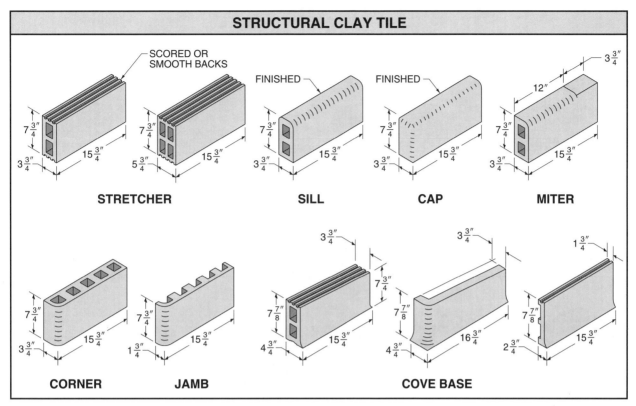

Figure 7-3. Structural clay tile is available in a variety of designs and shapes and is used in high-wear areas such as public kitchens and bathrooms.

Various materials, wall designs, and cavity fill materials affect the fire resistance ratings of concrete masonry unit walls. Fire resistance ratings are determined by a measurement of heat transmission as the rise in temperature on the side of the wall opposite the heat source. Plaster and other facing materials also affect the fire resistance ratings.

Concrete masonry units are cast in a variety of styles and shapes. See Figure 7-4. Concrete masonry units include stretcher, corner, header, bond beam, lintel, sill, column, jamb, etc. designs. Faces of concrete masonry units may be smooth, split with a rough finish, fluted, or ribbed.

Special Masonry Materials

In addition to brick, structural clay tile, and block, special masonry materials are used for a complete masonry installation. Special masonry materials include refractory materials and accessories such as expansion joints, control joints, reinforcing, ties, and sealants. Each of these accessories is noted in Division 4 of the CSI MasterFormat™ as applicable to a particular construction project.

Refractory Materials. A *refractory material* is a material that can withstand high temperatures without structural failure. Refractory materials include flue liners and refractory brick. As with many other masonry unit materials, flue liners and refractory brick are manufactured in standard sizes and shapes. See Figure 7-5. Availability of refractory materials is dependent on local suppliers. Estimators can find information concerning installation of refractory materials in areas such as fireplaces and furnace and boiler flues on drawings.

Accessories. Masonry materials expand and contract due to temperature and moisture variations. To minimize damage, expansion joints, control joints, specialized reinforcements, ties, and sealants are used in masonry construction.

An *expansion joint* is a separation between adjoining sections of a concrete slab to allow movement caused by expansion and contraction. Expansion joints such as soft joints or slip channels allow movement between masonry walls and where masonry walls abut adjoining floor and roof structures.

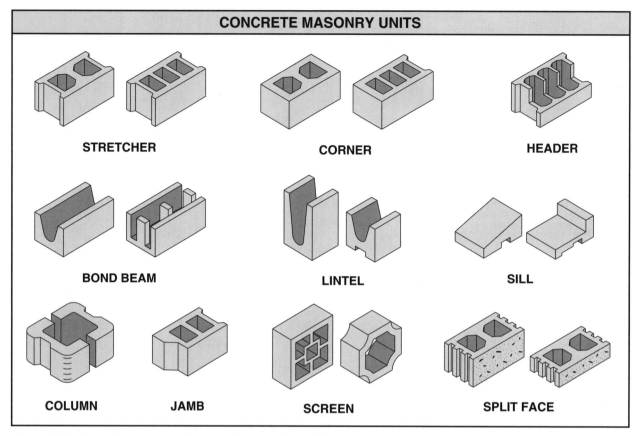

Figure 7-4. Concrete masonry units are widely used in building construction due to the variety of designs and finishes available.

Control joints are located in long straight walls to provide lateral stability across a joint and contain a wall seal where a crack may otherwise occur. Control joints may be made of rubber, plastic, or caulking.

Reinforcement is material embedded in another material to provide additional support or strength. Reinforcement normally consists of steel wire bent and shaped into various patterns and set in mortar joints to provide tensile strength to masonry construction. A *tie* is a device used to secure two or more members together. Ties are embedded in mortar joints to tie masonry walls to each other or to anchor veneer masonry walls to other structures.

A *sealant* is a liquid or semiliquid material that forms an air- or waterproof joint or coating. Sealants of flexible materials such as silicone or other elastomers are caulked into masonry joints to allow some movement without cracking. Accessory items are noted in the specifications, exterior elevations, or masonry detail drawings. Estimators must include these items in the overall masonry bid.

Mortars

Mortar is a bonding mixture consisting of lime, cement, sand, and water. Mortar is used to fill voids between aggregate or masonry units and reinforce a structure. Mortar selection for masonry construction is based on overall strength requirements, masonry units being joined, and the environmental conditions that the mortar must resist. Different mortars may be used for brick, stone, refractory, and corrosion-resistant applications. Coloring may be added to the mortar as required by the architect.

Masonry mortar includes Type M, Type S, and Type N mortar. Type M mortar has the highest compressive strength characteristics with an allowable compressive load strength of approximately 6100 lb per linear foot. Type S mortar has a compressive load strength of approximately 5200 lb per linear foot and Type N mortar has a compressive load strength of approximately 4800 lb per linear foot. The estimator should ensure that the proper type and color of mortar is used in each masonry application.

FLUE LINERS				
Liner	**Area***	**Dimensions****		
		A	**B**	**T**
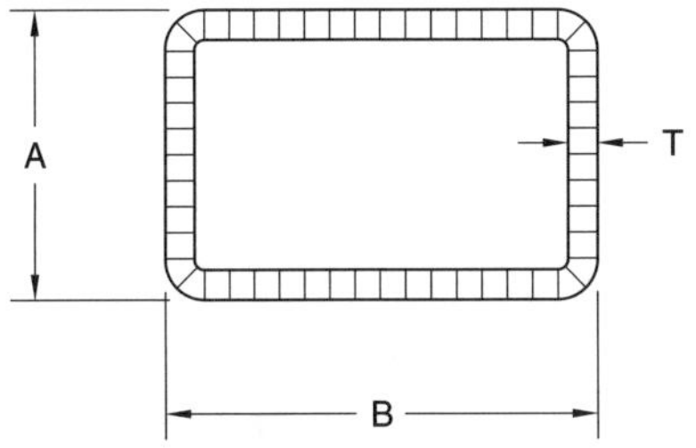 **RECTANGULAR (STANDARD)**	51	8½	8½	¾
	79	8½	13	⅞
	108	8½	18	1
	125	13	13	⅞
	168	13	17¾	1
	232	17¾	17¾	1¼
	279	20	20	1⅜
	338	20	24	1½
	420	24	24	1⅝
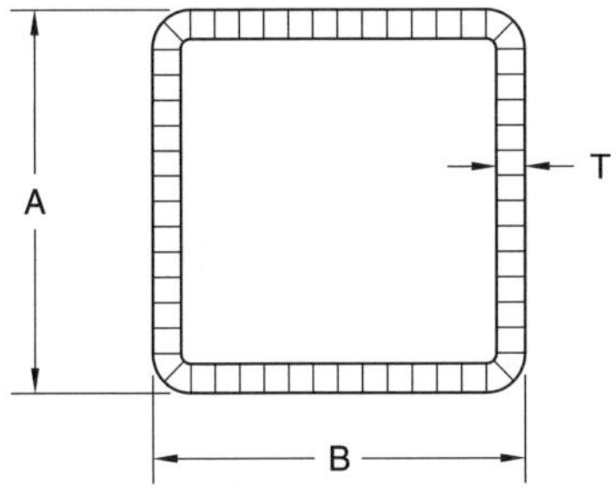 **RECTANGULAR (MODULAR)**	57	7½	11½	¾
	87	11½	11½	⅞
	120	11½	15½	1
	162	15½	15½	1⅛
	208	15½	19½	1¼
	262	19½	19½	1⅜
	320	19½	23½	1½
	385	23½	23½	1⅝
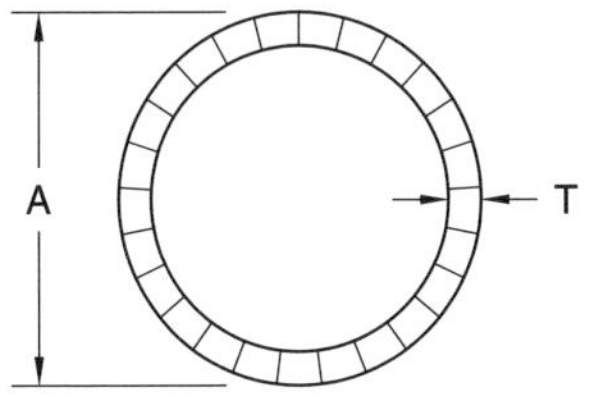 **ROUND**	47	8	—	¾
	74.5	10	—	⅞
	108	12	—	1
	171	15	—	1⅛
	240	18	—	1¼
	298	20	—	1⅜
	433	24	—	1⅝

* in sq in.
** in in.

Figure 7-5. Rectangular and round flue liners are refractory materials available for masonry construction.

MASONRY UNIT CONSTRUCTION METHODS

Construction methods for masonry walls and fireplaces depend on the wall design, materials used, height of the structure, environmental conditions, and exterior and interior finishes. The wall design may be hollow or solid, a veneer covering another structural member, or pre-laid curtain walls. Masonry units are commonly used with other construction methods such as wood framing, structural steel, or structural concrete.

Various equipment is required in masonry construction. For example, loaders are often necessary to move pallets of brick and block at the job site. Mixers are used to mix batches of mortar. Scaffolding is needed to facilitate the placement of brick and block. Protection and heat may be required to allow for proper set of mortar if cold weather is a possibility. See Figure 7-6. For high, unsupported walls, temporary bracing may be required to hold masonry walls in position until roof members and other wall supports are set in place.

Figure 7-6. In cold weather, protection and heat are required to allow for proper set of mortar.

Figure 7-7. Masonry corners act as guides for the body of masonry work on walls.

Masonry construction begins with a solid foundation of poured-in-place concrete. The foundation may be a concrete footing, foundation wall, or thickened portion of a slab-on-grade. A keyway or steel dowels are set in place to tie the masonry and concrete together. Work proceeds from each corner of a structure. Brick masons lay out the work on the footing to determine brick or block spacing. Corners are set in mortar with heights determined according to a mason's rule. A *mason's rule* is a rule graduated in increments equal to various mortar joint thicknesses. A mason's rule helps lay out courses of brick or block. A *course* is a continuous horizontal layer of masonry units bonded with mortar. See Figure 7-7. After corners are set in place, strings are stretched between corners to guide masonry work up to completion at the top of the structure. Muratic acid and water are commonly used after mortar has set to clear excess mortar from the face of the masonry work.

Masonry Walls

Exterior masonry walls may be all brick, all concrete masonry units, or a combination of brick and/or concrete masonry units with other building materials. Masonry walls may be structural bearing walls or a veneer built around and between structural members. Masonry units for exterior finish are shown on detail drawings, wall section views, and exterior elevation prints. See Figure 7-8.

Brick and concrete masonry units are also used for interior fire break walls and for exposed interior walls. Finish information for interior masonry walls is similar to exterior exposed masonry, including material, mortar joint color and finish, and bond pattern.

Where brick or block are used for an exterior wall, the exterior finish, including brick bond, face finish of concrete masonry units, mortar color, lintels, and stone patterns and types, is given in the specifications and on elevation views. Special masonry units may be fabricated specifically for a particular job. Special brick shapes and coursing information are included in the architectural elevation views. See Figure 7-9.

> ▲ *All masonry walls require temporary bracing that must remain in place until permanent bracing is constructed. The cost of bracing walls during construction must be included in the price of the masonry bid.*

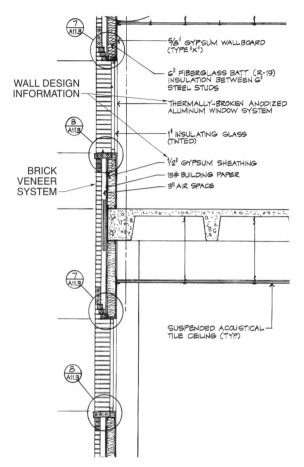

WALL DESIGN INFORMATION

5/8" GYPSUM WALLBOARD (TYPE "X")

6" FIBERGLASS BATT (R-19) INSULATION BETWEEN 6" STEEL STUDS

THERMALLY-BROKEN ANODIZED ALUMINUM WINDOW SYSTEM

1" INSULATING GLASS (TINTED)

1/2" GYPSUM SHEATHING

15# BUILDING PAPER

3" AIR SPACE

BRICK VENEER SYSTEM

SUSPENDED ACOUSTICAL TILE CEILING (TYP.)

Figure 7-8. Estimators consult detail drawings, wall section views, and exterior elevation prints for a complete understanding of masonry installations.

Bonds. A *bond* is the arrangement of masonry units in a wall by lapping them one upon another to provide a sturdy structure. Architects specify a brick bond and joint finish for a masonry structure. Common brick bonds include running, common, English, Flemish, etc. See Figure 7-10. Each bond uses a combination of brick positions such as stretchers, headers, soldiers, shiners, rowlocks, and sailors to create a design and provide structural support.

Solid Masonry Walls. Solid masonry walls consist of masonry materials only, with no other building material added. Solid masonry walls may be formed of one or two layers of brick tied together with metal ties, brick tied to the face of concrete masonry units, concrete masonry units only, or structural or decorative tile set onto concrete masonry. The estimator should check architectural detail drawings and symbols to accurately determine the solid masonry wall in each building area.

Cavity Walls. A *cavity wall* is a masonry wall with at least a 2″ void between faces. See Figure 7-11. The air space is provided to improve insulation and reduce thermal transmission. Cavity walls use metal ties to hold the two separate walls together to create a single structural unit.

Veneer Masonry Walls. Veneer masonry walls are used in frame, structural steel, and reinforced concrete construction. Ties are used to anchor the masonry veneer material to the structure. Masonry veneer walls are commonly built in place. Masonry veneer walls may also be made of prefabricated panels set in a manner similar to precast concrete panels. Prefabricated masonry veneer panels are laid with steel support frames and inserts, lifted into place, and attached to structural members. Estimators should also check plan drawings for other masonry veneer applications such as columns, pilasters, and interior partitions.

Fireplaces

Allowances are made in structural design, whether wood, metal, or masonry, for fireplace installation. A minimum distance of 2″ from structural members to fireplace masonry is required for fire safety reasons. This space may be filled with a noncombustible material such as fiberglass or rock wool insulation. *Rock wool insulation* is lightweight heat and sound insulating material made by blowing steam through molten rock or slag. A *hearth* is the fireproof floor immediately surrounding a fireplace.

Insulation is placed between concrete block and brick veneer walls to increase the thermal resistance of the wall.

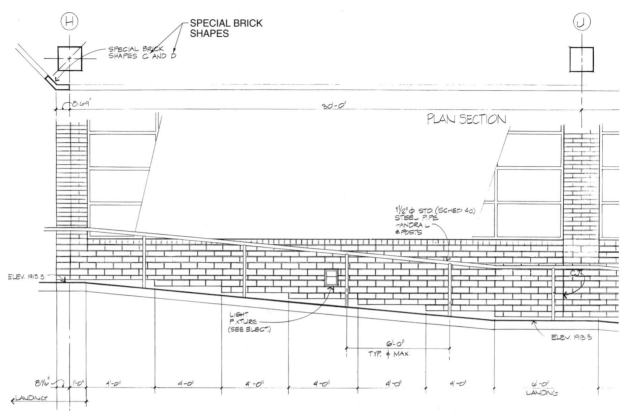

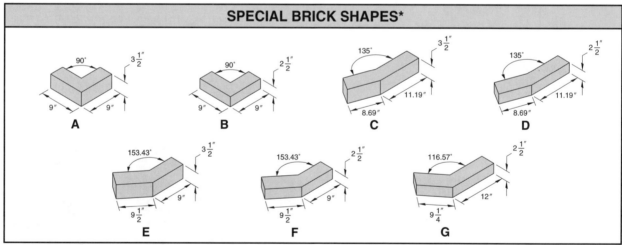

* special brick shapes fabricated specifically for a particular job

Figure 7-9. Special brick shapes are included in the architectural elevation views.

The hearth of the fireplace is laid first using firebricks. Fire clay mortar is used in firebrick applications. *Fire clay* is clay that is highly resistant to heat and does not deform. The sides and back of the fireplace opening are also laid with firebrick and fire clay. Dampers are set in place and smoke shelves are laid as designed by the architect and shown on detail drawings.

See Figure 7-12. A *smoke shelf* is a section of a chimney directly over a fireplace and below the flue. A smoke shelf prevents downdrafts from blowing smoke into the living area. Refractory materials such as flues are set in place and masonry work proceeds up to the chimney. Face brick or other fireplace facing materials are applied after all other fireplace work is set in place.

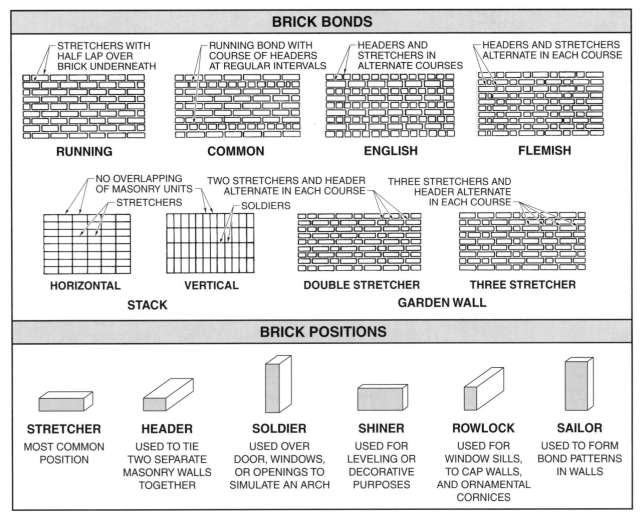

Figure 7-10. A bond is the arrangement of masonry units in a wall by lapping them one upon another to provide a sturdy structure.

MASONRY MATERIAL QUANTITY TAKEOFF

When estimating masonry material, an estimator commonly performs the takeoff from the bottom of the structure up, and from the outside of the structure in. Estimators begin by computing the foundation material, the exterior walls above grade, and finally the interior masonry work. After computing exterior material, the interior masonry material such as fireplaces, interior masonry walls, and hearths are taken off. In general, estimators should base masonry material costs on the delivered cost of the materials. The preliminary steps in masonry work include excavation and concrete work for footings and foundations. Some estimators may consider these items as part of the masonry cost of a structure.

Estimators must also include equipment costs in a masonry bid. Masonry equipment costs include costs for forklifts or other lifting equipment to move pallets of brick, block, and mortar material at the job site; scaffolding rental, erection, and removal; bracing of the masonry wall; and weather protection. Each of these contingency costs are considered by the estimator in the overall masonry cost.

The estimator should also include the cost of cleaning materials and labor because brick must be cleaned after it is laid. The specifications are checked to determine the cleaning materials and amount of cleaning required. Additional costs that may be included as a portion of the masonry work are costs for wood and metal forms or props for arched doorways, windows, or other special applications where temporary forms are necessary. Materials for metal lintels over doors, windows, fireplaces, dampers, and cleanout doors are usually computed last.

CAVITY WALLS

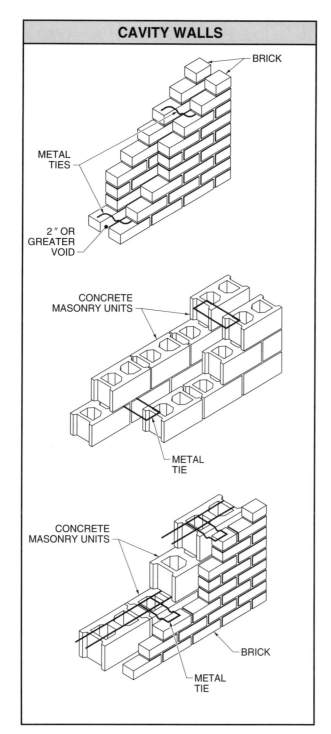

Figure 7-11. Cavity walls have a 2″ or greater void between faces.

▲ *Brick and block masonry units are priced by the unit, which is converted into a price per square foot. Openings of less than 2 sq ft are normally ignored because savings in units is normally offset by trimming and cutting.*

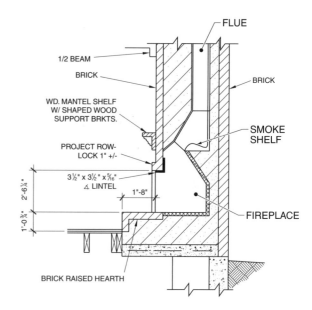

Figure 7-12. Specialized materials and construction methods are used in masonry fireplace construction.

Labor costs are normally based on a square foot cost. Standard labor tables or company historical data can provide information about the number of square feet of each masonry application that can be set in place by a particular labor crew. These labor hours are multiplied by the cost per crew hour to determine overall labor costs. Common waste factors are 2% to 5% for masonry units and 10% to 20% for mortar. These percentages are added after square foot calculations are completed.

Masonry Wall Takeoff

Masonry material takeoff begins by determining the area based on the number of square feet. To estimate the number of masonry units in a building, the estimator first computes the square feet of area of the structure and then subtracts the area not composed of masonry, including large door and window openings. These two steps yield the total area of masonry material. In estimating the quantity of face brick used as trim, the linear feet of trim is measured and multiplied by the number of bricks per linear foot.

Ledger Sheet. After calculating the area, an estimator using a ledger sheet refers to standardized tables to determine the number of masonry units per square foot. The material quantity is computed by multiplying the area in square feet by the number of masonry units per square foot or by using a masonry wall material table to determine the number of standard-size bricks required per square feet of wall. See Figure 7-13.

MASONRY WALL STANDARD SIZE FACE AND BUILDING BRICK*															
Wall†	Bonds														
	Running			Common (Header Course Every 7th Course)			English and English Cross** (Full Headers Every 6th Course)			Flemish (Full Headers Every 5th Course)			Double Headers (Alternating with Stretchers Every 5th Course)		
	Face Brick	Building Brick		Face Brick	Building Brick		Face Brick	Building Brick		Face Brick	Building Brick		Face Brick	Building Brick	
		8" Wall	12" Wall		8" Wall	12" Wall		8" Wall	12" Wall		8" Wall	12" Wall		8" Wall	12" Wall
1	6.16	6.16	12.32	7.04	5.28	11.44	7.19	5.13	11.29	6.57	5.75	11.91	6.78	5.54	11.70
5	31	31	62	36	27	58	36	26	57	33	29	60	34	28	59
10	62	62	124	71	53	115	72	52	113	66	58	120	68	56	117
20	124	124	248	141	106	229	144	103	226	132	115	269	136	111	234
30	185	185	370	212	159	344	216	154	339	198	173	358	204	167	351
40	247	247	494	282	212	458	288	206	452	263	230	477	272	222	468
50	308	308	616	352	264	572	360	257	565	329	288	596	339	277	585
60	370	370	740	423	317	687	432	308	675	395	345	715	407	333	702
70	432	432	864	493	370	801	504	360	791	460	403	834	475	388	819
80	493	493	986	564	423	916	576	411	904	526	460	953	543	444	936
90	555	555	1110	634	476	1030	648	462	1017	592	518	1072	611	499	1053
100	616	616	1232	704	528	1144	719	513	1129	657	575	1191	678	554	1170
200	1232	1232	2464	1408	1056	2288	1438	1026	2258	1314	1150	2382	1356	1108	2340
300	1848	1848	3696	2110	1584	3432	2157	1539	3387	1971	1725	3573	2034	1662	3510
400	2464	2464	4928	2816	2112	4576	2876	2052	4516	2628	2300	4764	2712	2216	4680
500	3080	3080	6160	3520	2640	5720	3595	2565	5645	3285	2875	5955	3390	2770	5850
600	3696	3696	7392	4224	3168	6864	4314	3078	6774	3942	3450	7146	4068	3324	7020
700	4312	4312	8624	4928	3696	8010	5033	3591	7903	4599	4025	8337	4746	3878	8190
800	4928	4928	9856	5632	4224	9152	5752	4104	9032	5256	4600	9528	5424	4432	9360
900	5544	5544	11,088	6336	4752	10,296	6471	4617	10,161	5913	5175	10,719	6102	4986	10,530
1000	6160	6160	12,320	7040	5280	11,440	7190	5130	11,290	6570	5750	11,910	6780	5540	11,700

* For other than $1/2$" joints, add 21% for $1/8$" joint, 14% for $1/4$" joint, 7% for $3/8$" joint. Subtract 5% for $5/8$" joint, 10% for $3/4$" joint, 15% for $7/8$" joint, and 20% for 1" joint
** Quantities also apply to common bond with headers in every sixth course
† in sq ft

Figure 7-13. Masonry wall material tables provide information concerning the number of standard-size masonry units required for various masonry construction designs.

For example, to calculate the number of face and building bricks required for a common bond 8" thick solid brick wall building with exterior dimensions of 24'-6" × 24'-6", a wall height of 10'-0", and two floor-to-ceiling doors each 9'-0" wide, apply the procedure:

1. Compute the area of the exterior walls. The area of each wall is 245 sq ft (24'-6" × 10'-0" = 245 sq ft).
2. Determine the total of the four walls. The total is 980 sq ft (4 × 245 sq ft = 980 sq ft).
3. Deduct the area of the openings. The area of the openings totals 180 sq ft (9'-0" × 10'-0" × 2 = 180 sq ft). The total area after deducting the area of the openings is 800 sq ft (980 sq ft − 180 sq ft = 800 sq ft).
4. Determine the number of bricks required. The number of bricks required is determined by multiplying 800 sq ft by the multiplier for an 8" thick solid brick wall or by referencing 800 sq ft on a masonry wall

material table. The number of face bricks required for a solid wall 8″ thick with a common bond is approximately 5632 bricks and the number of building bricks required is approximately 4224 bricks. Waste factors may be added to these totals and other adjustments made for mortar joint variations.

The number of concrete masonry units (CMUs) required is determined by calculating the square feet of wall area and multiplying this number by a standard multiplier based on the CMU used and the size of mortar joints. This gives the number of CMUs required for each wall. Corner and other special units are calculated separately.

Concrete masonry unit quantity takeoff may also be performed by determining the length of wall (in linear feet), dividing the length by the number of CMUs per course, and multiplying this subtotal by the number of courses in the overall wall height. See Figure 7-14. For example, a wall 32′-0″ long and 7′-6″ high is to be built of CMUs 16″ long and 8″ high. The wall length of 32′-0″ is divided by the CMU length of 1.33̄′ to determine that 24 CMUs are required per course. The wall height of 7.5′ is divided by the CMU height of .66̄′ to determine that 11 courses are required per wall. The CMUs in each course are multiplied by the number of courses to obtain a total of 264 CMUs per wall (24 CMUs per course × 11 courses = 264 CMUs per wall). This total is multiplied by a waste factor of 5% to obtain an overall total of 277.2 CMUs for the wall (264 × 1.05 = 277.2).

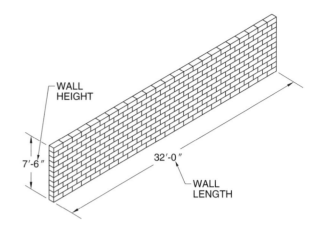

$$\frac{WALL\ LENGTH}{CMU\ LENGTH} = CMUs\ PER\ COURSE$$

$$\frac{32′-0″}{1.3\overline{3}′} = 24\ CMUs\ PER\ COURSE$$

CMUs PER COURSE × NUMBER OF COURSES = CMUs PER WALL

24 × 11 = **264 CMUs PER WALL**

Figure 7-14. Wall length and height information can be used to calculate the required number of concrete masonry units.

Mortar quantity is determined from standard mortar quantity tables based on varying wall thicknesses. See Figure 7-15. For example, for a cement-lime mortar, 2.40 sacks of cement, 2.40 sacks of lime, and .60 cu yd of sand lays about 1000 concrete blocks in a wall 12″ thick with ½″ joints. Additional tables are available for CMUs and other masonry materials.

The linear feet of reinforcing installed in mortar joints is based on spacing between courses of reinforcing and the linear feet of wall. Masonry below grade level is commonly laid with ½″ mortar joints. Specifications are checked to determine these variables and other reinforcing information.

Elevation views and floor plans are used to determine the number and types of lintels over openings and deductions for wall openings. Lintels are calculated on a per opening basis from door and window schedules. The total square feet of door and window opening areas is deducted from masonry walls to provide an accurate masonry material takeoff. Door, window, and opening dimensions are determined from floor plans, elevations, and door and window schedules.

Special units required for a concrete masonry wall are calculated separately from standard concrete masonry units.

MORTAR QUANTITY*						
	8″ Wall			12″ Wall		
Proportion	Cement Sacks	Lime Sacks	Sand**	Cement Sacks	Lime Sacks	Sand**
Cement Mortars						
Cement 1, Sand 2	5.00	—	.55	5.40	—	.60
Cement 1, Sand 2½	4.20	—	.55	4.40	—	.60
Cement 1, Sand 3	3.70	—	.55	4.00	—	.60
Lime Mortars						
Lime 1, Sand 2	—	3.80	.55	—	4.20	.60
Lime 1, Sand 2½	—	3.30	.55	—	3.60	.60
Lime 1, Sand 3	—	2.70	.55	—	3.00	.60
Cement-Lime Mortar						
Cement 1, Lime 1, Sand 6	2.20	2.20	.55	2.40	2.40	.60

* Quantities given are for foundation work below grade or other places where joints between tiers of bricks are filled with mortar
** in cu yd

Figure 7-15. Standard mortar quantity tables give mortar quantity and ingredient amounts based on masonry wall thickness.

Spreadsheet/Estimating Software. When using a spreadsheet, the appropriate cell for brick or block information may contain a standard formula that automatically performs necessary multiplication or refers to a standard chart. When using estimating software, entry of wall length and height dimensions into the appropriate item screen enables the software to automatically calculate brick and block materials, mortar amounts, equipment costs, reinforcing, and other items required by the estimator. See Figure 7-16. Various assemblies may be established in the database based on various wall bonds, types, and standard masonry construction. As with other assemblies, after this information is transferred from the database to the spreadsheet, changes and modifications appropriate for a specific job can be made, including addition of lintels and other accessory quantities.

Fireplaces

A standard table is used to determine bricks per foot of height for small chimneys. See Figure 7-17. Variations in the table are dependent on the size and number of flues in the chimney. Chimneys that have more than one flue lining must have 4″ of brickwork between the flues.

▲ *Fireplaces require a foundation of poured concrete or concrete block for support because of the weight of the chimney and fireplace.*

MASONRY
MATERIALS

WALL
INFORMATION

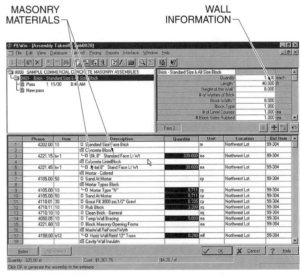

Figure 7-16. Estimating software simplifies masonry estimating by calculating a variety of masonry materials at one time based on basic wall information.

In fireplace masonry quantity takeoff, the estimator calculates surfaces of a fireplace as solid areas. Using this method, the number of bricks for the surface area is multiplied by the number of tiers of brick deep. Deductions are then made for the number of bricks displaced by all flues and openings, face brick, and firebrick lining.

For chimneys above the roof, masonry quantity takeoff is performed by determining the linear distance around the chimney and multiplying this value by the chimney height above the roof line. This results in the number of square feet of chimney above the roof line. Standard tables or square foot calculations are then used to determine the number of masonry units required.

In estimating the brick required for the chimney or fireplace, the easiest ledger sheet method is to calculate the cubic feet of chimney or fireplace from the plan drawings and deduct the cubic areas required for the flue lining and hearth. The estimator must also add the square feet of firebrick for the firebox. The number of firebricks is determined from standard face brick tables. Flue materials are determined based on the linear feet and type required. Dampers and lintels are taken off as single items from elevation views, detail drawings, and plan views.

SMALL CHIMNEY BRICKS PER FOOT OF HEIGHT		
Number of Flues	Size of Flue*	Number of Bricks**
1	8 × 8	26
2	8 × 8	44
3	8 × 8	63
1	8 × 12	31
2	8 × 12	52
3	8 × 12	74
1	12 × 12	38
2	12 × 12	60

* in in.
** per foot of height

Figure 7-17. Small chimney brick quantities are determined based on a standard table and the height of the chimney.

STONE MATERIALS

Stone is used in many construction applications, including veneer material, structural applications, and decorative placements. Three characteristics that affect the features of building stone are color, pattern, and texture. Stone color varies based on stone type. Stone color may be consistent throughout or vary within a particular stone. Patterns are variable and give special features to building stone. Texture varies from coarse to fine. Stone should be selected for use in a construction project based on moisture penetration, weatherability, price, availability, and color.

Stone

Three classes of stone (rock) are igneous, sedimentary, and metamorphic. *Igneous rock* is rock formed from the solidification of molten lava. Igneous rock has a high compressive strength. Igneous rock includes granite and traprock. *Granite* is extremely hard natural rock consisting of quartz, feldspar, and other minerals produced under intense heat and pressure. Granite varies in color from almost white to gray and white, is relatively hard, and retains a cut shape. Granite is commonly used for building stone, as a stone veneer material on walls, and for other decorative purposes. *Traprock* is fine-grained, dark-colored igneous rock. Traprock is used as a base under or between layers of asphalt paving.

Sedimentary rock is rock formed from sedimentary materials such as sand, silt, and rock and shell fragments. Sedimentary rock is weak in compressive strength but relatively high in shear strength. Sedimentary rock includes limestone and sandstone. *Limestone* is sedimentary rock consisting of calcium carbonate. Limestone is white to light gray in color and is relatively soft. Limestone scratches and cuts easily and has a relatively high level of water absorption compared to other stone materials. Common applications of limestone include dry walls, lintels, sills, and light decorative uses. *Sandstone* is sedimentary rock consisting of quartz held together by silica, iron oxide, and/or calcium carbonate. Sandstone has a granular texture and has been used for building walls but is now generally used as a decorative facing.

Metamorphic rock is sedimentary or igneous rock that has been changed in composition or texture by extreme heat, pressure, or chemicals. Metamorphic rock is high in compressive strength but low in shear strength. Metamorphic rock includes marble and slate. *Marble* is crystallized limestone. Marble is slightly harder than limestone and is usually used for decorative veneer. Marble is commonly available in pre-shaped squares or panels. Marble colors are highly varied and commonly show veins of different colors. *Slate* is fine-grain metamorphic rock that is easily split into thin sheets. Slate is used as roofing tile and flooring material.

STONE CONSTRUCTION METHODS

Many of the masonry construction methods that apply to brick and CMUs are also applicable to stone. Stone construction begins with a firm footing or a firm backing material for veneer stone. Stone members require equipment to set them in place due to their size. Mortar setting, scaffolding requirements, and weather protection are similar for brick, CMUs, and stone construction.

As with brick, a variety of bonds exist for stone. See Figure 7-18. A *course* is a continuous row of masonry bonded with mortar. *Ashlar* is masonry composed of squared building stones of various sizes. The estimator should consult specifications and elevation drawings to determine the stone bond.

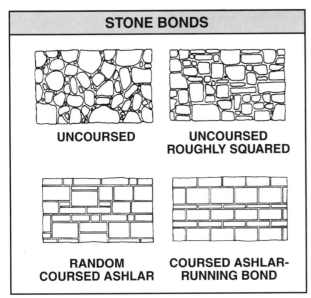

Figure 7-18. Stone bonds include uncoursed, uncoursed roughly squared, random coursed ashlar, and coursed ashlar-running bond.

Placement

For some applications, stone such as limestone may be placed dry and fitted without mortar. This method is usually applicable for low decorative walls and small retaining walls. For building applications, structural stone is set with mortar similar to brick and CMUs. Non-staining cement mortar is used for light-color stone. Additional metal ties are often required on structural stone applications. Care is taken to use metal that is non-staining, such as plated steel or zinc.

A variety of mastics, mortars, and clips may be used where decorative face stone panels are fastened to structural members. The estimator should check specifications and details for fastening information for face stone.

▲ *Stone that must be shipped from other countries or other parts of the country must be ordered by the estimator soon after being awarded the bid.*

STONE QUANTITY TAKEOFF

Estimators determine the structural or decorative use of the stone materials on a construction site. Structural stone members such as lintels and sills are listed on a ledger sheet, spreadsheet, or estimating software as individual items, with length, width, and depth noted. Stone such as ashlar or other wall materials are ordered by the ton. Estimators should consult with quarries to help determine the number of square feet of coverage expected per ton of various stone materials for structural uses.

Stone Wall Takeoff

The stone wall design affects the stone estimate. Structural walls require construction costs similar to laying concrete block or brick, including mortar, scaffolding, weather protection, and temporary bracing. Unit costs per ton of stone guide the estimator in costing this item.

For stone veneer walls, takeoff and cost estimates may be based on square foot or per piece costs. The decorative stone applied, thickness, location, and fastening methods all affect material, labor, and equipment costs. Standard industry information or company historical data is referenced for labor unit production levels for various stone veneer applications.

Stone Takeoff

Quarrying costs, quarry location, construction site location, and availability of stone materials all affect stone costs. Estimators should check in the geographic location of a project for the availability of stone required for a project. Some stone materials may require transportation over great distances between the quarry and the construction site. Estimators should obtain stone costs with transportation costs included to ensure an accurate estimate of stone material costs.

ESTIMATING SOFTWARE – CALCULATOR USE

Precision estimating software contains a calculator into which values and operators are entered to calculate various building material quantities. The calculator contains a single field that prefills with the value from the field where the cursor was placed before the calculator was opened. The calculator is used by following a standard procedure. See Figure 7-19.

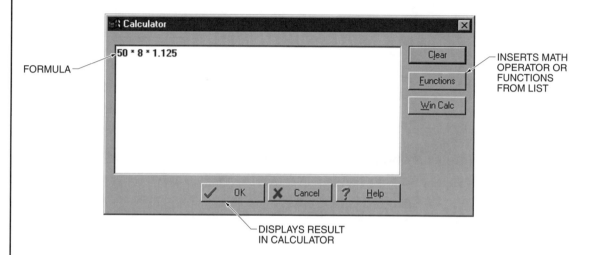

PRECISION ESTIMATING SOFTWARE – CALCULATOR USE

1. Perform a quick takeoff for item 4221.10 rw 1 Blk 12″ Standard Face Reg Wt masonry blocks by clicking the Quick Takeoff button and double-clicking through the group, phase, and item categories.
2. Close the Quick Takeoff window by clicking the close button.
3. Calculate the number of blocks needed for a 50′ × 8′ block wall by clicking on the takeoff quantity cell for the 12″ block and opening the calculator by clicking the calculator ( ) button from the toolbar menu. Click the clear button to delete the contents of the calculator if required.

FORMULA

INSERTS MATH OPERATOR OR FUNCTIONS FROM LIST

DISPLAYS RESULT IN CALCULATOR

4. Develop the formula. Use the calculation: wall length × wall width × # of blocks per sq ft. Enter 50 * 8 * 1.125 directly into the window or click the Functions button to insert a math operator or functions from the list.
   Note: The takeoff unit for this item is "each" so the formula developed must calculate the total number of blocks.
5. Click ENTER to display the result in the calculator.
6. Press OK to insert the result into the quantity cell and close the calculator.
   Note: The value generated by a formula or the calculator must always be in the units specified as the takeoff unit for the item. For example, if the takeoff unit is in cubic yards, the calculation must also return the value in cubic yards.

Figure 7-19. The estimating software calculator enables the estimator to quickly develop formulas to calculate various building material quantities.

Estimating

_____ **1.** A(n) masonry _____ joint is a continuous vertical joint in a masonry wall without mortar in which rubber, plastic, or caulking is installed.

_____ **2.** _____ is a masonry unit used for building or paving purposes.

_____ **3.** _____ brick is brick commonly used for an exposed surface.

_____ **4.** _____ is brick manufactured from clay containing properties that resist crumbling and cracking at high temperatures.

_____ **5.** _____ block is hollow opaque or transparent block made of glass.

_____ **6.** _____ clay tile is a hollow or solid clay building unit.

T   F   **7.** A concrete masonry unit (CMU) is a precast hollow or solid masonry unit made of portland cement and fine aggregate, with or without admixtures or pigments.

T   F   **8.** Masonry is noted in Division 5 of the CSI MasterFormat™.

T   F   **9.** A refractory material is a material that can withstand high temperatures without structural failure.

_____ **10.** A(n) _____ joint is a separation between adjoining sections of a concrete slab to allow movement caused by expansion and contraction.

_____ **11.** _____ is a bonding mixture consisting of lime, cement, sand, and water.

_____ **12.** A(n) _____ is a device used to secure two or more members together.

_____ **13.** A(n) _____ is a liquid or semiliquid material that forms an air- or waterproof joint or coating.

_____ **14.** A(n) _____ rule is a rule graduated in increments equal to various mortar joint thicknesses.

_____ **15.** A(n) _____ is a continuous horizontal layer of masonry units bonded with mortar.

_____ **16.** A(n) _____ is the arrangement of masonry units in a wall by lapping them one upon another to provide a sturdy structure.

T   F   **17.** A cavity wall is a masonry wall with at least a ¼″ void between faces.

T   F   **18.** A smoke shelf is a section of a chimney directly over a fireplace and above the flue.

_____ **19.** A(n) _____ is the fireproof floor immediately surrounding a fireplace.

_____ **20.** _____ clay is clay that is highly resistant to heat and does not deform.

Stone

_____   **1.** Igneous rock

_____   **2.** Granite

_____   **3.** Traprock

_____   **4.** Sedimentary rock

_____   **5.** Limestone

_____   **6.** Sandstone

_____   **7.** Metamorphic rock

_____   **8.** Marble

_____   **9.** Slate

A. Crystallized limestone

B. Sedimentary or igneous rock that has been changed in composition or texture by extreme heat, pressure, or chemicals

C. Sedimentary rock consisting of quartz held together by silica, iron oxide, and/or calcium carbonate

D. Fine-grained metamorphic rock that is easily split into thin sheets

E. Rock formed from the solidification of molten lava

F. Extremely hard natural rock consisting of quartz, feldspar, and other minerals produced under intense heat and pressure

G. Rock formed from sedimentary material such as sand, silt, and rock and shell fragments

H. Fine-grained, dark-colored igneous rock

I. Sedimentary rock consisting of calcium carbonate

CMU Calculations

_____   **1.** There are _____ 1.33′ CMUs per course.

_____   **2.** There are _____ courses.

_____   **3.** There are _____ CMUs per wall.

36′-0″
WALL LENGTH

Stone Bonds

_____   **1.** Coursed ashlar-running bond

_____   **2.** Random coursed ashlar

_____   **3.** Uncoursed

_____   **4.** Uncoursed roughly squared

Brick Positions

_____ **1.** Stretcher

_____ **2.** Header

_____ **3.** Soldier

_____ **4.** Shiner

_____ **5.** Rowlock

_____ **6.** Sailor

Brick Bonds

_____ **1.** English

_____ **2.** Flemish

_____ **3.** Stack, horizontal

_____ **4.** Stack, vertical

_____ **5.** Common

_____ **6.** Running

_____ **7.** Garden wall, double stretcher

_____ **8.** Garden wall, three stretcher

STRETCHERS WITH HALF LAP OVER BRICK UNDERNEATH — Ⓐ

RUNNING BOND WITH COURSE OF HEADERS AT REGULAR INTERVALS — Ⓑ

HEADERS AND STRETCHERS IN ALTERNATE COURSES — Ⓒ

HEADERS AND STRETCHERS ALTERNATE IN EACH COURSE — Ⓓ

NO OVERLAPPING OF MASONRY UNITS / STRETCHERS — Ⓔ

SOLDIERS — Ⓕ

TWO STRETCHERS AND HEADER ALTERNATE IN EACH COURSE — Ⓖ

THREE STRETCHERS AND HEADER ALTERNATE IN EACH COURSE — Ⓗ

Concrete Masonry Units

_____ **1.** Split face

_____ **2.** Column

_____ **3.** Screen

_____ **4.** Header

_____ **5.** Bond beam

_____ **6.** Sill

_____ **7.** Jamb

_____ **8.** Lintel

_____ **9.** Stretcher

_____ **10.** Corner

Structural Clay Tile

_____________ **1.** Cove base

_____________ **2.** Jamb

_____________ **3.** Miter

_____________ **4.** Corner

_____________ **5.** Stretcher

_____________ **6.** Sill

_____________ **7.** Cap

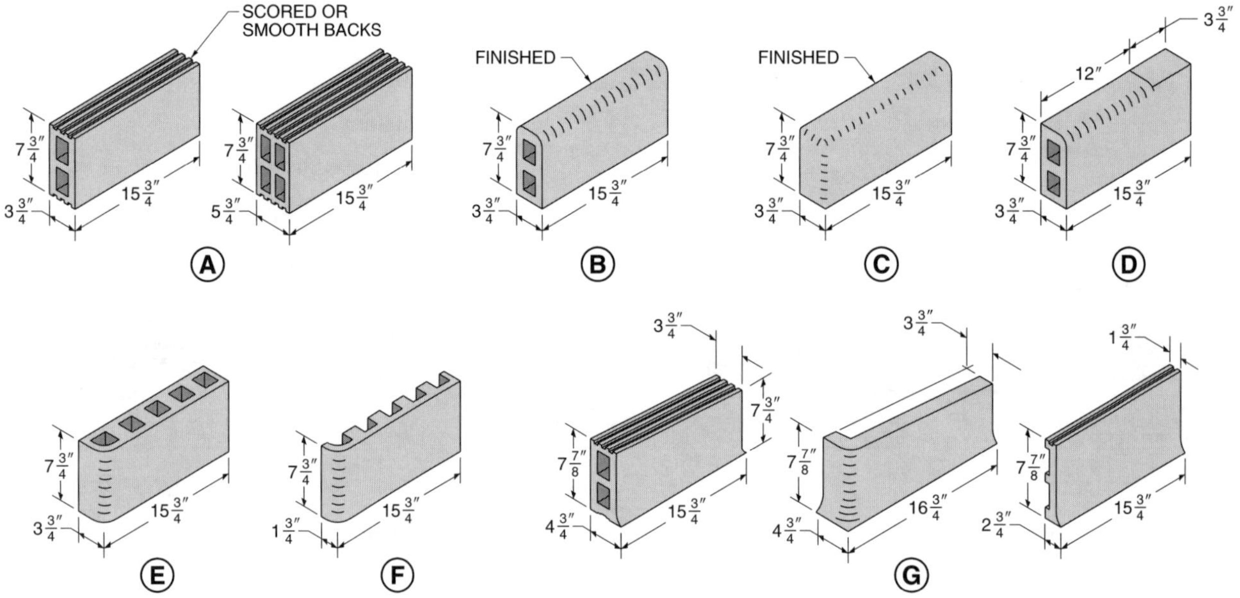

Brick

_____________ **1.** Economy

_____________ **2.** Engineer

_____________ **3.** Norman

_____________ **4.** SCR

_____________ **5.** Standard modular

BRICK			
Designation	Nominal Dimensions*		
	t	h	l
A 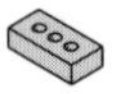	4	2²/3	8
B 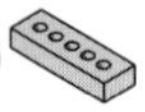	4	2²/3	12
C 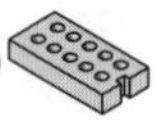	6	2²/3	12
D 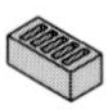	4	3¹/5	8
E 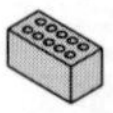	4	4	8

* in in.

Activities
Masonry

Name _____ Date _____ | **7** Chapter

Activity 7-1 – Ledger Sheet Activity

Refer to the cost data and Estimate Summary Sheet No. 7-1. Take off the number of bricks. Determine the total material and labor cost for the brick.

COST DATA			
Material	**Unit**	**Material Unit Cost***	**Labor Unit Cost***
Utility brick 4″ × 4″ × 12″ 3.00 bricks/sq ft	1000	918.50	819.50

* in $

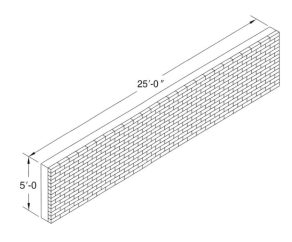

25'-0″

5'-0

ESTIMATE SUMMARY SHEET

Sheet No. _____7-1_____

Project: _____   Date: _____
Estimator: _____   Checked: _____

No.	Description	Dimensions			Quantity		Material		Labor		Total	
		H	**L**			**Unit**	**Unit Cost**	**Total**	**Unit Cost**	**Total**	**Unit Cost**	**Total**
	Utility brick 4″ × 4″ × 12″											
	Total											

Activity 7-2 – Spreadsheet Activity

Refer to the cost data and Estimate Summary Sheet No. 7-2 on the CD-ROM. Take off the square feet of concrete block for the foundation walls of a building 25′ × 70′ × 6′ high. Determine the total material and labor cost for the concrete block.

COST DATA			
Material	**Unit**	**Material Unit Cost***	**Labor Unit Cost***
Concrete block	sq ft	1.36	2.67

* in $

Activity 7-3 – Estimating Software Activity

Create a new estimate and name it Activity 7-3. Perform a quick takeoff for a masonry wall 35′ long and 8′ high having no openings. Take off items **4202.30 Modular Face Brick**, **4105.10 Mortar Type N**, and **4710.10 Clean Brick – General**. The bricks are 4″ wide and the wall consists of a single wythe (vertical course) of bricks. Print a standard estimate report.

Metals

S tructural steel construction uses a series of horizontal beams and trusses and vertical columns that are joined to create large structures with open areas. The three general classifications of structural steel construction are beam and column, long span, and wall bearing. Common steel shapes required for structural steel construction include wide flange, beam, light column, channel, angle, tee, bearing pile, zee, plate, flat bar, tie rod, and pipe column. Estimators rely on bids from a steel fabricator for costs for structural steel members.

STRUCTURAL STEEL MATERIALS

Structural steel construction is construction in which a series of horizontal beams and trusses and vertical columns are joined to create large structures with open areas. The various types and grades of steel used in structural steel construction include carbon, high-strength, high-strength low-alloy, corrosion-resistant high-strength low-alloy, and quenched and tempered alloy steel. See Figure 8-1. The most common steel used for structural steel construction is designated by the American Society for Testing and Materials (ASTM) as A36. A36 carbon steel has a yield strength of 36,000 lb/sq in. The uses and applications of the various types of steel depend on the engineering requirements for a particular structure.

Many steel shapes are required for structural steel construction. Common steel shapes include wide flange, beam, light column, channel, angle, tee, bearing pile, zee, plate, flat bar, tie rod, and pipe column. See Figure 8-2. A variety of letters, numbers, and symbols are used to show the different steel shapes on architectural and structural drawings. Estimators must be familiar with these symbols to accurately take off structural steel members. *Note:* While common notations are used for different types of steel shapes, different manufacturers, steel fabricators, and architects may use slightly different notations. Estimators should check with the manufacturer, fabricator, or architect when there may be the potential for error. All structural steel requires painting with a primer and sometimes requires a finish coat. The finish coat may be applied at the fabrication shop or at the job site.

A variety of steel designs are manufactured to meet all of the applications and loading requirements for columns, beams, girders, joists, braces, studs, runners, plates, and other building members. Variables in the designs include the steel composition, shape, thickness, weight, and length. Variances in the weight of the steel shape (in lb/ft) create differences between nominal size classifications and actual sizes of structural steel members. To prevent failure in case of fire, structural steel members are treated with several different spray cement mixtures and may be encased in concrete, masonry, or gypsum.

STRUCTURAL STEEL				
Steel	ASTM Designation	Minimum Yield Stress*	Form	Characteristics
Carbon	A36	36	Plates, shapes, bars, sheets and strips, rivets, bolts, and nuts	For buildings and general structures. Available in high-toughness grades
	A529	42	Plates, shapes, bars	For buildings and similar construction
High-Strength	A440	42 to 50	Plates, shapes, bars	Lightweight and superior corrosion resistance
High-Strength Low-Alloy	A441	40 to 50	Plates, shapes, bars	Primarily for lightweight welded buildings and bridges
	A572	42 to 65	Several types. Some available as plates, shapes, or bars	Lightweight, high toughness for buildings, bridges, and similar structures
Corrosion-Resistant High-Strength Low-Alloy	A242	42 to 50	Plates, shapes, bars	Lightweight and added durability. Weathering grades available
	A588	42 to 50	Plates, shapes, bars	Lightweight, durable in high thicknesses. Weathering grades available
Quenched and Tempered Alloy	A514	90 to 100	Several types. Some available as shapes, others as plates	Strength varies with thickness and type

* KSI (1000 lb)

Figure 8-1. The type of steel used for a structural steel project depends on the engineering requirements for the structure.

The initial fastening of structural steel members is commonly performed using bolts and nuts.

Beams

A *beam* is a structural member installed horizontally to support loads over an opening. Structural steel beams include wide-flange beams, American Standard beams, girders, and lightweight beams. Steel members are classified as beams and girders when they carry horizontal loads and are spaced greater than 4′ OC. A *girder* is a large horizontal structural member that supports loads at isolated points along its length. Girders carry the horizontal loading of beams and joists. Girders are commonly the heaviest horizontal members in a structure.

Shop drawings are often used for beam fabrication to show the beam size, type, length, cutout dimensions to allow for intersection with other structural steel members, hole location dimensions, and any required connecting angle sizes. See Figure 8-3. A plan view and elevation view may be necessary to give all required beam material information for takeoff of the necessary members.

> ▲ *A36 carbon steel is the most common steel used for structural steel buildings. A36 carbon steel is ductile (able to deform without fracturing) and can be welded or bolted.*

STANDARD STEEL SHAPES

Description	Pictorial	Symbol	Use
WIDE-FLANGE	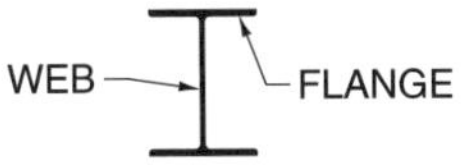 WEB — FLANGE	WF	24WF76
BEAM		I	12I29 x 18'-3"
LIGHT COLUMN	 NOMINAL DEPTH	M	8 x 8M 34.3
CHANNEL	 DEPTH	⊔	9 ⊔ 13.4
ANGLE	 LEGS	L	L 3 x 3 x $\frac{1}{4}$
TEE	 FLANGE — STEM	T	T4 x 3 x 9.2
BEARING PILE		BP	14 BP 73
ZEE		Z	Z6 x 3$\frac{1}{2}$ x 15.7
PLATE	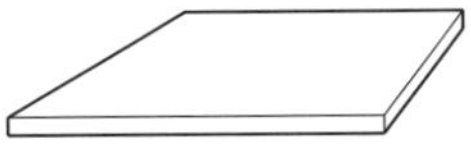	PL	PL 18 x $\frac{1}{2}$ x 2'-6"
FLAT BAR	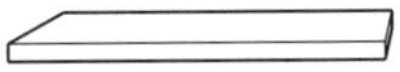	BAR	BAR 2$\frac{1}{2}$ x $\frac{1}{4}$
TIE ROD	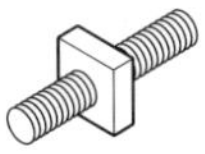	TR	$\frac{3}{4}$ Ø TR
PIPE COLUMN		○	○6Ø

Figure 8-2. Standard steel shapes used for structural steel construction are determined from symbols used on architectural and structural drawings.

I Beams. An *I beam* is a structural steel member with a cross-sectional area resembling the capital letter I. Information for I beams is given in a standard format. The sequence of this format is the nominal depth of the web, type of beam, weight per linear foot, and often the overall beam length. For example, a beam noted as 12I29×18'-3" indicates an I beam with a nominal web depth of 12", a weight of 29 lb/ft, and a length of 18'-3". American Standard beams are commonly referred to as I beams and designated on prints with the letters I or S. The nominal and actual size of American Standard beams are equal.

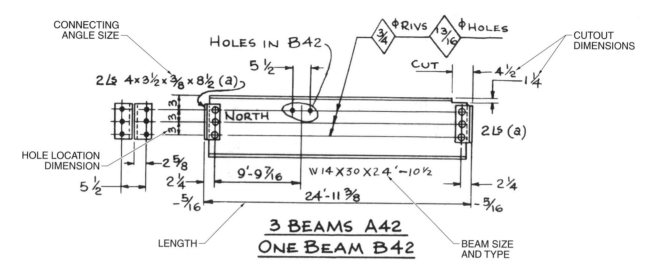

Figure 8-3. Shop drawings for structural steel members indicate fabrication information required for labor pricing calculations.

WF Beams. A *wide-flange beam* is a structural steel member with parallel flanges that are joined with a perpendicular web. The notation used on erection plans for wide-flange beams is W or WF. See Figure 8-4. Wide-flange beams have a nominal depth and an actual depth that varies depending on the manufacturer and flange and web thicknesses. Various lightweight beams are shown on erection plans with the letters B or JB for junior beams. Other miscellaneous-shaped beams are designated on prints with the letter M.

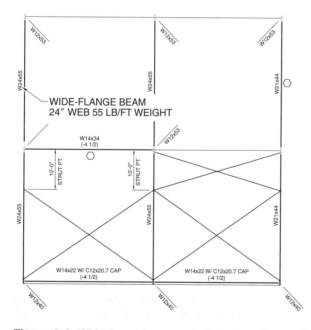

Figure 8-4. Wide-flange beams are shown on floor and erection plans with the symbol W or WF.

Columns

A *column* is a vertical structural member used to support axial compressive loads. Columns are the principal load-carrying vertical member in structural steel construction. Column shop detail drawings indicate column height, hole locations for the attachment of beams and bracing, base plate information such as plate thickness and size, and support angle location. See Figure 8-5.

American Standard (I) and wide-flange shapes are commonly used for steel columns. The size and design of the columns may be provided on a schedule and detail drawings. The loading requirements for the column determine column size and design. Architects and engineers use standard tables to indicate sizes of steel shapes for column applications. See Figure 8-6. Steel columns may also be made from round pipe or square tubing where loads are relatively light.

Joists

A *joist* is a horizontal structural member that supports the load of a floor or ceiling. Structural steel joists are lightweight beams spaced less than 4′ OC. Joists are used in all types of steel construction, including beam and column, long span, and wall bearing. Joists may be formed of a single structural member such as a C channel or built up as a trussed or open web joist.

A *purlin* is a horizontal support that spans across adjacent rafters. Purlins are placed to span between beams, columns, or joists to carry intermediate loads, such as roof decking or wall panels. Purlins placed horizontally to span between columns to carry wall panels are also referred to as girts. Purlins are formed from channel or zee shapes.

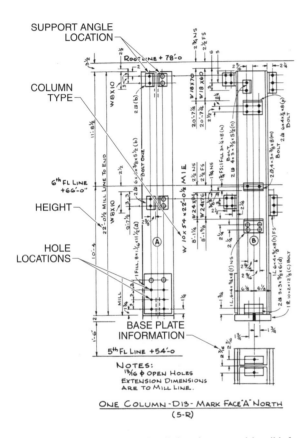

ONE COLUMN·D13· MARK FACE "A" NORTH
(5-R)

Figure 8-5. Column shop detail drawings provide all information required for fabrication and estimating of columns.

Open Web Joists. An *open web joist* is a framing member constructed with steel angles that are used as chords with steel bars extending between the chords at an angle. Open web joists are the most commonly used joists in structural steel construction. A *bar joist* is an open web joist with steel angles at the top and bottom of the joist and bars for the intermediate members. See Figure 8-7. Standard designation information for open web joists includes the nominal depth, span classification, type of steel, and size of chord material. For example, an open web joist with a designation of 25LJ10 has a nominal depth of 25″, is a longspan joist, is made of normal strength steel, and has chord material of #10 steel bar (1¼″ diameter).

WIDE-FLANGE SHAPES				
Designation	**Depth d***	**Flange**		**Web Thickness t_w***
		Width b_f*	**Thickness t_f***	
W18×71	18½	7⅝	¹³⁄₁₆	½
×65	18⅜	7⅝	¾	⁷⁄₁₆
×60	18¼	7½	¹¹⁄₁₆	⁷⁄₁₆
×55	18⅛	7½	⅝	⅜
×50	18	7½	⁹⁄₁₆	⅜
W18×46	18	6	⅝	⅜
×40	17⅞	6	½	⁵⁄₁₆
×35	17¾	6	⁷⁄₁₆	⁵⁄₁₆
W16×100	17	10⅜	1	⁹⁄₁₆
×89	16¾	10⅜	⅞	½
×77	16½	10¼	¾	⁷⁄₁₆
×67	16⅜	10¼	¹¹⁄₁₆	⅜
W16×57	16⅜	7⅛	¹¹⁄₁₆	⁷⁄₁₆
×50	16¼	7⅛	⅝	⅜
×45	16⅛	7	⁹⁄₁₆	⅜

* in in.

Figure 8-6. Column designation indicates depth, flange width and thickness, and web thickness.

OPEN WEB JOIST

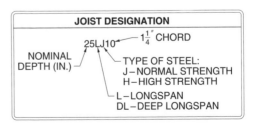

JOIST DESIGNATION

25LJ10

1¼″ CHORD

NOMINAL DEPTH (IN.)

TYPE OF STEEL:
J – NORMAL STRENGTH
H – HIGH STRENGTH

L – LONGSPAN
DL – DEEP LONGSPAN

Figure 8-7. Open web bar joists are identified by nominal depth, span, type of steel, and size of chord material.

Decking

Decking is light-gauge metal sheets used to construct a floor or deck form or used as floor or deck members. Corrugated metal decking forms are used to form floor or roof decks by fastening to structural members. See Figure 8-8. A plan view on the erection plans indicates the deck layout and opening locations. Floor plans may provide information concerning the number of floor or roof decking panels in each run. Detail drawings indicate methods for closing openings and bracing around columns and edge treatments.

Floor Decking. The uses of metal floor decking include providing a work platform during construction and a form and reinforcement for concrete slabs. A common method of creating a floor in a structural steel building is by attaching corrugated steel decking to the top of the joists.

The corrugated steel decking may be fastened to the joists with self-tapping screws or by welding. Corrugated decking is commonly topped with several inches of concrete. The depth of the concrete varies depending on the loads to be supported. The thickness of the concrete on the corrugated steel decking is given on erection plans or detail drawings.

Roof Decking. Metal roof decking provides a high strength-to-weight ratio that reduces the amount of dead load on a structure. Variations in roof decking design include the width and height of the corrugated ribs and the metal finish. Some steel roof decking is designed to create a finished surface with watertight joints between decking pieces. Others are designed to be covered with additional insulation and roofing materials.

Other Structural Members

In addition to columns, beams and girders, and decking, other structural steel members used in structural steel construction include structural load-bearing studs, channels, and other metal shapes used for various applications. All of these shapes are used in different applications for structural and finish metal work.

Studs. A *stud* is a vertical framing member in a wall. Cold-formed metal studs may be used in light construction applications in the same manner as wood studs for bearing walls in commercial and residential construction. Load-bearing metal studs and joists are most commonly made of 14 gauge, 16 gauge, or 18 gauge steel. Load-bearing metal studs include channel, C, and nailable studs. Common channel studs are available in widths of

$2\frac{1}{2}''$, $3\frac{1}{4}''$, $3\frac{5}{8}''$, $4''$, and $6''$ and depths of $1''$ and $1\frac{3}{8}''$. Common C studs are available in widths of $2\frac{1}{2}''$, $3''$, $3\frac{1}{4}''$, $3\frac{1}{2}''$, $3\frac{5}{8}''$, $4''$, $5\frac{1}{2}''$, $6''$, $7\frac{1}{2}''$, and $8''$ and depths of $1\frac{1}{4}''$, $1\frac{3}{8}''$, $1\frac{1}{2}''$, and $1\frac{5}{8}''$. Common nailable studs are available in widths of $3\frac{5}{8}''$ and $4''$ and depths of $1\frac{13}{16}''$ and $1\frac{15}{16}''$. Metal studs are sized in a manner similar to wood framing members. See Figure 8-9.

CORRUGATED FORMS FOR CONCRETE SLABS – NONCOMPOSITE			
Forms	Span*	Width**	Max Length*
$\frac{1}{2}''$	1 - 2	96	2 - 6
$\frac{9}{16}''$	1.5 - 3	30	40
$\frac{15}{16}''$	3 - 5	29	40
$1''$	3 - 5	28	30 - 40
$\frac{9}{16}''$	4 - 9	27	30 - 40
$2''$	7 - 12	24	30 - 40

* in ft
** in in.

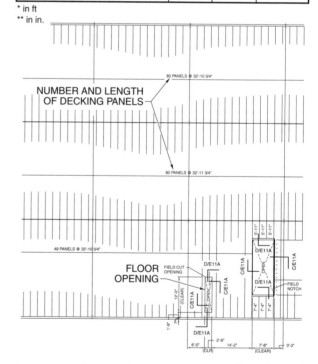

Figure 8-8. Floor and roof decking forms are manufactured in various designs, widths, and lengths that are indicated on floor plans.

METAL STUDS

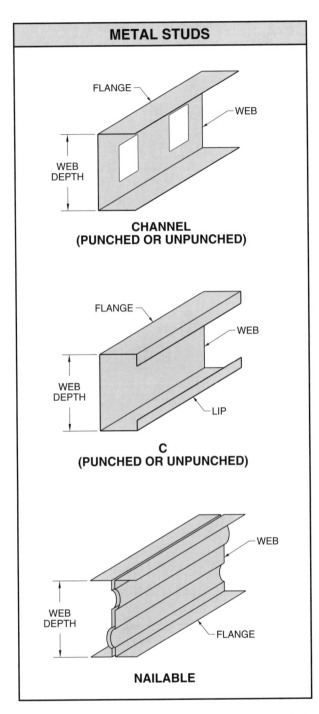

CHANNEL
(PUNCHED OR UNPUNCHED)

C
(PUNCHED OR UNPUNCHED)

NAILABLE

Figure 8-9. Metal studs of various gauges are used for load-bearing members similar to wood framing studs.

The most cost-efficient steel stairs use common materials and standard details and methods of job site assembly. Steel stair job site assembly should be concise and able to be accomplished efficiently with a minimum amount of labor and equipment.

Due to the various gauges of metal used for load-bearing metal stud applications, it is commonly necessary for the review and approval of the plans to be more rigorous than for a wood-framed wall. The estimator must check stud spacing and placement on each job because engineering of stud placement may vary from job to job. In some areas, the different gauges of metal studs and joists are installed by a drywall contractor or subcontractor rather than a structural steel subcontractor.

Channels. *Channel* is light-gauge metal members used for framing and supporting lath or drywall. Light-gauge channels are referred to as junior channels or C channels. Channels are shown on erection plans by the symbol [. Sizes of channels are given in the standard format of channel depth, flange width, and steel thickness. For example, a notation of [4×2×$^3/_8$ indicates a channel with a depth of 4″, a flange width of 2″, and a steel thickness of $^3/_8$″. Light-gauge steel framing of C or zee channel is used for purlins and girts. When used to span between columns or roof beams, C or zee channel dimensions are given on erection plan detail drawings.

Miscellaneous Metal. Other metal shapes used in a variety of applications include tees, bars, and zees. These members are used for bracing, joist, and truss construction. Structural tees are commonly made by cutting standard I beams, wide-flange beams, or smaller beams through the center of the web to form two tees.

Notation systems for tees give the flange width, nominal depth, and steel thickness. For example, a notation of T4×3×$^5/_{16}$ indicates a tee with a flange width of 4″, a nominal depth of 3″, and a steel thickness of $^5/_{16}$″. Steel bars are shown on prints by their thickness and width dimensions.

Zees are shown on drawings in the sequence of depth of the zee, flange width, and weight of steel per foot or steel thickness. For example, a note of Z4×3×3×$^1/_4$ indicates a Z-shaped steel member with a depth of 4″, equal flange widths of 3″, and steel thickness of $^1/_4$″.

Stairs

Metal stair systems are installed in high-traffic areas, for fire escapes, or as freestanding structures such as spiral stairs. Metal stair components are similar in terminology and dimensioning to wood stair components. Cast iron, steel, or aluminum may be used for metal stairs. When spiral metal stairs are installed, they are supported by a vertical pipe onto which treads are fastened. See Figure 8-10.

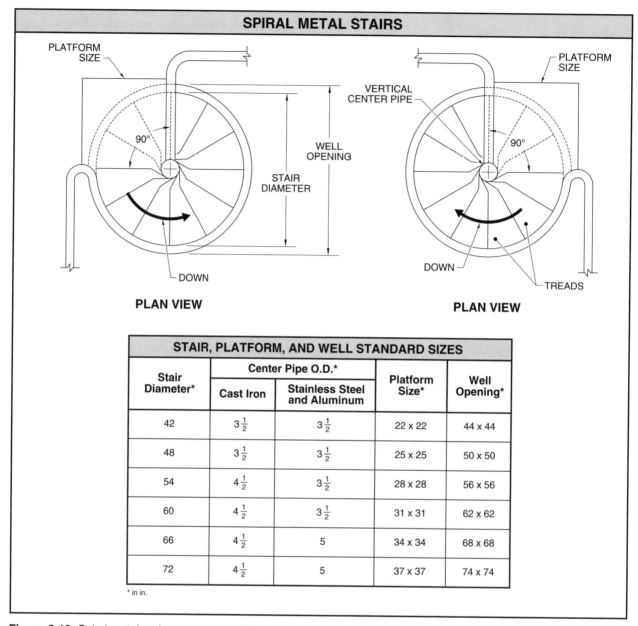

Figure 8-10. Spiral metal stairs are supported by a vertical center pipe onto which treads are fastened.

Treads. Metal stair treads may be cast iron, checked plate steel, expanded metal grating, C channel, angles, or bar grating depending on the tread application and location. Metal stair treads may also be covered or coated with an abrasive material to provide greater traction. Metal pans or other metal designs may be used with poured concrete stairs to provide a long-wearing, non-skid nosing cast into the concrete.

Handrails. Metal handrails are made of steel or aluminum piping designed to support a minimum of 200 lb at any point in any direction. Pipe is joined with mechanical connectors or by cutting and welding to form a

smooth surface. Pipe handrails and posts are commonly bolted in place with special brackets designed to fit the proper pipe diameter. Metal handrails and balusters may also be used with wood handrail components that are fastened to the top of the metal baluster/handrail frame.

STRUCTURAL STEEL CONSTRUCTION METHODS

Structural steel construction may use lightweight members to build industrial buildings and storage structures or large beam and truss assemblies to build skyscrapers and bridges. Structural steel may be used as an independent construction method or with other methods, such as masonry or reinforced concrete. Several common steps are used when building with structural steel.

In structural steel construction, a structural engineer determines the type of steel to be used and sizes and shapes of required steel members after all loads to be placed on a structure have been determined. Estimators take off the structural steel items to be fabricated according to the plans and possibly develop shop drawings. Various steel members are fabricated at a location away from the job site according to shop drawings. The structural steel components are transported to the job site where the members are unloaded (shaken out). *Shaking out* is the process of unloading steel members in a planned manner to minimize the moving of pieces during the erection process. Structural steel members are lifted into place with various lifting equipment. See Figure 8-11. Structural steel components are erected, braced, and fastened together to create the structural frame. This frame is covered with the materials for floors, walls, and roofing.

Figure 8-11. Structural steel construction uses a variety of prefabricated and on-site constructed steel components.

Varying engineering requirements and construction site conditions create the need for several methods of structural steel construction. Three general classifications of structural steel construction are beam and column, long span, and wall bearing. Erection plans provide information for placement of steel members for each structural steel construction method.

The proper handling, lifting, and placement of steel members is essential in each structural steel construction method. Safety is also a primary consideration during structural steel construction. Estimators

should check the specifications concerning the need for worker fall protection including a body harness and lanyard. Division 5 of the CSI MasterFormat™ should be consulted concerning worker tie-off, which may be required for all workers under some conditions.

Beam and Column

Beam and column construction is construction consisting of bays of framed structural steel which are repeated to create large structures. Beam and column construction is the most common structural steel construction method. See Figure 8-12. In beam and column construction, a series of foundations, footings, piers, or pilings are constructed. Columns are supported by the foundations, footings, piers, or pilings. Anchor bolts and base plates are prepared prior to setting columns in place. When concrete foundations and footings are placed, the anchor bolts for steel columns are set to specific dimensions for spacing and height according to the prints. Templates may be set on the anchor bolts and grouted in place to ensure exact elevations for steel column base plates. Base plates are welded onto small columns at a fabrication shop. Large column base plates are set separately. Columns are normally the first members erected for beam and column construction.

Figure 8-12. Beam and column construction uses structural steel beams and columns that distribute horizontal loads onto vertical columns.

Horizontal steel beams and girders are attached to the columns. Angles may be attached to columns to form a seat for the beams or they may be attached to beams that are bolted to columns. Tie rods, channels, and other bracing members are used to enhance stability. Joists and purlins are placed between the columns and beams to finish the structural portion of the construction.

Beams and girders are rigged and lifted into place before being bolted to columns or other girders. A *spandrel beam* is a beam in the perimeter of a structure that spans columns and commonly supports a floor or roof. Angles attached to the columns or beams enable the fastening of spandrel beams.

Bolts and nuts are commonly used for the initial fastening of columns and beams. Seat lugs or shear plates made of angles attached to the columns may be used to support beam and column connections during erection. For multi-member beams, splice plates may be used to hold the end of each beam in place. As construction progresses, guy wires, turnbuckles, sag rods, sway rods, and other cross-bracing members are added to plumb, level, and hold all members at the proper elevations. Bolting or welding of the connections of these members is required to complete the erection.

> ▲ *Structural steel accessories such as bolts, nuts, washers, and connection angles and plates can add a significant amount to the weight of a project. A factor of 10% may be added to the total weight to account for these accessories.*

Long Span

Long span construction is construction consisting of large horizontal steel members, such as girders and trusses, that are fastened together to create large girders and trusses. Long span construction is used for structures such as bridges and large arenas. In long span construction, a series of built-up girders and trusses are used for spanning large areas without the need for excessive intermediate columns or other supports. See Figure 8-13. Long span construction is commonly used for bridge construction. Large girders and beams span between bridge abutments, piers, and other supports. Decking of concrete or steel is supported by the large structural steel members. Detail drawings and shop drawings show the arrangement of the various angles, channels, and other members used to construct girders and trusses.

Wall Bearing

Wall bearing construction is construction integrating horizontal steel beams and joists into other construction methods such as masonry and reinforced concrete. Steel support members are used to span floors and roofs between masonry or reinforced concrete-constructed walls. Masonry or concrete walls support the vertical loads and structural steel beams and joists carry the horizontal loads. The installation of bearing base plates on the masonry or concrete walls provides proper load distribution at the points where the structural steel members rest on the walls.

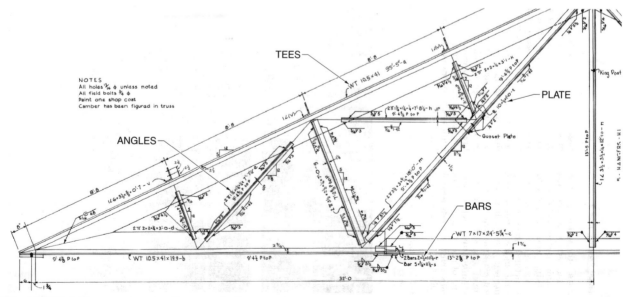

Figure 8-13. Long span construction uses built-up steel trusses and girders to span long distances.

Joists and Decking

Joists are commonly welded to their supporting members. Welding symbols are used to indicate the type of weld required. Detail drawings indicate the manner of joist attachment to the beams and girders. Print information for the installation of purlins and girts includes spacing and direction of placement. C channels or girts are most commonly bolted to their supporting members. Spacing and sizes of bolts are given in detail drawings.

After the structural frame is erected, fastened, and braced, various panel and decking members are attached to the frame. For structural steel buildings, the panel and decking members create floors, walls, and roofs on the framework to complete the structure. Erection plans provide information concerning the various panel and decking members.

Load Bearing Framing

Load bearing framing is a system of medium-gauge steel members designed by structural engineers to support live and dead loads. Load bearing studs are welded, fastened with self-tapping screws, or bolted to top and bottom plates or runners in a design similar to wood platform framing. Door and window openings are framed in the same manner as wood framing with heavy-gauge studs at either side of the openings and header and cripple framing. As the studs and runners are welded together, strapping or bracing is also attached to provide lateral strength. Exterior finish panel and decking materials may be fastened to the metal frame prior to setting in place.

STRUCTURAL STEEL QUANTITY TAKEOFF

Due to the size and shape of many structural steel pieces, it is common for a contractor to work with a steel fabrication shop for structural steel members. A structural steel fabricator works with the designer, architect, and structural engineer to design each piece of steel. Working drawings or shop drawings are developed that indicate the base plates, bearing members, types of welds, types of steel or other metal, locations for drilled or burned holes, and other special elements necessary for structural steel members.

Some installation costs should be considered during finalization of a bid for structural steel. The installation items include the size and type of crane that is required for setting and holding the structural steel members in place during erection, the height of the building and the need for additional lifting or safety equipment, the schedule of the job and its potential impact on erection accessibility, and the season of the year, which could affect labor costs.

Structural Steel Estimation

Estimators rely on bids from a steel fabricator for costs for structural steel members. Estimators at the fabrication shop determine costs for each element, including steel materials, labor costs, and transportation costs to the job site. See Figure 8-14. Structural steel fabrication estimators rely on company historical data to determine costs for fabrication labor. Steel and transportation prices depend on supply and distance to the job site. Structural steel fabrication estimators begin with the structural drawings to determine the number of each type of column, beam, girder, and other structural steel members needed. Designs for each include attachments for adjoining members, base plates, and bracing. Labor costs at the job site may be significantly increased when full tie-off is required by all workers erecting structural steel.

Figure 8-14. Structural steel members are calculated as individual units and grouped whenever possible.

Columns. Column locations are shown on erection plans according to a grid of letters and numbers. References to the letters and numbers identify each column. For example, a column located at the intersection of grid lines D and 2 is referred to as column D2. The design, size, and weight of each column may be noted at this intersection. For example, a notation of W12×53 indicates a column made of a wide-flange shape with a 12″ web weighing 53 lb per linear foot. A schedule of column information may also be provided to give design, web, and weight information about each column in relation to the plan grid.

For round pipe columns, the nominal inside diameter and schedule of the pipe is indicated. For example, a round pipe column shown on prints as 8 Sch 60 indicates an 8″ diameter pipe column with schedule 60 wall thickness. The outside dimensions of the tubing are given along with the steel wall thickness for square tubing columns. For example, a print notation of 3×3×¼ indicates a 3″ square tube column with ¼″ wall thickness.

Estimators count the number of columns of each type and enter the number onto a ledger sheet, spreadsheet, or estimating software cell. Some columns may be similar in design and can be grouped. Pricing for each column must include determination of overall height, bearing capacity, steel, base plate, and connections for intersecting beams or girders or additional columns for multistory buildings.

Beams and Girders. Beam and girder sizes are given on plan grid and elevation drawings. Sizes are noted along the gridlines on plan drawings. Lengths are obtained from the grid spacing dimensions.

In a manner similar to columns, a letter and number system is commonly used to identify each beam or girder. The letters and numbers are marked on the beam or girder at the fabrication shop and correspond to letter and number notations on the erection plan drawings. A schedule of beam sizes and types may be provided in addition to the drawings. Estimators must take material size and type and connection information into account when pricing structural steel beams. Estimators count the number of structural steel beams of each type and enter the number into a ledger sheet, spreadsheet, or estimating software cell.

Structural steel members and concrete masonry units may be used together in the design of a structure.

Joists. Joists are not normally identified with letters and numbers in the manner of columns and beams. The spacing of joists and the direction for their placement is noted on the erection floor plans. See Figure 8-15. The type of joist to be installed may also be noted by a manufacturer identification code, standard classification format, or fabrication shop code number. Open areas for stairwells or other access between levels of a structure are shown on the joist plan view. Open areas are indicated by dashed lines in an × pattern. Solid × lines between joists indicate crossbracing.

Estimators determine the number of each type of joist of a particular length and develop a joist schedule. Each portion of floor or roof decking may have many different types of joists depending on the span and loads to be supported. Estimators should take off and place steel joist orders with the fabrication shop well in advance of the projected construction schedule. Fabrication of steel joists may take time and could affect the project schedule if not ordered in a timely manner. Information forwarded to the fabrication shop for each type of joist should include the quantity, length (span), depth, top and bottom channel design, web design, and bearing and support design at each end of the joist.

Studs. Estimators determine stud spacing, length, gauge, and linear feet of wall necessary. Multipliers for stud spacing are similar to multipliers for other framing members. For example, a 1′ OC spacing requires 1 as a multiplier, a 16″ OC spacing requires .75 as a multiplier, and a 24″ OC spacing requires .50 as a multiplier. Overall wall height determines stud length. Linear feet of wall is used to determine linear feet of top and bottom runner. Estimators should add additional heavy-gauge studs at either side of window and door openings and additional runner material for headers and sills for window and door openings. Estimators determine the number of each stud classified by stud length and metal gauge. Totals for each length and gauge classification should include approximately 3% to 5% for waste, depending on the overall size of the project.

> *Structural steel is sold by the ton. The total number of tons required is determined by taking off the linear feet of each type of beam, girder, column, or joist required. The linear foot dimension is multiplied by the weight per foot and converted to tons to determine the total weight.*

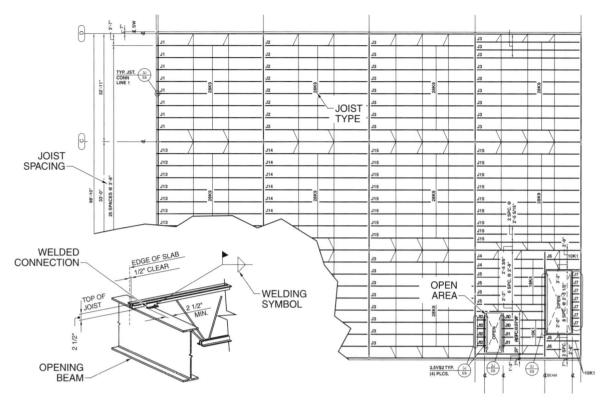

Figure 8-15. Erection floor plans provide estimators with information concerning joist types and spacing.

Decking

A roof panel plan of a steel floor or roof deck application is included in the erection plans. See Figure 8-16. A manufacturer schedule of the types and sizes of floor or roof decking may be provided to help take off the proper quantity of each type of panel. Floor decking can be cut to length and around unusual shapes with a cutting torch. Floor decking is fastened to bar joists with self-tapping screws or welding. Roof decking is attached to purlins or ceiling joists with self-sealing screws or clips. Estimators count the number of decking pieces of each type and enter the number into a ledger sheet, spreadsheet, or estimating software cell.

Stairs

Metal stairs are commonly fabricated off-site and delivered as a single unit to the construction site. Estimates for metal stairs of a common pan design to have treads filled with concrete use total run, total rise, and total width information from plan view and interior elevation drawings. Costs include fabrication labor, stringer materials, tread pans, and risers if required. There may be handrail costs included in the stair bid.

For spiral metal stairs, the total rise is used along with the stairwell opening and the required headroom. Several designs of prefabricated treads are available for use on spiral stairs and several types of center pipe and handrail systems are available. Estimators should determine if installation is part of the stair costs prior to submitting a bid. Stair information is entered into a ledger sheet, spreadsheet, or estimating software cell.

ESTIMATING SOFTWARE – LISTS

Lists are available throughout Precision estimating software to help make selections, provide ways to find information, and to sort the contents in alternate order. Lists are modified by applying a standard procedure. See Figure 8-17.

> ▲ *When estimating and taking off metal decking, the estimator should determine who is supplying decking terminations. Some terminations may be part of the decking while others may be part of the steel contract. The estimate must include all pieces required for proper construction.*

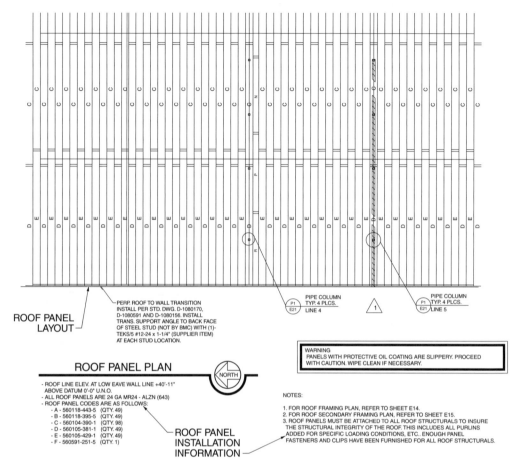

Figure 8-16. A roof panel plan provides information concerning the number of each type of roof panel as specified by a metal building manufacturer.

PRECISION ESTIMATING SOFTWARE – LISTS

Expanding and Collapsing Lists

1. Open the Quick Takeoff window.

2. Using the mouse, double-click the group to expand from the group level to the phase level. Double-click the phase to expand from the phase level to the item level. Double-click the Phase to collapse back to the phase level. Double-click the Group to collapse back to the item level.

Using the shortcut menu, right-click in the list to display the shortcut menu to expand or collapse the entire list. Select Collapse All or Expand All.

Using the keyboard, highlight the group and press the Right Arrow key to expand the list from the group level to the phase level. Highlight the phase and press the Right Arrow key to expand the list from the phase level to the item level. Highlight the phase and press the Left Arrow key to collapse the list back to the phase level. Highlight the phase and press the Left Arrow key to collapse the list back to the item level.

Sorting Lists

The Item list can be sorted by description or phase/item. To select the sort order, right-click in the item list. A shortcut menu displays. Select Description or Phase/Item. The list is immediately re-sorted.

Finding Items

In any list, the quickest way to get to a specific entry item is by typing its first few characters. The entry item displays at the top of the window, and the list moves to the closest match at the same level. If there is no match, the cursor moves to the end of the list. To find items in a list, choose the sort option that is most useful for the search. If necessary, use the Left Arrow key to collapse the list or the Right Arrow key to expand the list. Using correct capitalization, enter the first few numbers or characters of the entry.

Figure 8-17. Estimating software lists are used to help make selections, provide ways to find information, and to sort the contents.

Review Questions
Metals

Name _____ Date _____ **Chapter** 8

Estimating

_____ 1. _____ construction is construction in which a series of horizontal beams and trusses and vertical columns are joined to create large structures with open areas.

_____ 2. A(n) _____ is a structural member installed horizontally to support loads over an opening.

_____ 3. A(n) _____ is a large horizontal structural member that supports loads at isolated points along its length.

_____ 4. A(n) _____ beam is a structural steel member with a cross-sectional area resembling the capital letter I.

_____ 5. A(n) _____ beam is a structural steel member with parallel flanges that are joined with a perpendicular web.

T   F   6. A column is a vertical structural member used to support axial compressive loads.

T   F   7. A joist is a vertical structural member that supports the load of a floor or ceiling.

_____ 8. A(n) _____ is a horizontal support that spans across adjacent rafters.

_____ 9. A(n) _____ joist is a framing member constructed with steel angles that are used as chords with steel bars extending between the chords at an angle.

_____ 10. A(n) _____ joist is an open web joist with steel angles at the top and bottom of the joist and bars for the intermediate members.

_____ 11. _____ is light-gauge metal sheets used to construct a floor or deck form or used as floor or deck members.

T   F   12. A stud is a horizontal framing member in a wall.

T   F   13. Channel is light-gauge metal members used for framing and supporting lath or drywall.

_____ 14. _____ is the process of unloading steel members in a planned manner to minimize the moving of pieces during the erection process.

_____ 15. _____ construction is construction consisting of bays of framed structural steel which are repeated to create large structures.

_____ 16. A(n) _____ beam is a beam in the perimeter of a structure that spans columns and commonly supports a floor or roof.

_____ 17. _____ construction is construction consisting of large horizontal steel members, such as girders and trusses, that are fastened together to create large girders and trusses.

_____ 18. _____ construction is construction integrating horizontal steel beams and joists into other construction methods such as masonry and reinforced concrete.

_____ 19. _____ framing is a system of medium-gauge steel members designed by structural engineers to support live and dead loads.

T   F   20. Structural steel is sold by the ton.

Standard Steel Shapes

_____   **1.** Zee

_____   **2.** Plate

_____   **3.** Flat bar

_____   **4.** Tee

_____   **5.** Angle

_____   **6.** Channel

_____   **7.** Wide flange

_____   **8.** Beam

_____   **9.** Light column

_____   **10.** Tie rod

_____   **11.** Pipe column

_____   **12.** Bearing pile

WF	I	M
(A)	(B)	(C)

⊔	L	T
(D)	(E)	(F)

BP	Z	PL
(G)	(H)	(I)

BAR	TR	◯
(J)	(K)	(L)

Activity 8-1 – Ledger Sheet Activity

Refer to the cost data, Print 8-1, and Estimate Summary Sheet No. 8-1. Take off the square feet of metal decking. Determine the total material and labor cost for the metal decking.

COST DATA			
Material	**Unit**	**Material Unit Cost***	**Labor Unit Cost***
Metal decking, galvanized 18 gauge	sq ft	3.60	.77

* in $

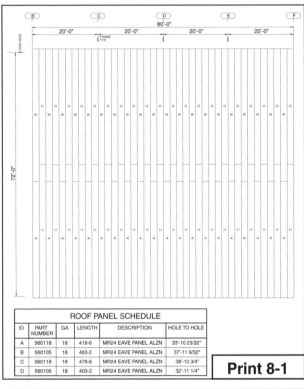

ROOF PANEL SCHEDULE

ID	PART NUMBER	GA	LENGTH	DESCRIPTION	HOLE TO HOLE
A	560118	18	418-6	MR24 EAVE PANEL ALZN	33'-10 23/32"
B	560105	18	463-2	MR24 EAVE PANEL ALZN	37'-11 9/32"
C	560118	18	478-6	MR24 EAVE PANEL ALZN	38'-10 3/4"
D	560105	18	403-2	MR24 EAVE PANEL ALZN	32'-11 1/4"

Print 8-1

ESTIMATE SUMMARY SHEET

Sheet No. __8-1__

Project: _____

Estimator: _____

Date: _____

Checked: _____

No.	Description	Dimensions			Quantity		Material		Labor		Total	
		L	W			Unit	Unit Cost	Total	Unit Cost	Total	Unit Cost	Total
	Metal decking											
	Total											

Activity 8-2 – Spreadsheet Activity

Refer to the cost data, Print 8-2, and Estimate Summary Sheet No. 8-2 on the CD-ROM. Take off the linear feet of structural steel beams W16×26, W16×31, W16×36, W24×55, W24×62, W24×76, and W27×84. Determine the total material, labor, and equipment cost for the structural steel beams.

COST DATA				
Material	**Unit**	**Material Unit Cost***	**Labor Unit Cost***	**Equipment Unit Cost***
Structural steel beams				
W16×26	lf	18.87	1.84	1.11
W16×31	lf	22.55	2.04	1.23
W16×36	lf	24.48	2.11	1.31
W24×55	lf	37.95	2.40	1.09
W24×62	lf	42.90	2.40	1.09
W24×76	lf	52.80	2.40	1.09
W27×84	lf	58.30	2.23	1.01

* in $

Activity 8-3 – Estimating Software Activity

Create a new estimate and name it Activity 8-3. Perform a quick takeoff for item **5055.00 5 Anchor Bolts (All Sizes)** and item **5510.05 10 Steel Lintels 3×3×¼**. Take off 500 anchor bolts and 20 lintels. Print a standard estimate report.

Wood and Plastics

Information about wood and plastic materials, wood treatments, wood coatings, and wood and plastic fastenings is given in Division 6 of the CSI MasterFormat™. Wood is used in construction for structural support members, temporary bracing, shoring, sheathing, decking, finish moldings, paneling, and casework. Estimators calculate the number of pieces, number of square feet, number of sheets, or the linear feet of wood depending on the material installed and the quantities required. Estimators rely primarily on interior finish elevation drawings, floor plans, schedules, and the specifications to take off wood finish members.

MATERIALS

Division 6 of the CSI MasterFormat™ includes basic information concerning wood and plastic materials, wood treatments, wood coatings, and wood and plastic fastenings. Wood products are available in a variety of types, sizes, shapes, and finishes. Wood is used in construction for structural support members, temporary bracing, shoring, sheathing, decking, finish moldings, paneling, and casework. Estimators calculate the number of pieces, number of square feet, number of sheets, or the linear feet of wood depending on the material installed and the quantities required.

Wood is used in many other applications on a construction site that are not included in Division 6 of the CSI MasterFormat™. These include concrete forming materials, bracing, and stakes (Division 3), temporary signage and temporary construction devices such as sawhorses and work platforms (Division 1), doors and windows (Division 8), finish wood flooring (Division 9), and casework of various hardwoods (Divisions 11 and 12). These are items that the estimator must in-clude in the overall bid but are not a portion of the wood information described in Division 6.

Plastics are being introduced into many new areas of construction. Division 6 of the CSI MasterFormat™ pertains to structural plastic materials used as a substitute for structural wood members and finish plastic materials such as laminates.

Wood and plastic materials contained in Division 6 of the CSI MasterFormat™, specifications, and bid documents include wood framing, heavy timber construction, sheathing, glulam members, finish paneling, shelving, millwork, window sills, and architectural woodwork including cabinetry, countertops, stairs, railings, finish trim members, and structural and finish plastic members. In many cases, Division 6 may also include information concerning prefabricated plastic laminate casework.

Structural Wood Members

A *structural wood member* is a wood component that provides support for live and dead loads to a structure.

A *live load* is the predetermined load a structure is capable of supporting. Live loads include moving and variable loads such as persons, furniture, and equipment. A *dead load* is a permanent, stationary load composed of all building material and fixtures and equipment permanently attached to a structure.

Wood is classified as hardwood and softwood. *Hardwood* is wood produced from broad leaf, deciduous trees such as ash, birch, maple, oak, and walnut. Hardwood is used in construction primarily for finish materials such as cabinetry and millwork. *Millwork* is finished wood materials or parts, such as moldings, jambs, and frames completed in a mill or manufacturing plant. *Softwood* is wood produced from a conifer (evergreen) tree. Softwood is used for structural framing members and trim components and includes pine and fir. Cedar is a softwood often used for exterior framing and finish applications.

Grading. *Grading* is the classification of various pieces of wood according to their quality and structural integrity. Wood is graded based on strength, stiffness, and appearance. *Lumber* is sawn and sized lengths of wood used in construction. High grades of lumber have few knots and defects. Low grades have many loose knots. Specifications indicate the grades required for various building applications. Grade marks are stamped on softwood lumber to provide information concerning grade. See Figure 9-1. Material costs increase for high grades of lumber.

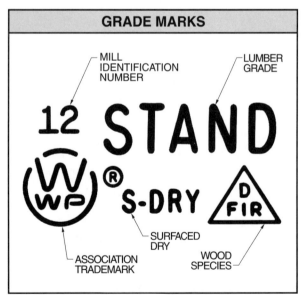

Figure 9-1. Lumber grade marks indicate lumber grade, wood species, mill information, etc.

Sizing. Structural wood framing member size is indicated as nominal size. *Nominal size* is a designated or theoretical size that may vary from the actual size. Nominal sizes are based on the dimensions of the rough lumber prior to planing. See Figure 9-2. For example, the actual dimensions of a 2″ × 4″ member are 1½″ × 3½″. Nominal wood framing members are cut to standard lengths of 8′, 10′, 12′, 14′, and 16′.

A *board* is a wood member that has been sawn or milled having a nominal size from 1″ × 4″ up to 1″ × 12″. *Dimension lumber* is a wood member precut to a particular size for the building industry, normally having a nominal size from 2″ × 4″ up to 4″ × 6″. *Timber* is a heavy wood member that has a nominal size of 5″ × 5″ or larger.

Glulam. *Glulam (glue-and-laminated)* is a structural member constructed by bonding several layers of lumber with adhesive. For timber construction, planks and beams of either a single piece of wood or glulam members are used. Glulam members are specified on plan drawings as to the type and grade of wood in the glulam members and the exact dimensions. In many cases, various geometric shapes are specified for glulam wood members. These members often require fabrication to specific shapes. See Figure 9-3.

Use of glulam members also allows architects and engineers to create curved wood members that are not possible with solid timber members. Applications for glulam members include floor and roof beams, columns, and trusses of all designs. A *beam* is a horizontal structural member installed to support a load over an opening. A *column* is a vertical structural support member. A *truss* is a manufactured roof or floor support member. Specific types, sizes, and grades of lumber are chosen for glulam members. Choices of lumber are based on appearance and structural requirements for span and loading.

Prefabricated Units. A *prefabricated unit* is a standalone module composed of many different building components. Structural wood members are joined into prefabricated units to form trusses and joists. A *joist* is a horizontal structural member that supports the load of a floor or ceiling. Various configurations allow for trussed wood members to span wide areas and withstand additional levels of live and dead loading according to the design used and the sizes and grades of lumber. See Figure 9-4. Metal webbing may be used in these designs for flat trusses and floor trusses with wood top and bottom members.

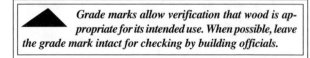

Grade marks allow verification that wood is appropriate for its intended use. When possible, leave the grade mark intact for checking by building officials.

STANDARD LUMBER SIZE

Type	Nominal Size*		Actual Size*	
	Thickness	Width	Thickness	Width
BOARD	1	2	$\frac{3}{4}$	$1\frac{1}{2}$
	1	4	$\frac{3}{4}$	$3\frac{1}{2}$
	1	6	$\frac{3}{4}$	$5\frac{1}{2}$
	1	8	$\frac{3}{4}$	$7\frac{1}{4}$
	1	10	$\frac{3}{4}$	$9\frac{1}{4}$
	1	12	$\frac{3}{4}$	$11\frac{1}{4}$
DIMENSION LUMBER	2	2	$1\frac{1}{2}$	$1\frac{1}{2}$
	2	4	$1\frac{1}{2}$	$3\frac{1}{2}$
	2	6	$1\frac{1}{2}$	$5\frac{1}{2}$
	2	8	$1\frac{1}{2}$	$7\frac{1}{4}$
	2	10	$1\frac{1}{2}$	$9\frac{1}{4}$
	2	12	$1\frac{1}{2}$	$11\frac{1}{4}$
TIMBER	5	5	$4\frac{1}{2}$	$4\frac{1}{2}$
	6	6	$5\frac{1}{2}$	$5\frac{1}{2}$
	6	8	$5\frac{1}{2}$	$7\frac{1}{2}$
	6	10	$5\frac{1}{2}$	$9\frac{1}{2}$
	8	8	$7\frac{1}{2}$	$7\frac{1}{2}$
	8	10	$7\frac{1}{2}$	$9\frac{1}{2}$

* in in.

Figure 9-2. Nominal board, dimension lumber, and timber sizes noted on plan drawings differ from actual sizes.

GLULAM MEMBER

Figure 9-3. Glulam members consist of laminated structural wood members used to span large open areas.

Wood Panels

Wood panels include plywood, oriented strand board, particleboard, and fiberboard. Wood panels are used for subflooring, wall sheathing, roof sheathing, shelving, and cabinetry. Wood panel products are commonly manufactured in 4′ × 8′ sheets of 32 sq ft each (4′ × 8′ = 32 sq ft). Wood panel products are also available in 4′ × 9′ and 4′ × 10′ sheets. Floors are covered with plywood or oriented strand board subflooring. Plywood and fiberboard are commonly used for wood wall sheathing materials. Roofs are commonly covered with plywood or oriented strand board roof sheathing. Laminated casework uses particleboard panels covered with plastic laminate. Casework may also use particleboard panels covered with hardwood veneer.

Plywood. *Plywood* is a wood panel manufactured from thin layers of lumber. The layers are glued together in an odd number of plies (most commonly 3 or 5) with alternating grain placement, resulting in cross-lamination. Plywood is graded according to appearance, size and number of repairs in the plies, and veneer grade. See Figure 9-5. Plywood is commonly available in thicknesses varying from ¹/₄″ to 1¹/₈″.

Oriented Strand Board. *Oriented strand board (OSB)* is a wood panel in which wood strands are mechanically oriented and bonded with resin under heat and pressure. OSB panels are comprised of strands of wood layered in a manner similar to plywood veneers, with alternating grain placement and cross-lamination. An odd number of plies is used similar to plywood. OSB panels are commonly available in thicknesses varying from ⁵/₁₆″ to 1¹/₈″.

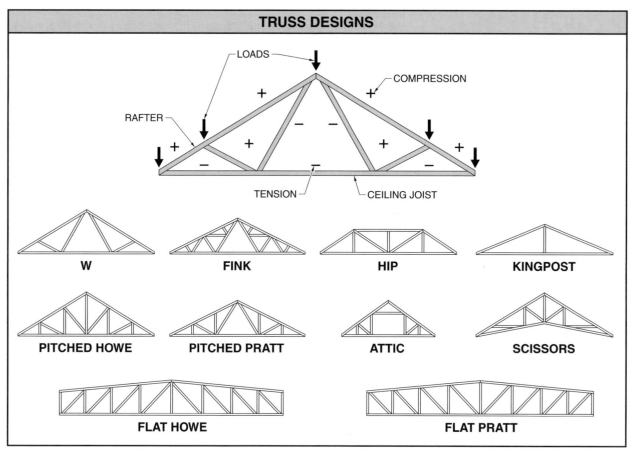

Figure 9-4. A variety of engineered designs create large-span trusses from framing lumber.

Particleboard. *Particleboard* is a wood panel constructed of wood particles and flakes that are bonded together with a synthetic resin. Particleboard has no grain and produces a smooth finish, suitable for underlayment and various cabinet components. Three grades of particleboard are low-density, medium-density, and high-density. Low-density particleboard panels weigh less than 37 lb per cu ft and are of limited architectural use. Medium-density particleboard panels weigh from 37 lb to 50 lb per cu ft and are the most commonly used. High-density particleboard panels weigh over 50 lb per cu ft and are used where high strength and hardness are required. Particleboard is available in thicknesses varying from $\frac{1}{4}''$ to $1\frac{1}{8}''$.

Fiberboard. *Fiberboard* is a wood panel manufactured from fine wood fibers mixed with binders. Panels are formed with heat and pressure. Fiberboard panels are used with or without veneers and surfacing materials. Densities are measured in the same manner as particleboard. The visual edge quality of fiberboard is excellent when compared to the poor visual edge quality of particleboard. Fiberboard is available in thicknesses varying from $\frac{3}{8}''$ to $\frac{3}{4}''$.

Finish Wood Members

A *finish wood member* is a decorative and nonstructural wood in a structure. Hardwood and softwood are used as finish wood members. Most paneling, cabinetry, and stair components are made of hardwood, with millwork made from softwood and hardwood. Millwork includes trim moldings, window and door frames, and stairs. Wood ornaments specified on plan drawings also include mantels, pediment heads, and decorative and structural columns.

> *Wood roof trusses are available from manufacturers in many different designs. Members are connected with wood or metal gussets, and are glued and/or power nailed in a jig to ensure uniformity.*

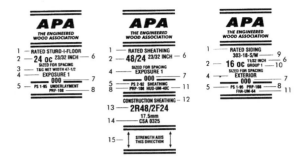

1 Panel grade
2 Span Rating
3 Tongue-and-groove
4 Exposure durability classification
5 Product Standard
6 Thickness
7 Mill number
8 APA's Performance Rated Panel Standard

9 Siding face grade
10 Species group number
11 HUD/FHA recognition
12 Panel grade, Canadian standard
13 Panel mark – Rating and end-use designation, Canadian standard.
14 Canadian performance-rated panel standard
15 Panel face orientation indicator

PLYWOOD VENEER GRADES

GRADE	CHARACTERISTICS
A	Smooth, paintable. Not more than 18 neatly made repairs, boat, sled, or router type, and parallel to grain, permitted. May be used for natural finish in less damanding applications.
B	Solid Surface. Shims, circular repair plugs and tight knots to 1″ across grain permitted. Some minor splits permitted.
C PLUGGED	Improved C veneer with splits limited to ⅛″ width and knotholes and borer holes limited to ¼″ x ½″. Admits some broken grain. Synthetic repairs permitted.
C	Tight knots to 1½″. Knotholes to 1″ across grain and some to 1½″ if total width of knots and knotholes is within specified limits. Synthetic or wood repairs. Discoloration and sanding defects that do not impair strength permitted. Limited splits allowed. Stitching permitted.
D	Knots and knotholes to 2½″ width across grain and ½″ larger within specified limits. Limited splits are permitted. Stitching permitted. Limited to interior (Exposure 1 or 2) panels.

Figure 9-5. Plywood panel grades, span rating, veneer grade, etc., are shown by codes stamped on each sheet.

Hardwood lumber for finish use is graded according to appearance. Hardwood lumber is graded as first and second (FAS), No. 1 common, No. 2 common, and No. 3 common. First and second is the highest appearance grade. An FAS grade is usually required for hardwood trim materials that have a natural or stained finish. No. 1 common is a low-cost shop grade of hardwood. No. 1 common boards are a minimum of 3″ wide and 4′ long. No. 2 common boards are the same size as No. 1 common boards, but yield a minimum of 50% clear face cuttings a minimum of 3″ wide and 2′

long. No. 2 common boards are used for some flooring and paneling applications. No. 3 common boards have similar dimensions to No. 1 and No. 2 common boards, with minimum yields of 33⅓% clear face cuttings for No. 3A common boards and 25% clear face cuttings for No. 3B common boards. No. 3 common is a utility grade hardwood used for crates and other rough applications.

Cabinetry. A *cabinet* is an enclosure fitted with any combination of shelves, drawers, and doors usually used for storage. Cabinetry on a job site may be specified as standard prefabricated cabinets or as custom-built cabinets. The specifications and drawings may indicate the use of standard cabinets available in stock sizes. The architect may also specify custom-built and finished cabinets. Cabinets are shown on plan, elevation, and detail drawings in width/height/depth notation. For example, a notation of 24/36/12 indicates a cabinet 24″ wide, 36″ high, and 12″ deep.

Standard prefabricated cabinets are commonly constructed of particleboard and plywood with a wood or plastic laminate veneer. Custom-built cabinets are commonly made of particleboard, plywood, wood veneer, plastic laminate, and hardwood or softwood materials. Architects provide drawings detailing the various materials and design for custom-built cabinets.

Millwork. Millwork is finished wood materials or parts, such as moldings, jambs, and frames completed in a mill or manufacturing plant. Millwork is cut and molded from hardwood or softwood and is available unfinished, primed, or prefinished. Standard sizes and shapes are used by mills when cutting and shaping millwork. Baseboard, casing, chair rail, handrail, base shoe, crown, panel, cove, quarter round, and outside and inside corner moldings are available in a wide variety of standard sizes and shapes. See Figure 9-6. Softwood molding with a primed finish is normally specified for applications where millwork is to have a painted finish. Hardwood molding is normally specified for applications where millwork is to remain a natural wood finish.

> ▲ *Wood panels should be properly stored, handled, and installed to ensure top performance. Always protect the edges and ends of panels. Whenever possible, store panels where protected from the elements. Panels stored outside should be stacked on a level platform supported by full-width 4″ × 4″ lumber blocks. Never leave panels in direct contact with the ground.*

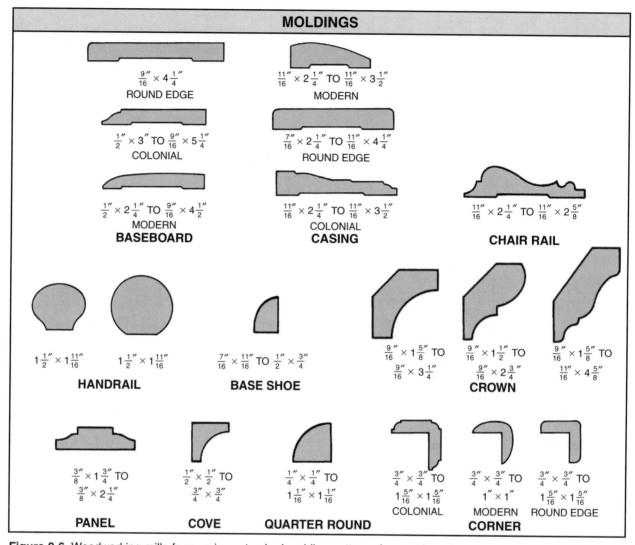

Figure 9-6. Woodworking mills form various standard molding patterns from cedar, white pine, and hardwoods.

Paneling. Wood paneling is available in a wide variety of prefinished and unfinished styles and types of wood. Wood paneling is formed from ¼″ or ¾″ thick plywood panels covered with a hardwood veneer such as birch, oak, walnut, or mahogany. The surface may be plain or textured and prefinished or unfinished.

Stair Components. A large number of specialized stair components are included in a completed staircase. See Figure 9-7. A *stringer (carriage)* is a support for stairs or steps. A *tread* is the horizontal surface of a step. A *riser* is the piece forming the vertical face of a step. Structural wood members of 2″ thick lumber are used for stringers to support the treads and risers. Boards 2″ × 10″ or 2″ × 12″ are used in lengths that are multiples of two feet for stringers. Treads are built of 2″ thick lumber for a rough stair and 1″ or greater thick-

ness of oak or other hardwood for a finished stair. The material for treads on a rough stair is usually 2″ × 10″ or 2″ × 12″ stock. Risers and open and closed stringers are made of ¾″ thick softwood or hardwood lumber, depending on the finish. An *open stringer* is a finish member that has been cut out to support the open side of a stairway. A *closed stringer* is a finish member installed at the meeting of a staircase and wall.

A *baluster* is the upright member that runs between the handrail and the treads. Balusters support the handrail and may be made from softwood or hardwood, depending on the finish. A *handrail* is a rail that is grasped by the hand for support in using a stairway. Handrails are built of 2″ nominal thickness lumber for rough stairs and milled hardwood for finished stairs. Special components such as goosenecks, starting and landing newel posts, and cove moldings are shown on architectural detail drawings of wood stairs. A *gooseneck* is a curved or bent section of the handrail. A *starting newel post* is the main post supporting the handrail of a stair at the bottom of a stairway. A *landing newel post* is the main post supporting the handrail at a landing. A *landing* is the platform that breaks the stair flight from one floor to another. A *cove molding* is the trim material often used to cover the joint between the tread and the stringer and the joint between the tread and the top of the riser. *Nosing* is the projection of the tread beyond the face of the riser.

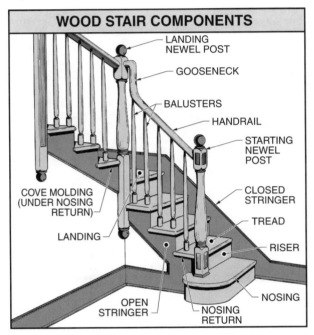

Figure 9-7. Wood stairs are made from a variety of components, many of which are provided precut from woodworking mills.

Plastics

Various structural framing members made from recycled plastic material are available. This material is used in applications where decay resistance and weatherproof qualities are required, such as exterior decks and fences. Plastic is also used in cabinet construction as a surfacing material for horizontal and vertical surfaces. Plastics may be used as structural members or as laminates.

Structural. *Structural plastic* is plastic used for exterior, light bearing applications. Plastic that is shaped to common lumber sizes may also be used for fencing or other decorative exterior applications such as trellises, etc. Plastic framing members are available in sizes similar to nominal size structural wood framing members. Plastic is warp, split, and weather resistant, and is commonly fastened together with screws.

Laminates. *Plastic laminate* is a sheet material composed of multiple layers of paper material and resins bonded together under high heat and pressure. Plastic laminate is commonly referred to as high-pressure or low-pressure laminate, depending on the grade. High-pressure plastic laminate is used where heavy wear qualities are needed and is commonly referred to as horizontal grade laminate. Low-pressure plastic laminate is used for vertical application or where post-forming of the laminate is required. Plastic laminate is available in sheets of $^1/_{32}''$ or $^1/_{16}''$ thickness. Solid-core laminates maintain their surface color throughout the entire thickness of the laminate. Plastic laminate is commonly applied with contact cement to particleboard as a finish material. Plastic laminate is available in widths from 16″ to 60″ and lengths of 8′ to 12′. Plastic laminate is available in many patterns and designs.

WOOD CONSTRUCTION

Wood may be used for structural or finish construction. *Structural wood construction* is the use of wood members as the means of support for all imposed loads of a structure. *Finish wood construction* is the use of wood members for finish applications.

Structural Wood Construction

Structural wood construction includes frame construction and the use of prefabricated trusses. In frame construction, individual wood members are assembled and fastened into structural frames on the job site. Prefabricated trusses are engineered to span various distances and support specific loading. Trusses are commonly built off-site, transported to the job site, and set in place. Frame construction includes platform, balloon, and timber framing. Each frame construction method requires different wood framing members used in different applications.

Platform Framing. *Platform (western) framing* is construction in which single or multistory buildings are built one story at a time with work proceeding in consecutive layers or platforms. Wood members are used for joists, headers, and plates. See Figure 9-8. A joist is a horizontal structural member that supports the load of a floor or ceiling. A *header* is a horizontal framing member at the top of a window or door opening. A *plate* is a horizontal support member in a frame wall, usually 2″ material. Wood members installed vertically include studs, posts, and other structural support members. A *stud* is a vertical support member in a frame wall that is spaced at regular intervals to support loads and provide a surface for finish wall material attachment. A *post* is a vertical support member such as a column or pillar.

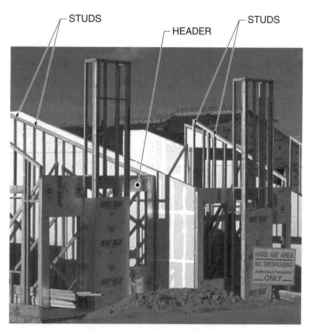

Figure 9-8. Platform framing is construction in which single or multistory buildings are built one story at a time with a variety of wood members.

Blocking is a piece of wood fastened between structural members to strengthen the joint, provide structural support, or block air passage. Blocking, permanent and temporary bracing, and backing are additional wood uses in platform framing. Wood sheathing is attached to the frames to add lateral strength. Platform framing is the most common framing method.

Balloon Framing. *Balloon framing* is construction in which one-piece studs extend from the first floor line or sill to the top plate of the upper story. The primary difference between balloon framing and platform framing is that the studs for multistory balloon-framed load-bearing walls extend the full height of the walls from foundation to roof. Wood members used in balloon framing are similar to those used in platform framing, with only the lengths of the members being different. Balloon framing is used in some applications to mini-

mize the effects of lumber shrinkage and to create open-framed areas in multistory buildings.

Timber Framing. *Timber framing* is construction in which large, laminated wood members (glulam) or one-piece timbers are used to form large open areas. See Figure 9-9. Glulam or timber framing members are spaced in a wide manner to create large open areas. Areas between timber framing members are fitted with glass or covered with thick wood planking of sufficient thickness to span between posts and beams without deflection.

California Redwood Association

Figure 9-9. Timber framing uses large, laminated wood members or one-piece timbers to form large open areas.

Finish Wood Construction

Finish wood construction is comprised of items built of wood or wood products and applied to the interior or exterior surfaces of walls, partitions, floors, or ceilings to provide decorative or functional finishes. Many components of finish wood construction are prefabricated and delivered to the job site in a precut and prefinished condition. Care is taken at the job site to install these materials near the completion of the work when climatic conditions and on-site workers cannot damage the finished materials. Due to the moisture content of some finished wood, certain installations require finish wood materials to be delivered in time to adjust to the interior climatic conditions of the final installation space prior to cutting and fitting into final position. Finish wood construction includes cabinetry, millwork, paneling, and stair components.

Cabinet units are commonly prefabricated at a cabinet shop and delivered to the job site ready for installation. Some custom-built cabinets may be built in-place. Countertops are built of particleboard covered with plastic laminate, veneer, or ceramic tile, or made of solid surface material. Dimensions and finishes for cabinets are based on standard sizes or custom sizes determined from the specifications, plan details, interior elevation drawings, job site measurements, and architectural notes.

Millwork, including baseboard, crown, chair rail, and other specialized moldings, is measured, cut, and fitted in place at the job site. Wood paneling is installed by fastening to structural members or special framing. Dimensions for finished wood stairways are shown on print drawings but are commonly measured at the job site when structural work is completed. Stair components, including risers, stringers, treads, balusters, and handrails, are commonly fabricated at a woodworking shop, shipped to the job site, and installed. Two or three stringers are normally required for a stairway. Additional stringers may be required on very wide stairs.

QUANTITY TAKEOFF

Methods of specifying lumber quantities vary according to their use. For example, structural wood framing members for platform and balloon framing are ordered according to the individual number of pieces and priced based on board feet. Timber components are specified according to each piece and dimension. Sheathing materials are calculated according to the number of square feet to be covered. The amount of millwork necessary is stated in linear feet. Cabinetry is noted by the dimensions, number of doors, number of drawers, and finish of each cabinet. Countertop calculations are based on the number of square feet. Wood paneling is calculated as the number of sheets.

Structural Wood Members

Platform framing wood members include sill plates, floor joists, floor sheathing, wall plates, wall studs, headers, bracing and blocking materials, wall sheathing, ceiling joists, roof rafters, roof trusses, and roof sheathing. Balloon framing includes these same members. Timber framing wood members include glulam members, timbers, and planking.

The species, grade, stress rating, and moisture content of lumber to be used is noted in the specifications.

See Figure 9-10. Lumber measurements are stated as thickness, width, and length. Lumber length is based on a multiple of 2′, normally beginning at 8′ in length and extending to 16′ or 18′. Structural lumber length is always rounded up to the closest multiple of 2′. Lumber quantities are often stated in board feet when calculating pricing for large projects. A *board foot* is a unit of measure for lumber based on the volume of a piece 12″ square and 1″ thick. Board feet is calculated as thickness (in in.) multiplied by width (in in.) multiplied by length (in ft). This total is divided by 12. For example, a board that is 2″ thick, 4″ wide, and 16′ long contains $10.6\overline{6}$ bf ($[2″ \times 4″ \times 16′] \div 12 = 10.6\overline{6}$ bf). Lumber prices are often stated in cost per board foot or thousand board feet (Mbf).

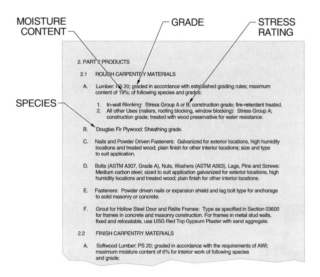

Figure 9-10. Specifications provide information concerning species, grade, stress rating, and moisture content of lumber.

When using a ledger sheet, the dimensions are taken from the architectural drawings and used to determine the number of framing members. When using a spreadsheet or estimating software, the calculation of the number of framing members is similar to ledger sheet estimating in that the dimensions are taken from the architectural drawings. Depending on the software program, individual calculations may be necessary to determine the number of framing members or the program may automatically determine the number of framing members needed based on spacing and dimension information. When using a spreadsheet, the number of framing members is entered into the proper spreadsheet cell. When using estimating software, the number of framing members is entered into the takeoff quantity column on the row corresponding to the applicable framing member. In estimating software, material and labor costs are calculated electronically based on information retrieved from a previously-stored database containing cost information.

Sill Plates. A *sill plate* is a support member (usually a treated 2″ × 4″ or 2″ × 6″) laid flat and fastened to the top of a foundation wall. Sill plates provide a nailing base for floor joists or studs. Sill plates are calculated by determining the linear feet of the foundation walls to be covered and the nominal size of the sill plates. The linear feet of sill plates is determined by calculations from the foundation and floor plans. The nominal size of sill plates is shown by an architectural note or detail drawings. For example, an L-shaped foundation wall 23′-11″ × 32′-3″ requires 56′-2″ (56.16̄′) of sill plates. See Figure 9-11. An architectural note may indicate 2″ × 4″ lumber for this application. The number of board feet is determined by multiplying 56.16̄′ by 2″ × 4″ and dividing by 12. The 56.16̄′ wall requires 37.44 bf ([56.16̄′ × 2″ × 4″] ÷ 12 = 37.44 bf). The total number of board feet is multiplied by the cost of the lumber per board foot to determine the total cost of the sill plates.

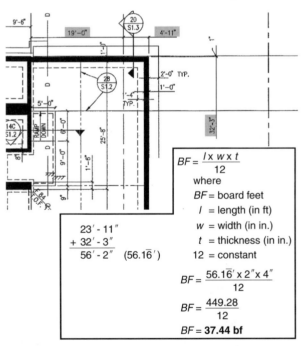

Figure 9-11. Sill plate linear feet are determined from foundation and floor plan drawings.

The linear feet of sill plates is calculated for spreadsheet and estimating software in the same manner as the ledger sheet. The quantity is taken from the architectural drawings or the linear feet can be determined

using a digitizer. A *digitizer* is an electronic input device that converts analog data to digital form. A digitizer enables the user to take measurements from prints and input the data directly into a computer program. When using a spreadsheet, the linear feet of sill plates is entered into the proper spreadsheet cell. When using estimating software, the linear feet of sill plates is entered into the takeoff quantity column on the row corresponding to the sill plates of a given dimension, in this case 2″ × 4″. Material and labor costs are calculated electronically based on information retrieved from a previously-stored database containing cost information. See Figure 9-12.

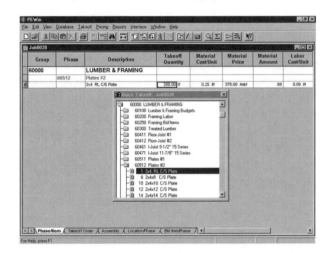

Figure 9-12. The linear feet of wood sill plate material is entered into the takeoff quantity column when using estimating software.

Floor Joists. The type and number of floor joists required is calculated by determining the length and on center (OC) spacing from the floor plan and the nominal size from the specifications, an architectural note on the floor plan, or detail drawings. Floor joist length depends on the distance between structural supporting members such as the sill plates and beams. Sufficient length is allowed for a certain amount of overlap where required. For example, a 14′ floor joist is used when the distance between the outside of a sill plate and the center of the supporting beam is 13′-4″.

When using a ledger sheet, the number of floor joists required is calculated by determining the length (run) for a given type of floor joist and the OC spacing of the run. See Figure 9-13. A multiplier is used to determine the actual number of joists required depending on the OC spacing. The multiplier is one when joists are spaced 1′-0″ OC. For example, an 18′ run of joists requires 18 joists spaced 1'-0" OC (18' × 1 = 18 joists). The multiplier is .75 when joists are spaced 16" OC. For example, an 18' run of joists requires 14 joists spaced 16" OC (18' × .75 = 13.5 joists rounded to 14). The multiplier is .5 when joists are spaced 2' OC. For example, an 18' run of joists requires 9 joists spaced 2' OC (18' × .5 = 9 joists). Additional floor joists are added or subtracted around stairwell openings, the perimeter of the structure, and at cantilevers. Additional joist material is also included for blocking. The number of board feet of floor joists is determined and the total board feet is multiplied by the cost of the lumber per board foot to determine the material cost.

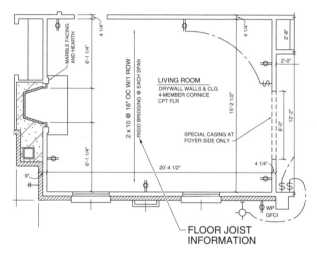

Figure 9-13. Wood floor joist calculations are determined from information provided on floor plans.

The calculation of the number of joists using a spreadsheet or estimating software is similar to calculations using a ledger sheet. The quantity is taken from the architectural drawings or a digitizer may be used. When using a spreadsheet, the number of joists is entered in the proper spreadsheet cell. When using estimating software, the quantity is entered into the take-off quantity column on the row corresponding to the applicable floor joist size and type. In estimating software, material and labor costs are calculated electronically based on information retrieved from a previously-stored database containing cost information.

Girders and Beams. A *girder* is a large horizontal structural member used to support loads at isolated points along its length. A beam is a horizontal structural member installed to support a load over an opening. Dimensions for wood girders and beams are determined from floor plans and detail drawings. The lumber and design of any pre-engineered and trussed components are

shown in detail drawings and the specifications. The cost per unit for prefabricated members is obtained from the supplier based on the engineering specifications. Wood girders and beams built and fabricated on the job site are priced according to the lumber dimensions and grade.

Girders, beams, and pre-engineered wood members may be considered one-time items when using a spreadsheet or estimating software. When using a spreadsheet, material dimensions and costs are entered into the appropriate cell. When using estimating software, material dimensions and costs are entered on a one-time item pull-down menu that allows for the insertion of a description of the girder or beam, the quantity, and cost. This information is integrated into the estimate sheet. The information may also be added to the system database for future use if desired by the estimator.

Wall Framing Members. Wall framing members include the individual wood pieces used to provide structural support and partitioning. Wall framing members include plates, studs, headers, sills, blocking, bracing, and cripples. A plate is a horizontal support member in a frame wall, usually a 2" × 4". A stud is a vertical support member in a frame wall that is spaced at regular intervals to support loads and provide a surface for finish wall material attachment. A header is a horizontal framing member at the top of window or door framing. A sill is a horizontal framing member placed under window openings. Blocking is a piece of wood fastened between structural members to strengthen the joint, provide structural support, or block air passage. A *cripple* is a short wall stud spanning from a plate to an opening sill or header. Each wall framing member requires a different method for material estimating.

The number of wall studs required is determined by wall length and a multiplier based on the stud spacing.

Plates, studs, sills, blocking, and cripples are usually the same width. The number of linear feet of wall plates and lumber size are calculated from the wall length dimensions on the floor plan and notes on the detail drawings or specifications. Totaling the wall lengths and multiplying by 2 for single top plated walls or 3 for double top plated walls gives the required linear feet of plate material. For example, a rectangular building 24'-6" × 13'-0" (24.5' × 13') with no interior partitions has a total of 75 lf of walls ([24.5' × 2] + [13' × 2] = 75'). If the walls are all single-plated, the total linear feet of plate material is 150' (75' × 2 = 150'). The linear feet of wall is multiplied by 3 for a double top plate for a total of 225' (75' × 3 = 225'). When using a spreadsheet, the takeoff quantity is entered into the proper cell. When using estimating software, the quantity is entered into the proper row corresponding to the plate size. A digitizer may also be used to determine linear feet of wall plates. Plate material costs are commonly calculated by prices based on board feet.

Material quantity for headers over windows and doors is determined by taking the width of each opening, adding 2" for jamb installation and 3" to allow for 1½" of bearing on each end of the header. This value is multiplied by 2 for double headers. For example, a 3'-0" wide swinging door has a 3'-2" opening (including jamb allowance) plus an additional 3" for a total of 3'-5". A 3'-5" single solid header is needed for this opening. This value is multiplied by 2 for a double header (3'-5" × 2 = 6'-10"). An 8' piece of header material is ordered for a built-up double header. A note on the floor plan, elevation drawings, or specifications gives the lumber grade, type, thickness, and width of header material to determine whether they are single or double headers. This amount is entered into a ledger sheet, spreadsheet, or estimating software quantity cell.

The method for estimating the number of wall studs depends on the stud spacing. The proportional relationship of spacing of studs to each foot determines the multiplier for estimating. For 16" OC studs, the ratio of wall length to studs is $^{12}/_{16}$ ($^{3}/_{4}$). The wall length is multiplied by $^{3}/_{4}$ (.75) to calculate the number of studs. For 24" spacing, the ratio is $^{12}/_{24}$ ($^{1}/_{2}$). The wall length is multiplied by $^{1}/_{2}$ (.5) to determine the number of studs. For example, a 75' wall with studs spaced 16" OC requires 56.25 studs (75' × .75 = 56.25 studs). Adding 5% for waste equals a total of 59.06 studs (56.25 studs × 1.05 = 59.06 studs). This number is rounded up to 60 studs. Other elements to consider when determining the number of studs needed per wall include additional studs required at intersecting walls and studs that may be deleted at large door or window openings. One additional stud is added to the calculated amount to end the wall framing at a corner.

Balloon framing requires a 1" × 4" ribbon at the second-floor level. A *ribbon* is a supporting ledger applied horizontally across studs as a support for joists. To find the length of the ribbon, add the linear feet of the two outside walls that carry the ends of the joists.

Additional wall framing members are added for window sills, blocking, header framing, intersecting wall framing, and bracing. These additional members are added based on the number of door and window openings and other framing requirements. The estimator should check these structural items when determining the total number of wood framing members entered into the quantity figures.

Calculations for an entire wall assembly may be made in a single entry when using estimating software. For standard wall constructions, the estimator can enter information into a database including plates, studs, blocking, let-in bracing, temporary bracing material, drywall, molding, and paint. See Figure 9-14. By entering or accessing standard information, including wall width, stud spacing, single or double top wall plates, and finish information, the estimator can make a single entry to take off all material for a certain number of linear feet of standard wall. This includes wood framing members, surface finish materials such as gypsum drywall, fasteners, and labor. The number of linear feet of wall can be calculated traditionally or digitized into the estimating software program.

Ceiling Joists and Trusses. Ceiling joist length is determined from the location of bearing walls on the floor plan. The quantity and type of material for ceiling joists is calculated depending on the spacing, which is the same as for floor joists and wall studs. The ceiling joist quantity is entered into a ledger sheet, spreadsheet cell, or estimating software quantity cell in a manner similar to floor joists.

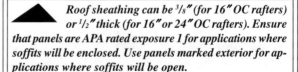

Roof sheathing can be ³/₈″ (for 16″ OC rafters) or ¹/₂″ thick (for 16″ or 24″ OC rafters). Ensure that panels are APA rated exposure 1 for applications where soffits will be enclosed. Use panels marked exterior for applications where soffits will be open.

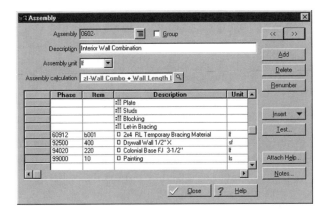

Figure 9-14. Calculations for entire wall assemblies may be made in a single entry when using estimating software.

The number of roof trusses required is also calculated according to the truss OC spacing. For example, for a structure 24'-6" long with roof trusses spaced 2'-0" OC, 24'-6" is multiplied by .5 to obtain 12.25 trusses (24.5' × .5 = 12.25 trusses). This number is rounded up to 13 trusses.

Wood Panels

The number of wood panels required for framing applications is determined based on area calculations. Each 4' × 8' sheet of paneling covers 32 sq ft. When using a ledger sheet, the number of sheets of wood sheathing required for an area is determined by dividing the total square feet to be covered by 32. Approximately 7% is added for waste material to determine the total number of sheets.

Flooring. For floors, the number of square feet to be covered is calculated based on floor plan dimensions. The type and thickness of wood sheathing product to be installed is indicated on the floor plan with an architectural note or shown on detail drawings and specifications. For example, a wood floor area of 60' × 250' requires 15,000 sq ft of wood sheathing. This number is divided by 32 to determine the number of sheets required (15,000 sq ft ÷ 32 = 468.75 sheets). Approximately 7% is added for waste for a total of 501.56 sheets (468.75 sheets × 1.07 = 501.56 sheets). This number is rounded up for a total of 502 sheets. Deductions may be made for large floor openings such as stairwells.

When using a spreadsheet, the quantity taken off is entered into the appropriate cell. When using estimating software, the length and width dimensions of the area to be sheathed are entered into the item form. This data may be entered by reading the print drawings or by a digitizer. The estimating software program automatically calculates the square footage including material, labor, and waste factors.

Walls. The square foot method is also used for calculating the amount of wall sheathing required. Estimators determine the total square feet of coverage, the sheathing materials, and calculate the number of sheets needed.

Exterior wall dimensions are totaled to determine the linear feet of the wall. The total linear feet is divided by 4 to estimate the number of sheets of wall sheathing for an 8' high wall because wall sheathing is 4' wide. The estimator may make allowances in the square feet of wall sheathing required for large or numerous wall openings. For example, a 24'-6" × 13'-0" rectangular building with 8' high walls has 75 lf of wall ([24'-6" × 2] + [13'-0" × 2] = 75 lf). Seventy-five divided by 4 equals 18³/₄ sheets (75' ÷ 4 = 18.75 sheets) of 4' wide wall sheathing. This number is rounded up to nineteen sheets. The square foot method may also be used to determine the wall sheathing material required. This method is similar to floor and roof sheathing.

For non-panelized sheathing, such as individual boards or other material that requires square footage calculations, the length of the walls requiring sheathing is measured and multiplied by the height to which the sheathing is applied. The area of all openings such as windows and doors is deducted. Thirty percent is added for waste when boards are placed diagonally and 20% is added for waste when boards are placed horizontally.

Roofs. Roof sheathing calculations are based on the roof and rafter length, which is based on the pitch of the roof. Roof length is located on the floor plans. Rafter length is determined by the run of the roof shown on the floor plan and the pitch shown on exterior elevation drawings. Steeply-pitched roofs result in longer rafter lengths and additional sheathing requirements.

When using a ledger sheet, rafter length is determined prior to beginning square foot sheathing calculations. Rafter length is determined mathematically or from printed tables. Mathematical rafter length calculations equal the square root of the sum of the squares of the rafter run and rise. See Figure 9-15. Rafter length multipliers become greater as the pitch becomes steeper. The estimator should consult typical wall section drawings or detail drawings to determine overhang distance to be added to rafter and sheathing totals. When using a ledger sheet or spreadsheet, the rafter length and number are entered into the proper column and row. When using estimating software, formulas may be established for rafter items to automatically calculate rafter length based on run and pitch information. See Figure 9-16.

The square feet of sheathing required for a roof is determined by multiplying rafter length by the length of the roof. The square feet of roof area is divided by 32 to determine the number of 4′ × 8′ sheets of roof sheathing required. For example, a building 24′-6″ wide and 13′-0″ deep (24.5′ × 13′) has a gable roof on a 3:12 pitch (12.36″ per run foot). The rafter length is determined by multiplying 12.36″ (run per foot) by 6.5′ (one half of the roof) and dividing by 12 to obtain 6.70′ (12.36″ × 6.5′ ÷ 12 = 6.7′). The building length is multiplied by the rafter length to find the roof area (24.5′ × 6.7′ = 164.15 sq ft). This value is multiplied by 2 to find the total area for the front and back of the roof (164.15 sq ft × 2 = 328.3 sq ft). Five percent is added for waste for a total of 344.72 sq ft (328.3 sq ft × 1.05 = 344.72 sq ft). The total is divided by the coverage of one piece of sheathing (32 sq ft) to find the number of pieces required (344.72 sq ft ÷ 32 = 10.77 sheets). This number is rounded up to eleven sheets of roof sheathing.

When using a spreadsheet, the total square feet or number of sheets of sheathing is entered into the proper cell. When using estimating software, a formula may be entered to automatically calculate the sheathing amount based on pitch and rafter length calculations. The estimator may use rafter length information based on various pitches in a standardized formula. Each time the roof sheathing item is selected from an item listing, the computer program asks for specific information concerning roof pitch, length, and width. The quantity of roof sheathing necessary is automatically calculated and entered into the takeoff quantity column corresponding to the item. Rafter and sheathing calculation is an assembly that could be established to automatically enter a variety of material and labor quantity information in estimating software.

Finish Wood Members

Finish wood members include cabinetry, millwork, paneling, and stair components. Quantity takeoff methods for finish wood members include subcontractor bids, linear feet calculations, piece counts, and angular calculations based on the material to be determined. Estimators rely primarily on interior finish elevation drawings, floor plans, schedules, and the specifications to take off finish wood members.

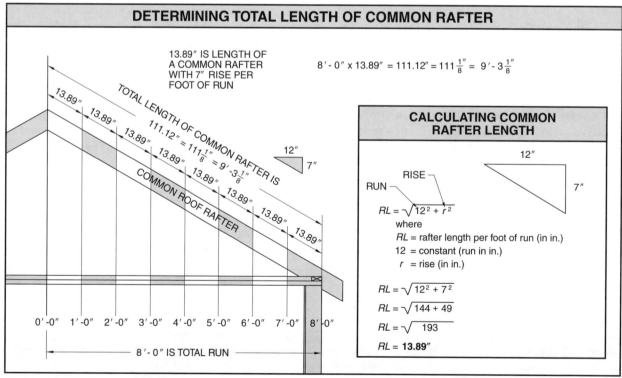

Figure 9-15. Estimators determine the length of roof rafter material based on run and pitch calculations.

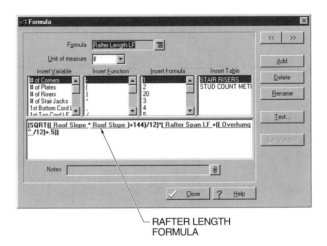

— RAFTER LENGTH
FORMULA

Figure 9-16. Formulas in estimating software programs allow roof material calculations to be performed based on run and pitch information.

Cabinetry. Interior elevation drawings and related schedules provide location, dimension, and finish information for cabinetry and casework. See Figure 9-17. Architectural notes and codes on interior elevation drawings relate to a schedule for cabinetry and casework. The schedule shows width, height, and depth of each cabinet. A schedule of each cabinet is provided to a cabinet shop for shop-built cabinets. A price for these cabinets is entered into the bid. For job-built cabinets, material calculated includes particleboard, plywood, and exterior finish material.

When using a spreadsheet, shop-built cabinet subcontractor prices are entered in the proper cell. For job-built cabinets, the number of sheets of particleboard, plywood, and exterior finish material is entered into the proper spreadsheet cell where costs per sheet are calculated based on actual costs. When using estimating software, assemblies may be established for certain cabinet types. This allows all material and labor quantities to be entered into the bid automatically based on size and style information.

Millwork. Millwork is estimated based on linear feet. The number of trim moldings such as baseboard, casing, chair rail, and crown molding are estimated by the linear foot. The total linear feet of these moldings is determined from wall length dimensions on the floor plan. A tabulation is made for each type of molding required. A cost per foot provides the material and labor costs.

Baseboard molding is calculated in linear feet as the total perimeter of each area to be based. Deduc-

tions are made where many doors or large openings exist. The perimeter of all rooms is added and the amount of baseboard is determined. For example, a 24'-6" × 13'-0" room (24.5' × 13') has 75 lf of wall ([24.5' × 2] + [13' × 2] = 75 lf). A minimum of 75'-0" of baseboard, chair rail, and crown molding is required for the room (with no allowances for doors or windows). Some additions may be necessary for fitting at the corners. In applications where special trim moldings are noted, they are estimated individually for length and type.

Casing length depends on the width and height of each door or window opening. For a 3'-0" wide by 6'-8" high door, casing is installed on each side (6'-8" × 2 = 13'-4"). Normally, a 7'-0" length is needed at each side to allow for fitting at each top corner. A 4'-0" piece of casing is required across the top of the door opening to allow for fitting (depending on the width of the casing).

The linear feet of each type of molding is taken off from the drawings and entered into the proper row and column on a ledger sheet or spreadsheet. Print takeoff or a digitizer may be used to enter linear foot information into the proper item information cell when using estimating software. See Figure 9-18.

Paneling. Sheets of paneling are installed with a particular grain orientation, which may be horizontal, vertical, or angular. Each sheet covers 4' perpendicular to the grain orientation and 8' parallel with the grain orientation. The linear feet of coverage perpendicular to the grain orientation is determined and divided by 4 to calculate the number of sheets of paneling in an area with wall coverage of less than 8'. For example, a rectangular room measuring 14' × 52' with 8' high ceilings has a total of 132 lf of wall space ([14' × 2] + [52' × 2] = 132 lf). With paneling installed with a vertical grain, the number of paneling sheets is determined by dividing 132 lf by 4, resulting in 33 sheets (132 lf ÷ 4 = 33 sheets). Five percent waste is added, resulting in 34.65 sheets (33 sheets × 1.05 = 34.65 sheets). This number is rounded up to the next full sheet, so 35 sheets are ordered. Cost per sheet is based on the type and grade of paneling as quoted by the supplier. Allowances may be made for large door openings.

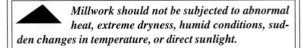
Millwork should not be subjected to abnormal heat, extreme dryness, humid conditions, sudden changes in temperature, or direct sunlight.

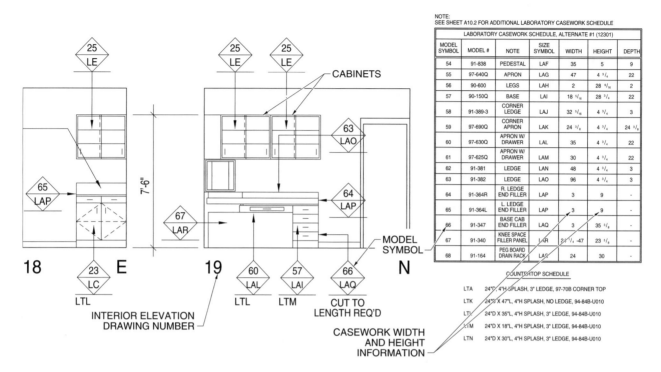

Figure 9-17. Interior elevation drawings and related schedules provide location, dimension, and finish information for cabinetry and casework.

Stair Components. Estimating stair components requires information obtained from interior elevation drawings, plan views, and specifications concerning finishes, the number of treads and risers, balusters, handrails, volutes, goosenecks, bullnosed treads and risers, newel posts, and stringers. See Figure 9-19. Rough stairs are not always shown in detail on the plans. The estimator must calculate stringer length and tread width before doing a quantity takeoff. Without detail and interior elevation drawings, the estimator calculates the rise and the run of the stairs.

Variables in stair calculations involve the total rise, total run, type of wood to be installed, and final finish. The actual number of treads, risers, balusters, volutes, goosenecks, newel posts, and bullnosed treads and risers are counted from the architectural drawings when possible.

The linear feet of handrail and stringers are calculated based on the hypotenuse of a triangle with the two legs being the total run and the total rise of the stair. To calculate the length of the hypotenuse, the two legs are squared and added. The square root of this value is the length of the hypotenuse. For example, a stair with a total run of 6′ and a total rise of 4′ requires stringers and handrails approximately 7′-3″ long ($6^2 \times 4^2 = 52′$, $\sqrt{52} = 7.21′$). Eight feet of material is purchased to allow for cutting and fitting.

To find the tread and riser height and the number of risers, the total floor to floor change in height elevation is divided by the desired riser height (commonly 7½″). If the division does not result in a whole number, choose the closest whole number and divide the total height by the whole number. The result of the division is the actual unit rise. The treads are always one less in number than the risers. The tread width is found by dividing the number of treads into the total run of the stairway. Sometimes there is only a limited distance available. Using this distance as a limit, the same process is used as for calculating riser height.

When the total run of the stairs is not limited by the conditions of the building, the tread width is found by using a ratio established by the local building code. One ratio is tread width plus riser height equals a total of 17″ to 18″. For example, if the riser is 7¼″, the tread is 9¾″, 10¾″, or some measurement in between.

Costs are based on a per piece cost for special stair components counted by the piece, such as balusters, volutes, and goosenecks. Costs are based on a linear foot cost for handrails and a cost per board foot for stringers, rough treads, and rough risers. The quantities determined are entered into the proper row and column on a ledger sheet or spreadsheet. When using estimating software, the quantities for each item are entered into the item takeoff cell. Stairs may be entered into the database by the estimator as a common assembly. Formulas may be built for stair items to assist in determining stringer and handrail lengths and number of risers and treads based on total rise and total run information. Stair work is a portion of the work that is commonly performed by a specialty contractor. This is an item that may be subcontracted and included in the overall final bid.

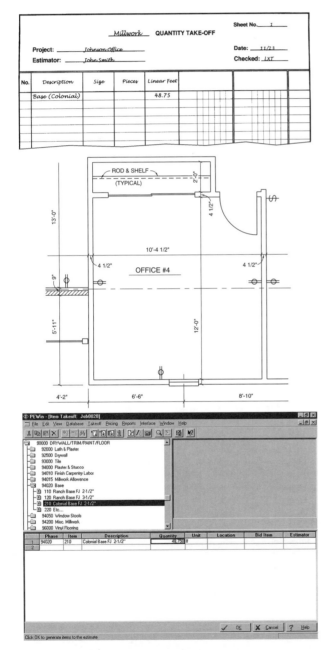

Figure 9-18. The number of linear feet of moldings is entered into the proper item information cell in estimating software.

Plastics

The primary location of structural plastics on architectural drawings is exterior deck areas and fencing components. High- and low-pressure plastic laminate sheets are applied in cabinetry areas. Linear foot, piece, and square foot calculations are required for plastic materials.

Structural. Structural plastics are relatively new on construction sites. Information concerning structural plastic material is shown on plan drawings and elevations and in specifications. Estimating for structural plastic is performed in the same manner as structural wood members, with the number of pieces depending on spacing of studs, joists, and plates. Costs for structural plastic materials are calculated based on a per piece cost or a cost per board foot calculation.

Laminates. Plastic laminate cost is determined by the number of square feet to be covered and the laminate material used. This applies to horizontal and vertical surfaces and high- and low-pressure laminate. Where patterned laminate is used rather than a solid color, quantity calculations must ensure that the pattern can be applied continuously without turning or reorienting the direction or design of the pattern. Costs for plastic laminate vary depending on the thickness, quality, manufacturer, finish, possible requirement of solid core laminate, and pattern. Costs are given per square foot of coverage.

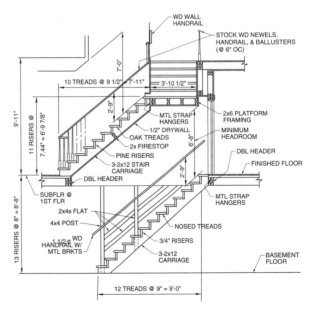

Figure 9-19. Interior elevation drawings and plan views show finish wood stair components and measurements.

ESTIMATING SOFTWARE – ASSEMBLY TAKEOFF

An *assembly* is a collection of items needed to complete a particular unit of work. Assemblies enable the estimator to take off multiple items with a single operation. Assemblies also enable the estimator to obtain a cost per unit for a group of items.

Assemblies may be database assemblies or one-time assemblies. A *database assembly* is an assembly that is stored in the database and can be used in any estimate based on that database. A *one-time assembly* is an assembly that is stored with the estimate and cannot be used in other estimates. Assembly takeoff is performed in Precision estimating software by following a standard procedure. See Figure 9-20.

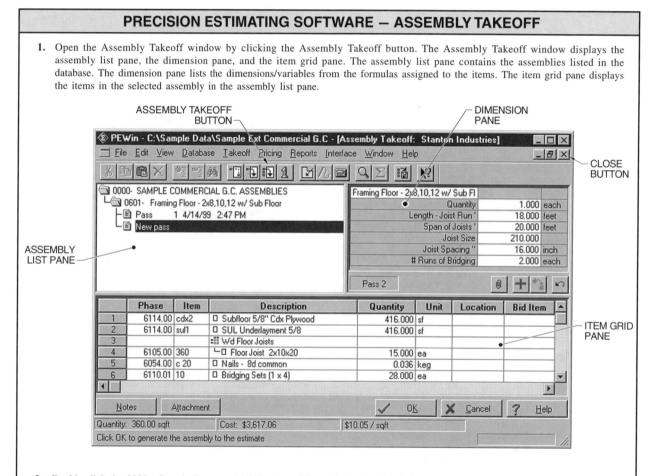

PRECISION ESTIMATING SOFTWARE – ASSEMBLY TAKEOFF

1. Open the Assembly Takeoff window by clicking the Assembly Takeoff button. The Assembly Takeoff window displays the assembly list pane, the dimension pane, and the item grid pane. The assembly list pane contains the assemblies listed in the database. The dimension pane lists the dimensions/variables from the formulas assigned to the items. The item grid pane displays the items in the selected assembly in the assembly list pane.

2. Double-click the 0000 - Sample Commercial G.C. Assemblies and then double-click assembly 0601- Framing Floor - 2 × 8, 10, 12 w/ Sub Floor. The items in the assembly display in the item grid and the variables from the formulas assigned to those items display in the dimension pane. *Note:* Some quantities in the dimension pane prefill from the last time the assembly was used.

3. Enter the dimensions for the assembly. Enter 1 for the Quantity, 18 for Length - Joist Run, 20 for Span of Joists, 210 for Joist Size, 16″ for Joist Spacing, and 2 for # Runs of Bridging. Click ENTER. *Note:* If there were two identical joist runs, a quantity of 2 would have been entered for this assembly. Also note that after the last dimension is input and ENTER is pressed, takeoff quantities are automatically generated to the item grid.

4. Click OK to send the items to the spreadsheet.

5. Close the Assembly Takeoff window by clicking the close button in the upper right-hand corner of the Assembly Takeoff window. *Note:* The items taken off using this assembly are viewed by sorting the spreadsheet in assembly order by clicking the assembly tab at the bottom of the spreadsheet. The estimate is closed by clicking the close button in the upper right-hand corner of the spreadsheet screen.

Figure 9-20. Assemblies enable the estimator to take off multiple items with a single operation.

Review Questions
Wood and Plastics

Name _____ Date _____

Chapter 9

Estimating

_____ 1. Division _____ of the CSI MasterFormat™ includes basic information concerning wood and plastic material.

_____ 2. A(n) _____ wood member is a wood component that provides support for live and dead loads to a structure.

_____ 3. _____ is wood produced from broad leaf, deciduous trees.

_____ 4. _____ is wood produced from a conifer (evergreen) tree.

_____ 5. _____ is finished wood material or parts completed in a mill or manufacturing plant.

_____ 6. A(n) _____ load is the predetermined load a structure is capable of supporting.

_____ 7. A(n) _____ load is a permanent, stationary load composed of all building material and fixtures and equipment permanently attached to a structure.

_____ 8. _____ is the classification of various pieces of wood according to their quality and structural integrity.

_____ 9. _____ is sawn and sized lengths of wood used in construction.

T   F   10. A board is a wood member that has a nominal size from $1'' \times 4''$ up to $1'' \times 12''$.

T   F   11. Dimension lumber is a wood member that has a nominal size from $1'' \times 4''$ up to $4'' \times 6''$.

T   F   12. Timber is a heavy wood member that has a nominal size of $5'' \times 5''$ or larger.

_____ 13. A(n) _____ is a horizontal structural member installed to support a load over an opening.

_____ 14. A(n) _____ is a vertical support member.

_____ 15. A(n) _____ is a manufactured roof or floor support member.

_____ 16. A(n) _____ unit is a stand-alone module composed of many different building components.

_____ 17. _____ is a wood panel manufactured from thin layers of lumber.

_____ 18. _____ is a wood panel in which wood strands are mechanically oriented and bonded with resin under heat and pressure.

_____ 19. _____ is a structural member constructed by bonding several layers of lumber with adhesive.

_____ 20. _____ is a wood panel constructed of wood particles and flakes that are bonded together with a synthetic resin.

_____ 21. _____ is a wood panel manufactured from fine wood fibers mixed with binders.

_____ 22. A(n) _____ wood member is a decorative and nonstructural wood in a structure.

T   F   23. A cabinet is an enclosure fitted with any combination of shelves, drawers, and doors usually used for storage.

T   F   **24.** Structural plastic is plastic used for interior, light bearing applications.

_____   **25.** Plastic _____ is a sheet material composed of multiple layers of paper material and resins bonded together under high heat and pressure.

_____   **26.** _____ framing is construction in which single or multistory buildings are built one story at a time with work proceeding in consecutive layers or platforms.

_____   **27.** A(n) _____ is a horizontal framing member at the top of a window or door opening.

_____   **28.** A(n) _____ is a horizontal support member in a frame wall, usually a 2″ material.

_____   **29.** A(n) _____ is a vertical support member in a frame wall that is spaced at regular intervals to support loads and provide a surface for finish wall material attachment.

_____   **30.** A(n) _____ is a vertical support member such as a column or pillar.

_____   **31.** _____ is a piece of wood fastened between structural members to strengthen the joint, provide structural support, or block air passage.

_____   **32.** _____ framing is construction in which one-piece studs extend from the first floor line or sill to the top plate of the upper story.

_____   **33.** _____ framing is construction in which large, laminated wood members or one-piece timbers are used to form large open areas.

_____   **34.** A(n) _____ is an electronic input device that converts analog data to digital form.

_____   **35.** A(n) _____ is a large horizontal structural member used to support loads at isolated points along its length.

Truss Designs

_____   **1.** Attic

_____   **2.** Scissors

_____   **3.** Flat Howe

_____   **4.** Flat Pratt

_____   **5.** Pitched Howe

_____   **6.** Pitched Pratt

_____   **7.** Hip

_____   **8.** Kingpost

_____   **9.** W

_____   **10.** Fink

Moldings

_____   1. Casing
_____   2. Chair rail
_____   3. Quarter round
_____   4. Base shoe
_____   5. Crown

_____   6. Corner
_____   7. Cove
_____   8. Handrail
_____   9. Baseboard
_____   10. Panel

Ⓐ   Ⓑ   Ⓒ   Ⓓ   Ⓔ

Ⓕ   Ⓖ   Ⓗ   Ⓘ   Ⓙ

Wood Stair Components

_____   1. Landing
_____   2. Open stringer
_____   3. Closed stringer
_____   4. Landing newel post
_____   5. Starting newel post
_____   6. Gooseneck
_____   7. Balusters
_____   8. Tread
_____   9. Riser
_____   10. Handrail
_____   11. Nosing

Lumber Grade Marks

_____ **1.** Mill identification number

_____ **2.** Association trademark

_____ **3.** Lumber grade

_____ **4.** Surfaced dry

_____ **5.** Wood species

Plywood Panel Codes

_____ **1.** Exposure durability classification

_____ **2.** Mill number

_____ **3.** Span rating

_____ **4.** Siding face grade

_____ **5.** Species group number

_____ **6.** Product standard

_____ **7.** Panel grade

_____ **8.** HUD/FHA recognition

_____ **9.** Thickness

_____ **10.** Performance rated panel standard

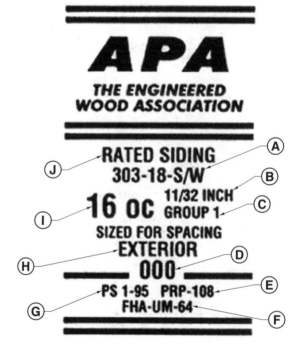

Activities
Wood and Plastics

Name _____ Date _____

Activity 9-1 – Ledger Sheet Activity

Refer to Print 9-1 and Quantity Sheet No. 9-1. Take off the linear feet of interior wall plates, number of studs spaced 16″ OC, and number of 4′ × 8′ sheets of plywood floor sheathing for the Biotech Study. Assume a double top and a single bottom plate. Round all plates, studs, and sheets to the next highest whole unit. Add one stud at each wall intersection. Do not make allowances for door openings.

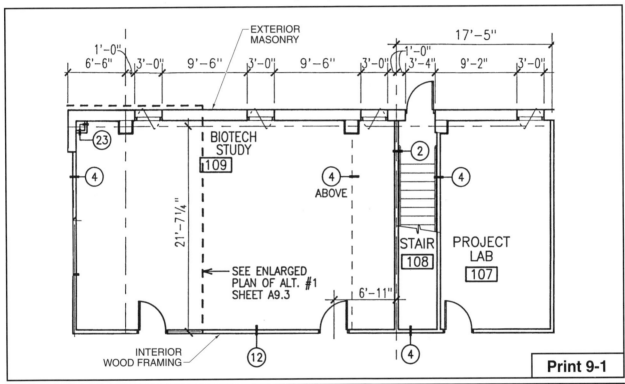

Print 9-1

QUANTITY SHEET

Sheet No. ____9-1____

Project: _____
Estimator: _____

Date: _____
Checked: _____

No.	Description	Dimensions				Unit		Unit		Unit		Unit
		L	W									
	Biotech study											
	Interior wall plates											
	Studs											
	Plywood floor sheathing											

Activity 9-2 – Spreadsheet Activity

Refer to Print 9-2 and Quantity Sheet 9-2 on the CD-ROM. Take off the total number of studs spaced 2′ OC, 4′ × 8′ sheets of plywood floor sheathing, and the linear feet of plates for Office 127, Project Lab 128, and Project Lab 129. Assume a double top and a single bottom plate. Round all studs, sheets, and plates to the next highest whole unit. Add one stud at each wall intersection. Do not make allowances for door openings.

Activity 9-3 – Estimating Software Activity

Create a new estimate and name it Activity 9-3. Take off items **6101.00 100 Plates 2×6 RL**, **6105.00 220 Floor Joist 2×8×12**, **6106.00 100 Studs 2×6×8**, and **6114.00 cdx2 Subfloor ⁵⁄₈″ cdx Plywood** for a room 12′ × 24′ with a 3′-0″ door. Add one stud at each intersection of walls and one stud at each side of the door opening. The studs and joists are 16″ OC. Print a standard estimate report.

Exterior Protective and Thermal Systems

10

Chapter

Division 7 of the CSI MasterFormat™ includes information concerning thermal protection, moisture protection, roof covering materials, siding panels, fire and smoke protection, and joint sealants. Information concerning exterior finish system materials is given on exterior elevation drawings, specifications, detail drawings, mechanical plans for rooftop units, vents, and exhausts, and roof and floor plans. Thermal systems are designed to control temperatures that affect the comfort of building occupants, deter condensation, and reduce heat transmission to improve energy use within the structure. Thermal system materials are installed after structural and framing members are in place. Estimators refer to specifications and detail drawings to determine the types of insulation, water vapor, and fire protection coatings installed in each area.

EXTERIOR FINISH MATERIALS AND METHODS

Exterior finish material information for takeoff and estimating is taken from exterior elevation drawings, specifications, and floor plans. These sources include all materials used for weather and thermal protection on roofs and walls. See Figure 10-1. Division 7 of the CSI MasterFormat™ includes information concerning thermal protection, moisture protection, roof covering materials, siding panels, fire and smoke protection, and joint sealants.

Roofing

A *roof* is the covering for the top exterior surface of a building or structure. Roofing includes all material placed on top of a building or structure to provide protection from environmental elements such as wind, rain, snow, etc. Many roofing materials are used depending on appearance, cost, roof pitch, temperature variation, and local climatic conditions such as annual rainfall. Roofing may be bituminous, elastomeric, or metal, and consist of shingles, tiles, roof pavers, and other roof accessories.

Figure 10-1. Exterior finish materials include thermal protection, moisture protection, roof covering materials, siding panels, fire and smoke protection, and joint sealants.

Bituminous. *Bituminous roofing* is roofing comprised of layers (plies) of asphalt-impregnated felt or fiberglass material that is fastened to the roof deck and mopped with hot coal tar to create a waterproof surface. Bituminous roofing is commonly referred to as built-up or hot tar roofing. Gravel, slag, or a mineral top sheet is applied to the roof surface to finish the application and provide protection for the felt/tar layers.

Elastomeric. *Elastomeric roofing* is roofing made of a pliable synthetic polymer. An elastomeric roofing system is comprised of large sheets that are laid in place and sealed at the joints. See Figure 10-2. The sheets are made of chlorinated polyethylene (CPE), ethylene propylene diene monomer (EPDM), polyvinyl chloride (PVC), or other chemical compounds. A vapor barrier and rigid roof insulation are installed on the surface of the roof deck. Elastomeric sheets are rolled out across the entire surface of the roof. Joints are sealed with a solvent that joins the sheets into a single unit. The entire surface may be covered with gravel after all joints are sealed.

Figure 10-2. Elastomeric roofing is made of pliable synthetic polymer sheets that are laid in place and sealed at the joints.

Metal. *Metal roofing* is roofing made of steel, aluminum, copper, and various metal alloys. Metal roofing materials may be formed into shingles, corrugated sheets, or sheet metal strips. Rigid insulation may be applied to the top of the roof deck and covered with building paper. *Building paper* is felt material saturated with tar to form a waterproof sheet. Sheets of prefinished or ornamental metal are set in place with the long dimension placed parallel to rafters running from the bottom of the roof to the top. Seams along the sides of the metal sheets are fastened in a variety of designs. The seams may be flat, ribbed, or standing. See Figure 10-3. Metal roof sheets are fastened to the roof deck with clips or self-tapping screws. Overlapped joints are coated with a waterproof joint sealant.

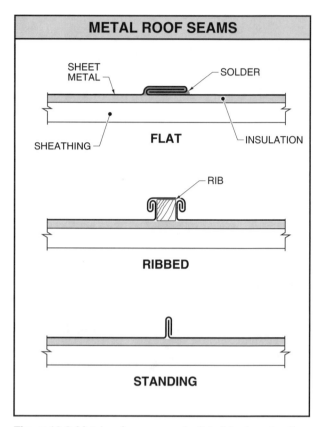

Figure 10-3. Metal roof seams may be flat, ribbed, or standing.

Shingles. A *shingle* is a thin piece of wood, asphalt-saturated felt, fiberglass, or other material applied to the surface of a roof or wall to provide a waterproof covering. A *composition shingle* is an asphalt or fiberglass shingle coated with a layer of fine mineral gravel. Composition shingles are available in a wide range of styles, colors, and thicknesses. Composition shingles include three-tab, laminated, and architectural shingles. The quality of composition singles is measured by weight. Heavy shingles are of higher quality than light shingles.

A *shake* is a hand-split wood roofing material. *Exposure* is the amount of a shingle or shake that is seen after installation. Exposure of wood shingles or shakes varies with their width and thickness and the roof slope. Wood shingles and shakes appear on elevation drawings as closely-spaced horizontal lines.

Tiles. *Roofing tile* is clay, slate, or lightweight concrete members installed as a weather-protective finish on a roof. Roofing tile is manufactured in several different shapes. See Figure 10-4. The most common shape of roofing tile is Spanish tile that is arched with a flat area that has holes for attachment to a roof deck. Roofing tiles are much heavier than other roofing materials and commonly require stronger roof support systems. Roofing tiles are laid in successive rows, with each tile covering a portion of the previous row.

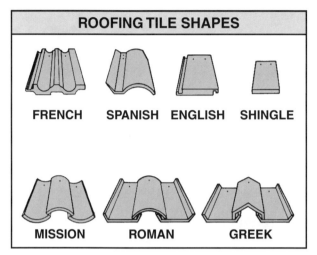

Figure 10-4. Roofing tile consists of clay, slate, or lightweight concrete members installed as a weather-protective finish on a roof.

Roof Pavers. A *roof paver* is a flat, precast concrete block that is set in place on a roof deck to provide a walking surface and to protect waterproofing materials. Roof pavers are placed on top of elastomeric roofing systems with support devices placed between the pavers and the roof surface to prevent punctures and leaks. Roof paver locations may occasionally be shown on roof plans or mechanical drawings for rooftop units.

Roof Accessories. Roof accessories include vents to provide air flow and removal of smoke, hatches to allow access to roof surfaces, and curbs around the perimeter of roof surfaces. Specifications and roof plans indicate the placement and types of roof accessories required. Items such as gutters and downspouts are normally included in CSI MasterFormat™ Division 7600 with other sheet metal items.

Walls

Building walls may be finished in a wide variety of finishes, including exposed concrete, brick, stone, and glass. Exterior wall finish materials included in Division 7 of the CSI MasterFormat™ include exterior insulation and finish systems, sheet metal, and various siding materials including metal, hardboard, cement, vinyl, and wood. Cement siding is similar in manufacture and design to precast concrete panels, but is smaller and lighter.

Exterior Insulation and Finish Systems. An *exterior insulation and finish system (EIFS)* is an exterior siding material composed of a layer of exterior sheathing, insulation board, reinforcing mesh, a base coat, and a textured finish. See Figure 10-5. Insulation board is made of extruded or molded expanded polystyrene. Insulation board provides thermal insulation and flexibility in the plaster structure to minimize cracking. A base coat approximately ¼″ thick containing acrylic polymers and portland cement is applied to the surface of the insulation board. The base coat is troweled onto the surface of the insulation board and reinforced with one or two layers of an open weave glass fiber reinforcing mesh. After the base coat has set, the textured finish of acrylic resins is troweled or sprayed onto the structure to create the desired finish.

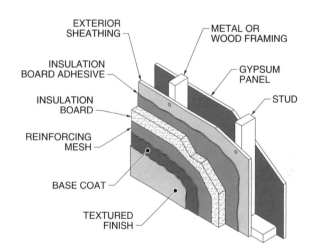

Figure 10-5. An EIFS is an exterior siding material composed of a layer of exterior sheathing, insulation board, reinforcing mesh, a base coat, and a textured finish.

Sheet Metal. A common exterior finish for small metal buildings is the application of sheet metal siding panels. Erection plans and elevation drawings indicate the types and applications for sheet metal siding panels. See Figure 10-6. Information includes the direction of application, finish trim members at the roof line and corners, and possibly the manufacturer name and identifying code number(s) and color(s) for the panels. Sheet metal wall panels are attached to purlins or girts with self-tapping and self-sealing screws. A *purlin* is a horizontal support member that spans across adjacent rafters. A *girt* is a horizontal bracing member placed around the perimeter of a structure. Exterior wall panels may also be used for soffit and canopy coverings.

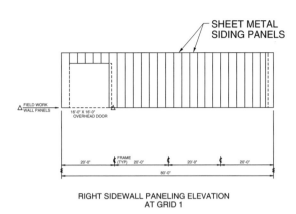

RIGHT SIDEWALL PANELING ELEVATION
AT GRID 1

Figure 10-6. Sheet metal siding panels are shown on erection plans and exterior elevation drawings.

Siding. Wood siding is available in long, narrow pieces of tempered hardboard or solid wood, 4′ × 8′ sheets, or wood shingles or shakes. Tempered hardboard siding is 4″ to 12″ wide, ½″ thick, and is available in smooth or textured finishes. Tempered hardboard siding is not susceptible to warping and twisting as is solid lumber. Tempered hardboard siding may be applied horizontally or at an angle. Solid wood siding includes bevel, shiplap, and clapboard siding. See Figure 10-7. *Bevel siding* is siding that has a tapered cross-section. Bevel siding is 8″ to 12″ wide and tapered in thickness from approximately ¾″ at the bottom to ⅛″ at the top. Bevel siding may be applied horizontally or at an angle. *Shiplap siding* is siding that has rabbeted edges that allow for a fitted overlap between edges. A *rabbet* is a square or rectangular groove cut in the edge or side of a board or plank. Shiplap siding is designed to allow for the overlapping of pieces without increasing the thickness of the wall covering. Shiplap siding is 4″ to 12″ wide and ¾″ thick. Shiplap siding may be applied horizontally, vertically, or at an angle. *Clapboard siding* is siding that has a consistent thickness and square edges. Clapboard siding is 4″ to 12″ wide and ¾″ thick. Clapboard siding may be applied horizontally or at an angle. Red cedar and redwood are the most common lumber used for solid wood siding because they are resistant to weather damage and decay.

Wood siding applied horizontally is attached to exterior walls by starting at the bottom of the wall and proceeding to the top. Each row overlaps the previous row by approximately 1″. Wood siding may also be applied vertically with battens covering the joints or at an angle for a decorative appearance. A *batten* is a narrow strip of wood used to cover the joints between boards.

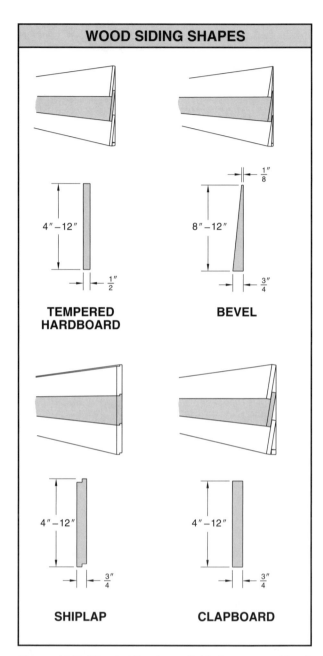

Figure 10-7. Wood siding is available in tempered hardboard and solid wood. Solid wood siding includes bevel, shiplap, and clapboard siding.

Sheet siding is commonly manufactured in 4′ × 8′ sheets ⅜″ to ⅝″ thick. Sheet siding is made of tempered hardboard or plywood with a special grade of glue and face veneer. Sheets are flat or patterned to simulate wood grain, reverse board and batten, stucco, and shiplap siding. Plywood sheets are manufactured with a special exterior glue and marked as exterior grade. A face veneer of cedar or some other water-resistant wood is applied to plywood sheet siding.

Wall siding wood shingles and shakes are made of red cedar. Wood shingles and shakes are flat or tapered in thickness from approximately 1¼″ at the bottom to ½″ at the top. Widths vary from 3″ to 14″ and lengths include 16″, 18″, 24″, and 32″. Wood shingles and shakes are applied over a layer of asphalt-impregnated building paper. Each piece is applied separately, with successive shingles or shakes overlapping previous shingles or shakes to close horizontal and vertical joints.

Metal and vinyl exterior siding provides the look of horizontal lap or vertical wood siding. Metal siding is made of aluminum or prepainted steel. Metal and vinyl siding are available with smooth finishes or woodgrain textures and in a variety of colors. Metal and vinyl siding panels are approximately ¹/₁₆″ thick, 12′-6″ long, and 8″, 10″, or 12″ wide. Each panel may have one or more horizontal offsets in the face of the siding to make one panel appear as several smaller pieces. See Figure 10-8.

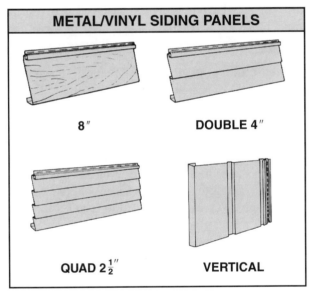

Figure 10-8. Interlocking siding panels made of metal or vinyl are available for use as an exterior wall finish.

EXTERIOR FINISH QUANTITY TAKEOFF

The primary sources of information for the estimator concerning exterior finish system materials include exterior elevation drawings, specifications, detail drawings, mechanical plans for rooftop units, vents, and exhausts, and roof and floor plans. Exterior elevation drawings provide information concerning roofing material and siding types and quantities. Specifications give detailed information concerning architectural specifications for all building materials and any manufacturer references for items such as fireproofing materials and sealants. Detail drawings show applications of flashing and roof accessories. Mechanical plans show roof vents, roof vent flashing materials, and other exterior penetration finishes. Roof and floor plans may provide additional quantity information for exterior wall finishes. Most exterior finish material quantities, including roofing and siding, are determined by the number of square feet of coverage required.

Roofing Takeoff

The method of estimating and quantity takeoff for roofing materials depends on the roofing applied. Bituminous, elastomeric, composition shingles, wood shakes or shingles, and sheet metal each require slightly different calculations. For all roofs, estimators should take into account roof accessibility and job scheduling. In some geographic locations, climatic conditions may cause roof labor application costs to vary depending on the time of year. The estimator should also note the pitch of the roof and the overall height of the building. For steep-sloped roofs, additional labor considerations may be required for fall protection and additional application time.

Roof pavers are precast concrete blocks that are placed on top of elastomeric roofing systems to provide a walking surface and protect the waterproofing material.

Georgia-Pacific

When estimating composition shingles, one extra course of shingles is required at the eaves because a starter course is required underneath the first full course.

Roofing material quantities are determined based on the total number of square feet of roof area or number of pieces for sheet metal roofs. See Figure 10-9. When using a ledger sheet, the roofing material quantities are entered on the ledger sheet. When using a spreadsheet, the calculated number of square feet of roof coverage or number of pieces is entered in the proper cell. When using estimating software, length and width dimensions are entered and the software calculates roofing materials based on individual item usage or by assemblies. Individual items such as roof accessories are counted and entered into the proper ledger sheet, spreadsheet, or estimating software cell.

Bituminous and Elastomeric. Bituminous and elastomeric roofing for flat or slightly pitched roofs is taken off according to the number of square feet of roof area to be covered. The number of roof accessories and edge treatment can affect labor costs. See Figure 10-10. Additional allowances may be required for many roof openings and irregular or lengthy edge treatments that require additional labor time for proper sealing and flashing. Costs per square foot of roof coverage are obtained from industry standard information or company historical data.

Sheet Metal. Information necessary for the estimator relating to sheet metal roofing includes the width of the metal sheets, the seams used at the edges, the sheets specified by the architect, and the square feet of coverage. The length of run of each piece of sheet metal roofing must be compared to the length of metal roofing materials available to determine the amount of seams necessary. The total length of the roof is divided by the width of a sheet metal roof sheet to determine the number of sheets and the number of seams necessary. See Figure 10-11.

Shingles and Tiles. Roof area quantity takeoff for shingles and tiles begins with the calculation of the number of squares of coverage. A square equals 100 sq ft. For sloped roofs, the roof width is multiplied by the length of the rafters to determine the total square footage. Rafter length is determined from floor plan dimensions and slope information on exterior elevation drawings. A waste factor of 10% is added for composition roof shingles. A waste factor of 20% is added for slate shingles. A waste factor of 18% is added for tile. The total is divided by 100 to determine the number of squares. The length of any ridges, valleys, and hips covered is also added.

Composition roof shingles are commonly packaged in bundles containing 33 sq ft. Three bundles or rolls of roofing material cover one square. Each bundle of composition roof shingles covers 33 lf of ridge. For example, material for a gable roof 39′ wide with 12′ rafters is estimated by multiplying 39′ by 12′ to obtain 468 sq ft (39′ × 12′ = 468 sq ft). A total of 468 sq ft is required for one side of the roof. This value is multiplied by 2 to obtain 936 sq ft (468 sq ft × 2 = 936 sq ft) for the entire roof.

Figure 10-9. Roofing material quantities such as shingles and underlayment are determined based on the total number of square feet of roof area.

 When taking off roof material, overall dimensions (including overhangs) are referenced.

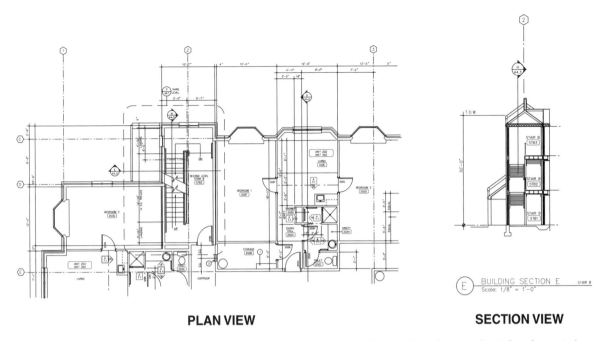

PLAN VIEW   **SECTION VIEW**

Figure 10-10. Building plan and section views are used to determine the number of square feet of roof area to be covered.

The total square feet of roof is multiplied by a waste factor of 10% to obtain a total of 1029.6 sq ft (936 sq ft × 1.1 = 1029.6 sq ft). This is rounded up to 1030 sq ft. This value is divided by 100 to obtain 10.3 squares required (1030 sq ft ÷ 100 = 10.3 squares). This is rounded up to 11 squares. For composition shingles, an additional 2 bundles of shingles are added to cover the 39′ of ridge. Eleven squares are ordered plus 2 bundles of roof shingles. For other roofing materials, a similar amount of ridge covering material is added.

To estimate the proper quantity of wood shingles or shakes, determine the pitch of the roof, find the required exposure to be used on the roof, and calculate the number of wood shingles or shakes required. In some areas, due to the weather conditions, up to three layers of wood shingles may be needed.

Wood shingles are usually bought by the square or in bundles. Four bundles make up one square (100 sq ft). For wood shingles on a roof slope of 4 to 12 and steeper, the standard exposures are 5″ for 16″ shingles, 5½″ for 18″ shingles, and 7½″ for 24″ shingles.

For hand-split shakes, the standard exposure is 7½″ for 18″ shakes, 10″ for 24″ shakes, and 13″ for 32″ shakes. The minimum roof slope for hand-split shakes is 4″ rise per foot of run. To take off the amount of wood shingles for various roof pitches, an estimator adds the appropriate exposure factor percentage to the total roof area to be covered. See Figure 10-12.

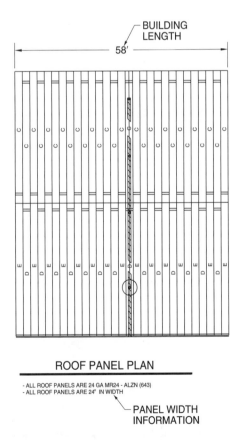

ROOF PANEL PLAN

- ALL ROOF PANELS ARE 24 GA MR24 - ALZN (643)
- ALL ROOF PANELS ARE 24″ IN WIDTH

PANEL WIDTH INFORMATION

Figure 10-11. Metal roof panels are taken off based on the overall building length and the width of the panel material used.

WOOD SHINGLE EXPOSURE FACTORS	
Roof Pitch	**Roof Area Addition***
4:12	5
5:12	8½
6:12	12
8:12	20

* in %

Figure 10-12. Wood shingle exposure factor percentages increase with roof pitch.

Once the amount of wood shingle material is determined, allowances are made for starter courses and double courses at eaves. One square of hand-split shakes provides about 120 lf of starter course. Approximately one square of cedar shingles and two squares of hand-split shakes are allowed for every 100 lf of valley. At standard exposure, it takes approximately 2½ lb of nails per square of cedar shingles and hand-split shakes.

Roof Pavers and Roof Accessories. Roof plans provide information concerning the number of square feet of coverage for roof pavers. Estimators use length and width calculations to determine the number of pavers required. Roof accessories are calculated on an item by item basis. The quantity of each roof accessory is determined from roof plans, specifications, detail drawings, or a schedule provided by the architect. Installation costs are based on standard labor tables or company historical data.

Wall Takeoff

Exterior wall covering materials are calculated according to the square feet of coverage required. Deductions are made for large wall openings such as doors and windows. Wall height and accessibility are key elements in accurate labor costing. Provisions may be necessary for rental, erection, and dismantling of scaffolding or other worker lifts for work crews to reach buildings several stories high.

> ▲ *Architectural drawings provide section and detail views of exterior insulation and finish systems. The specifications often indicate the specific product required and the manufacturer application and installation procedures. Some exterior insulation and finish systems require installation by a contractor that is approved and certified by the system manufacturer.*

Exterior Insulation and Finish Systems. Exterior insulation and finish system (EIFS) wall coverings include takeoff of insulation board, reinforcing mesh, and various coats of cementious materials. These materials are taken off according to the square feet of coverage required, resulting in the number of sheets of insulation board, rolls of reinforcing mesh, and gallons of coating materials. Architects may use EIFS materials for the creation of elaborate geometric designs. Estimators should include additional labor costs where complex layout and execution of designs are required. The assembly takeoff function in estimating software allows for quick takeoff of EIFS materials and labor. See Figure 10-13. The assembly function in estimating software leads the estimator through a series of dimension questions that determine labor and material quantities. A final review of all quantities should be performed to ensure all job site conditions have been considered.

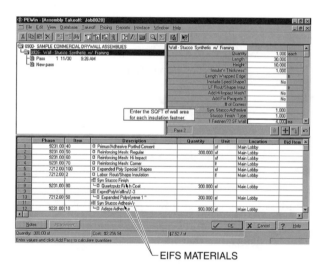

EIFS MATERIALS

Figure 10-13. The assembly takeoff function in estimating software allows for quick takeoff of EIFS materials and labor.

Sheet Metal. As with sheet metal roofing, the length of available metal siding panels is compared to the required wall height. The amount of linear feet of wall to be covered is divided by the width of each metal panel to determine the number of panels required. Certain metal trim members or anchor clips may be required by the manufacturer. See Figure 10-14. Specifications and manufacturer notes are checked to ensure all items and components are included for metal siding panel installation.

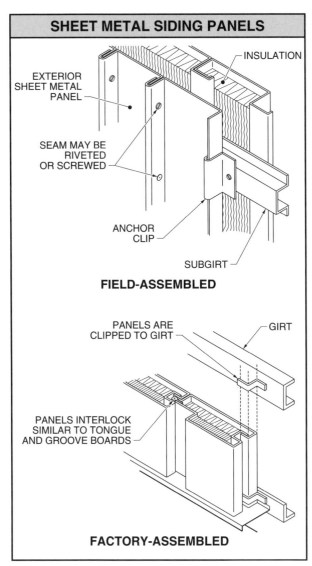

SHEET METAL SIDING PANELS

INSULATION

EXTERIOR
SHEET METAL
PANEL

SEAM MAY BE
RIVETED
OR SCREWED

ANCHOR
CLIP

SUBGIRT

FIELD-ASSEMBLED

PANELS ARE
CLIPPED TO GIRT

GIRT

PANELS INTERLOCK
SIMILAR TO TONGUE
AND GROOVE BOARDS

FACTORY-ASSEMBLED

Figure 10-14. Sheet metal siding panels may be field-assembled using rivets or screws or factory-assembled using an interlocking mechanism.

Siding. All siding types and styles appear on exterior elevation drawings with various symbols and architectural notes. Manufacturer codes and identification numbers may appear on the elevation drawings and in the specifications.

The areas of walls covered with siding are calculated by the square foot and divided into areas of 100 sq ft to determine the number of squares of material required. For example, an exterior wall 35′ long and 9′ high has 315 sq ft (35′ × 9′ = 315 sq ft). A waste factor of 5% is added for a total of 330.75 sq ft (315 sq ft × 1.05 = 330.75 sq ft). Approximately 3$^1/_3$ squares of siding are needed to cover the wall (330.75 ÷ 100 = 3.31 squares). Four squares of siding are ordered for the job.

Wood siding is packaged and delivered to the job site in bundles containing one square each. Metal siding is bundled in one-square packages with expanded foam backing board, or in two-square packages without foam backing board. Vinyl siding bundles contain two squares of siding. In addition to the siding, the amount of trim members and soffit material is determined from the elevation drawings.

THERMAL SYSTEM MATERIALS AND METHODS

Thermal systems are generally designed to perform within ambient temperature limits. Thermal systems are designed to control temperatures that affect the comfort of the building occupants, deter condensation, and reduce heat transmission to improve energy use within the structure. Thermal systems may also be used to add structural strength, support surface finishes, and reduce water vapor and noise transmission.

Insulation

Insulation is material used as a barrier to inhibit thermal and sound transmission. The North American Insulation Manufacturers Association (NAIMA) has adopted a uniform method of rating the effectiveness of all thermal system insulation when installed according to manufacturer instructions. The resistance of a specific thickness and type of insulation is indicated by an R value. *R value* is a unit of measure of the resistance to heat flow. Higher R values indicate higher heat flow resistance. For example, insulating material with an R value of 12 (R-12) offers ¾ as much resistance as insulating material having a value of R-16. The resistance of any thickness of material is equal to its resistivity per inch multiplied by its total thickness.

Waterproofing and dampproofing materials are sprayed on below grade walls to prevent the passage of moisture.

Materials. Insulating materials are categorized based on their structure and form. Insulation structures include cellular, granular, and fibrous. *Cellular insulation* is insulation composed of small individual cells separated from each other. Cellular insulation materials include polystyrene and polyurethane. *Granular insulation* is insulation composed of small nodules which contain voids or hollow spaces. Granular insulation materials include vermiculite, perlite, and cellulose. *Fibrous insulation* is insulation composed of small diameter fibers. Fibrous insulation materials include rock wool, slag wool, and glass (fiberglass). Each of these materials has its own density, resistance (R value), water vapor permeability, and dimensional stability properties. See Figure 10-15.

Insulation is manufactured in various forms for specific uses. Insulation forms include loose fill, flexible blankets and sheets, semirigid blankets and batts, rigid board, blocks, and sheets, tapes, spray-on fibers and cements, and foams.

Applications. The type and amount of thermal insulation used depends on the geographic location and intended use of the building. For example, a parking garage has different insulation requirements than a medical facility. Architects use standard tables and charts to determine minimum R values for common building usage in various zones of the U.S. See Figure 10-16.

 *The specific waterproofing and dampproofing requirements must be verified before taking off these items because the cost between waterproofing and dampproofing can vary greatly.*

Water Vapor and Fire Protection

Waterproofing is the treatment of a material so it is impervious to water. Waterproofing prevents the passage of water through the walls and floors of a building. *Dampproofing* is the treatment of a material so it is moisture-resistant. Dampproofing prevents the passage of moisture, but is not capable of withstanding hydrostatic pressure. *Hydrostatic pressure* is pressure exerted by a fluid at rest. The choice of the proper waterproofing or dampproofing system depends on the hydrostatic building conditions. Applications for waterproofing and dampproofing include below grade walls, under floors, under the walking surfaces of roofs, walls, and floors above grade, balconies, concrete canopies, pools, and around floor and roof drains.

Waterproofing and dampproofing systems include membranes, hydrolithic coatings, concrete admixtures, bentonite materials, and sheet metal. Membranes include built-up bituminous layers and rubber or PVC sheets with sealed joints. Hydrolithic coatings include materials such as tar, plastics, plaster, or cement that is troweled or sprayed onto building surfaces. Concrete admixtures include various chemical compounds added to concrete during placement to create an impervious surface. Bentonite is a natural compound that expands when exposed to moisture, allowing it to seal joints when water is present. Sheet metal includes coated aluminum and galvanized sheets used for roof valleys and ridges, gutters, downspouts, and flashing.

INSULATION MATERIAL PROPERTIES				
Material	**Density***	**Resistance****	**Water Vapor Permeability†**	**Dimensional Stability**
Polystyrene	.8 – 2.0	3.8 – 4.4	1.2 – 3.0	None
Polyurethane	2.0	5.8 – 6.2††	2.0 – 3.0	0% – 12% change
Vermiculite	4.0 – 10.0	2.4 – 3.0	High	None
Perlite	2.0 – 11.0	2.5 – 3.7	High	None
Cellulose	2.2 – 3.0	3.2 – 3.7	High	Settles 0% – 20%
Rock or slag wool	1.5 – 2.5	3.2 – 3.7	100	None
Fiberglass	.6 – 1.0	3.16	100	None

* in lb/cu ft
** R value
† in perms per in.
†† aged unfaced or spray applied

Figure 10-15. Insulation materials are selected for an application based on their density, resistance, water vapor permeability, and dimensional stability.

RECOMMENDED MINIMUM INSULATION R VALUES			
Zone	Ceiling	Wall	Floor
1	19	11	11
2	26	13	11
3	26	19	13
4	30	19	19
5	33	19	22
6	38	19	22

Figure 10-16. Recommended minimum R values relate to climatic conditions in specific geographic zones.

Fire protection materials include a variety of chemical compounds that are sprayed, troweled, or caulked into place. Some panel materials, such as calcium silicate and slag fiber, are formed into sheets and applied as boards to provide fire resistance. These materials resist high heat levels and retain their integrity by not allowing the flow of air, thus inhibiting the spread of smoke and fire.

Flashing. *Flashing* is any material installed at the seam or joint of two building components to inhibit air, water, or fire passage. Flashing includes base, cap, concealed, and exposed flashing. Flashing normally consists of job-formed and preformed sheet metal materials. *Base flashing* is a sheet metal strip installed at the lowest meeting point of a vertical and horizontal surface. Base flashing is a continuation of roofing membranes that are turned upward at the edge. *Cap flashing* is a sheet metal member wrapped around the top of a wall or other vertical projection to provide moisture protection. Cap flashing covers base flashing. *Concealed flashing* is flashing hidden from the interior and exterior of a building. Concealed flashing is made of sheet metal, plastic, or other materials. *Exposed flashing* is flashing open to view and most commonly made of metal.

Joint Sealants. *Joint sealant* is any material installed between two members to seal the seam between the members. Construction joints require a certain amount of movement due to variations in climate and building materials that cause different rates of expansion and

contraction. A solid substrate ensures a properly-sealed joint. *Substrate* is the underlying surface of a material. A primer may be applied to ensure proper adhesion of the substrate and the joint sealant. A joint filler or backer rod may be used to control the depth of the sealant and permit proper installation of the sealant. A bond breaker may also be applied to prevent adhesion of the sealant to surfaces that would reduce the performance of the sealant.

Common elastomeric sealants include acrylic materials, polysulfide, polyurethane, and silicone. Acrylic materials are one-part solvent-release compounds that cure by evaporation of solvents. Polysulfide, polyurethane, and silicone are one- or two-part chemical compounds that cure by reaction with moisture or oxygen in the air. Variables in these materials include the percent of solids, the curing process and characteristics, primers, application temperature ranges, hardness, and setting time.

THERMAL SYSTEM QUANTITY TAKEOFF

Insulation, waterproofing, and dampproofing materials are installed after structural and framing members are in place. Insulation installation must be coordinated and scheduled with other trades to ensure accessibility to required work areas. Estimators refer to specifications and detail drawings to determine R values required and the types of insulation and water vapor and fire protection coatings installed in each area. Calculations are based on cubic feet, square feet, or liquid measure, depending on the insulation and application. The square feet of coverage required is also applicable to thermal protection such as waterproofing and dampproofing. Linear feet calculations are used for various exterior trim members, flashing, and joint sealants.

Insulation Takeoff

Estimators use different takeoff methods depending on the insulation form. Insulation forms include rigid boards, blocks, and sheets, loose fill, flexible blankets and sheets, and foam.

Rigid insulation material quantities are calculated by the area to be covered, similar to floor or wall sheathing or siding. Length is multiplied by height or width to calculate the number of square feet of coverage required. See Figure 10-17.

The most common loose fill insulation material is dry granules or fibers. Loose fill insulation is poured in place or blown in place. Loose fill insulation quantities are measured in cubic feet or cubic yards of material to be poured or blown in place. Estimators determine the square feet of area to be covered and the depth or thickness to determine loose fill insulation quantities. For example, a ceiling area $45' \times 30'$ is to be insulated with $8''$ of blown fiberglass insulation. The cubic feet of insulation required is calculated by multiplying the length by width by depth to obtain 891 cu ft of insulation ($45' \times 30' \times .66' = 891$ cu ft). This value is divided by 27 to obtain 33 cu yd (891 cu ft $\div$ 27 = 33 cu yd) of insulation.

Flexible insulation for frame walls is calculated in square feet. Estimators take off the square footage of surface area to be insulated and deduct large openings such as windows and doors. Flexible insulation is manufactured in blankets or sheets that are $1''$, $2''$, $3''$, or $4''$ thick and designed to fit between studs spaced $16''$ or $24''$ OC. Each sheet is precut to $24''$, $48''$, or $96''$ long or may be obtained in blanket form in a continuous roll. Wall area is calculated in the same manner as for sheathing by multiplying wall length by wall height. When using estimating software, flexible insulation is an item that should be included in wall assemblies for insulated walls.

> ▲ *A high-quality insulating material should resist moisture, be inflammable, and not harbor pests, sustain fungus, or change volume. In addition, high-quality insulating material should not deteriorate or lose other properties due to aging or exposure to temperature extremes.*

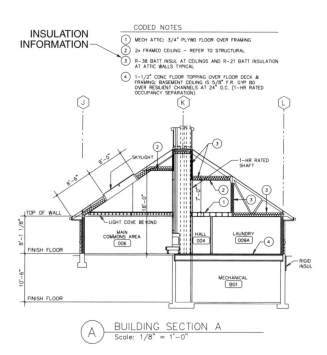

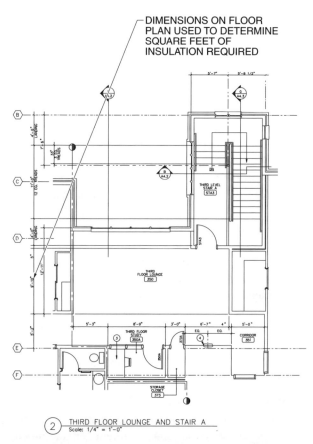

Figure 10-17. Elevation and plan views are used to determine the square feet of areas to be insulated.

For masonry walls, the takeoff process is the same as for frame walls. Flexible blanket insulation is often applied to the interior of masonry walls. Blanket insulation is stapled to furring strips and fills the space between the brick and interior finish materials.

Foam insulation is purchased in liquid form and is applied using compressed air or gas. Estimators should check manufacturer specifications concerning the amount of coverage or volume per gallon for various foam insulation materials. The total coverage area is divided by the coverage or volume per gallon to determine material quantities.

Water Vapor and Fire Protection Takeoff

Foundation plans, floor plans, elevation drawings, roof plans, specifications, and detail drawings are reviewed to check for locations of waterproofing, dampproofing, and fire protection materials. Special items that are taken off as individual units include smoke vents and joint gaskets to provide fire protection where pipes penetrate floors, roofs, and walls. Other waterproofing, dampproofing, and fire protection materials are taken off by the square or linear foot.

Floors. Concrete slab-on-grade floors commonly require waterproofing and dampproofing materials under the slab. The square feet of the slab coverage area is determined from dimensions on foundation plans and floor plans. Detail floor sections indicate the moisture protection and insulation used. See Figure 10-18. For above-grade floors, insulation may occasionally be specified on the underside of the floor for sound or thermal protection. Joint sealant quantities and types of materials for joints in pavement or concrete floors are determined by the number of linear feet of joints to be sealed.

Roofs. Estimators use floor and roof plans along with detail drawings to take off other items such as flashing and sheet metal roof specialties such as gutters, downspouts, and roof vents. Flashing is taken off by the number of linear feet of flashing required. Gutter and downspout quantities are also determined by the linear foot. Roof vents and other roof specialties are counted as individual units.

Walls. Waterproofing and dampproofing for below-grade foundation walls is shown on foundation plans and exterior wall section drawings. Membrane or rigid materials for this application are taken off by the number of square feet of coverage required. Cementious or other waterproofing coatings are taken off as the number of square feet of coverage per gallon of material.

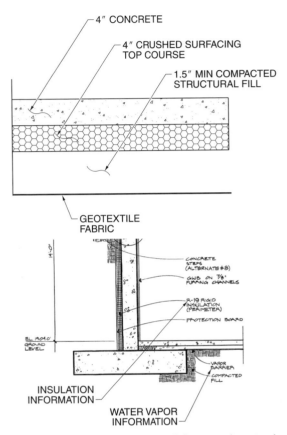

Figure 10-18. Estimators use detail floor sections to determine the moisture protection and insulation used.

ESTIMATING SOFTWARE – ASSEMBLY ITEM ADDITION

In Precision estimating software, adding, deleting, or substituting items in an assembly may be necessary for unique applications. Any changes made to an assembly apply to the current estimate only and do not affect the assembly in the database. Assembly item addition is performed by following a standard procedure. See Figure 10-19.

> ▲ *Warm air is capable of holding more moisture than cold air. Condensation forms when warm, moist air contacts a cold surface. Insulation material loses its insulation value if condensation is allowed to form on the insulation. To prevent a loss of insulation value, a vapor barrier is placed on the warm side of the insulation material.*

PRECISION ESTIMATING SOFTWARE – ASSEMBLY ITEM ADDITION

1. Create a new estimate. Name the estimate Parker Ind – Phase 2.

2. Open the Assembly Takeoff window by clicking the Assembly Takeoff button. Double-click 0000 – Sample Commercial G.C. Assemblies and then double-click Assembly 0601 – Framing Floor – 2 × 8, 10, 12 w/Sub Floor. Enter the dimensions for the assembly. Enter 18 for Length – Joist Run, 20 for Span of Joists, 210 for Joist Size, 16 for Joist Spacing, and 2 for # Runs of Bridging.

3. Click the Add Pass (⊞) button.

4. Display the item list by right-clicking in the assembly list pane and choosing List Items from the shortcut menu.

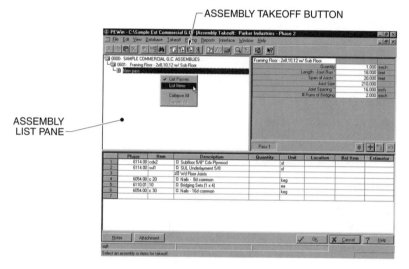

5. Add an item to the assembly by drilling down through 7000.00 Thermal and Moisture Protection, 7211.20 Insulation: Sound Blanket, and selecting 30 Sound Blanket 16″ × 3-⅝″.

6. Input the variables for the assembly addition by right-clicking the item quantity field in the item grid for the sound blanket and selecting SF L×H from the shortcut menu. Enter 1 for Quantity, 50 for Length, and 10 for Height for the dimensions for the floor in the SF L×H window.

7. Send the items to the spreadsheet by clicking the Add to Qty button and clicking OK.

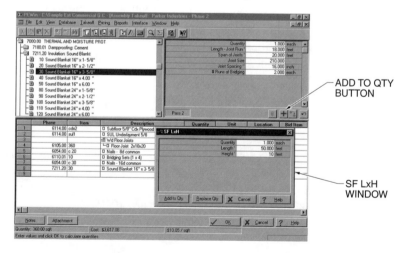

8. Close the Assembly Takeoff window. *Note:* Sound Blanket 16″ × 3-⅝″ has been added to the framing assembly.

Figure 10-19. In estimating software, adding, deleting or substituting items in an assembly may be necessary for unique applications.

Review Questions
Exterior Protective and Thermal Systems

Chapter 10

Name _____ Date _____

Estimating

_____ 1. A(n) _____ is the covering for the top surface of a building or structure.

_____ 2. _____ paper is felt material saturated with tar to form a waterproof sheet.

_____ 3. A(n) _____ is a thin piece of wood, asphalt-saturated felt, fiberglass, or other material applied to the surface of a roof or wall to provide a waterproof covering.

_____ 4. A(n) _____ is a hand-split wood roofing material.

_____ 5. Roofing _____ is clay, slate, or lightweight concrete members installed as a weather-protective finish on a roof.

_____ 6. A roof _____ is a flat, precast concrete block that is set in place on a roof deck to provide a walking surface and to protect waterproofing materials.

_____ 7. A(n) _____ system is an exterior siding material composed of a layer of exterior sheathing, insulation board, reinforcing mesh, a base coat, and a textured finish.

_____ 8. A(n) _____ is a horizontal support member that spans adjacent rafters.

_____ 9. A(n) _____ is a horizontal bracing member placed around the perimeter of a structure.

T   F   10. A batten is a narrow strip of wood used to cover the joints between boards.

T   F   11. R value is a unit of measure of the resistance to heat flow.

_____ 12. _____ is material used as a barrier to inhibit thermal and sound transmission.

_____ 13. _____ insulation is insulation composed of small individual cells separated from one another.

_____ 14. _____ is the treatment of a material so it is impervious to water.

_____ 15. _____ is the treatment of a material so it is moisture-resistant.

_____ 16. _____ pressure is pressure exerted by a fluid at rest.

_____ 17. _____ is any material installed at the seam or joint of two building components to inhibit air, water, or fire passage.

_____ 18. _____ is the underlying surface of a material.

_____ 19. _____ roofing is roofing made of a pliable synthetic polymer.

_____ 20. _____ is the amount of a shingle or shake that is seen after installation.

Metal Roof Seams

_____   **1.** Flat

_____   **2.** Ribbed

_____   **3.** Standing

Metal/Vinyl Siding Panels

_____   **1.** Vertical

_____   **2.** Quad 2½″

_____   **3.** Double 4″

_____   **4.** 8″

Roofing Tile Shapes

_____   **1.** Mission

_____   **2.** English

_____   **3.** French

_____   **4.** Spanish

_____   **5.** Roman

_____   **6.** Greek

_____   **7.** Shingle

Activity 10-1 – Ledger Sheet Activity

Refer to Print 10-1A, Print 10-1B, and Quantity Sheet No. 10-1. Take off the squares of composition shingles including hips. Calculate the building roof perimeter as square to obtain the material estimate. Do not make deductions for the cupola or entrance gable.

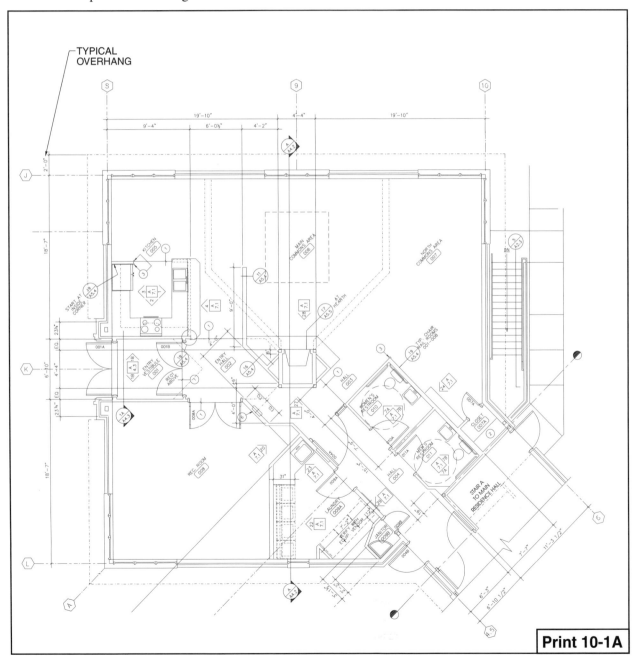

Print 10-1A

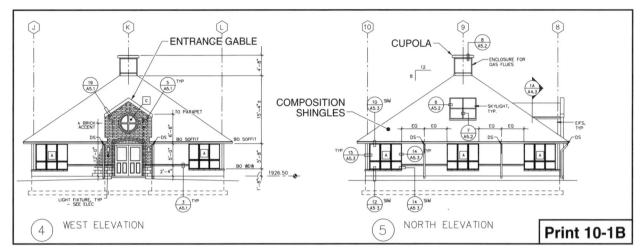

QUANTITY SHEET

Sheet No. _____10-1_____

Project: _____
Estimator: _____

Date: _____
Checked: _____

No.	Description	Dimensions				Unit		Unit		Unit		Unit
		L	W	D								
	Composition shingles											

Activity 10-2 – Spreadsheet Activity

Refer to the cost data, Print 10-2, and Estimate Summary Sheet No. 10-2 on the CD-ROM. Take off the square feet of elastomeric roofing, number of walk pads, and linear feet of flashing at the top of the parapet wall. Determine the total material, labor, and equipment cost for the elastomeric roofing, number of walk pads, and flashing.

COST DATA				
Material	**Unit**	**Material Unit Cost***	**Labor Unit Cost***	**Equipment Unit Cost***
Elastomeric roofing, 50 mils, reinforced	sq ft	3.08	.81	.15
Walk pads, concrete, 2″ thick	sq ft	1.53	1.64	—
Flashing	lf	.36	.61	—

* in $

Activity 10-3 – Estimating Software Activity

Refer to Print 10-3 and Specifications 10-3 on the CD-ROM. Open a new estimate and name it Activity 10-3. Perform an item takeoff for Insulation – Loose Fill Zonolite Insulation for the ceiling of the building. Use the perimeter of the building dimensions. Assume 6″ of insulation. Use the calculator provided with Precision estimating software.

Doors and Windows

Division 8 of the CSI MasterFormat™ includes information concerning door and window materials and methods. Metal and wood doors are described according to their fire rating, core materials, and finishes. Specialty doors include fire, access, coiling, folding and sliding, overhead, and entrance doors. Items considered in takeoff for doors include frames, doors, and hardware. Materials for window construction include various metals, woods, plastics, and glass. Window frames and glazing are commonly set in place after construction of structural members. Information necessary for window quantity takeoff includes the grade, thickness, and type of glass to be used at each location, the type of each window, and the number and size of the glass panes.

DOORS

Door and window materials and methods are included in Division 8 of the CSI MasterFormat™. Division 8 of the CSI MasterFormat™ includes basic information concerning doors, entrances and storefronts, windows, skylights, hardware, glazing, and curtain walls. Metal and wood doors are described according to their fire rating, core materials, and finishes. Estimators review door information in floor plans, specifications, and elevation drawings concerning the type of door and frame, door finish, and hardware. Metal door finishes include primer and finish paint coatings. Wood door finishes include veneers of various types and qualities and the application of various finish materials such as moldings, panels, and lights. The specifications commonly contain a door schedule providing detailed information about all doors on a construction project. See Figure 11-1. Letters or numbers on the door schedule correspond to doors specified on floor plans or elevation drawings.

Standard door sizes are stated in a sequence of door thickness, width, and height. The most common swinging and sliding door thicknesses are $1^3/_8''$ and $1^3/_4''$. Door widths are sized in 2″ increments beginning at a width of 1′-6″. Common swinging door heights are 6′-8″ and 7′-0″. Swinging door information also indicates the hand of the door. The hand of a door pertains to the direction of door swing. Swinging doors may be right-hand, left-hand, right-hand reverse, or left-hand reverse. Details for special door applications provide information about thickness, width, and height for special door applications. Details for doors may also indicate the installation of transoms where applicable. Common doors include flush, panel, louvered, view, and full view. See Figure 11-2.

> ▲ *All exit doors serving hazardous areas with occupancy levels of 50 people or more must swing in the direction of exit travel.*

DOOR SCHEDULE

Number	Size	Type	Remarks
1	2′-8″ × 6′-8″ × 1³/₈″	fl hollow core	—
2	2′-8″ × 6′-8″ × 1³/₈″	fl hollow core	—
3	3′-0″ × 6′-8″ × 1³/₄″	solid core	—
4	2′-8″ × 6′-8″ × 1³/₈″	fl hollow core	—
5	2′-6″ × 6′-8″	bi-fold	2′-6″ opening
6	2′-6″ × 6′-8″ × 1³/₈″	fl hollow core	—
7	2′-8″ × 6′-8″ × 1³/₄″	fl solid core	—
8	3′-0″ × 6′-8″	bi-fold	3′-0″ opening
9	3′-6″ × 6′-8″	bi-fold	3′-6″ opening
10	5′-0″ × 6′-8″	alum sliding	5′-0″ opening
11	2′-4″ × 6′-8″ × 1³/₈″	fl hollow core	—
12	2′-8″ × 6′-8″ × 1³/₈″	fl hollow core	—
13	4′-0″ × 6′-8″	alum sliding	12′-0″ unit
14	2′-8″ × 6′-8″	alum storm	—
15	2′-8″ × 6′-8″ × 1³/₄″	fl solid core	—
16	2′-6″ × 6′-8″	bi-fold	2′-6″ opening
17	2′-6″ × 6′-8″	alum sliding	5′-0″ opening
18	2′-6″ × 6′-8″ × 1³/₈″	fl hollow core	—
19/20	2′-6″ × 6′-8″	bi-fold	2′-6″ opening
21	2′-4″ × 6′-8″ × 1³/₈″	fl hollow core	2′-4″ opening
22/23	2′-6″ × 6′-8″ × 1³/₈″	fl hollow core	—
24	4′-0″ × 6′-8″	2 bi-fold	4′-0″ opening
25	4′-0″ × 6′-8″	2 bi-fold	4′-0″ opening
26	2′-6″ × 6′-8″	bi-fold	2′-6″ opening
27	2′-8″ × 3′-0″ × 1³/₈″	fl hollow core	crawl space
28	16′-0″ × 7′-0″	ovhd 4-panel	cedar faced

Figure 11-1. The specifications normally contain a door schedule providing detailed information about all doors on a construction project.

Custom metal doors, frames, and hardware require long lead times and should be ordered directly after approval of the door schedule.

Flush doors contain no lights and are covered with metal or wood on both sides. A *light* is a pane of glass or translucent material in a door or window. Panel doors are comprised of a series of rails and stiles forming frames for panels. A *rail* is a horizontal member of a door, window frame, or cabinet face frame. A *stile* is a vertical member of a door, window frame, or cabinet face frame. Louvered doors contain a frame with vents allowing air passage. View doors contain a framed opening with a glass or plastic insert. Full view doors are made primarily of glass with a glazed frame.

Metal Frames and Doors

Metal frames and doors are commonly referred to as hollow metal. Specifications for hollow metal frames and doors are available from the National Association of Architectural Metal Manufacturers (NAAMM). Various metal frames and doors are installed for commercial and residential applications according to frequency of use, traffic patterns, fire ratings, and climatic and temperature conditions.

Metal Frames. Hollow metal door frames are designed to receive a variety of doors, including metal and wood doors. Metal door frame gauge varies depending on the design of the door to be supported and the required fire rating. A variety of metal door frame anchors are available based on the wall into which the frame is installed. See Figure 11-3. Metal door frame anchors include loose T, loose wire, wood stud, core board, through bolt, adjustable loop, standard floor knee, adjustable floor knee, wedge clip, steel channel, steel stud, and extended frame with base anchors. Metal door frame anchors may be screwed, snapped, or welded into position. The different metal door anchors are used to secure metal door frames to a variety of interior and exterior wall materials.

In addition to width, height, and door thickness, variables in metal door frame design include jamb depth, rabbet, soffit, face, stop depth, and backbend. See Figure 11-4. The open area between the metal frame and the supporting structure may be required to be filled with grout or another material as specified by the architect.

> ▲ *Metal doors are used most often in commercial construction. Metal doors are available in 1³/₈″ and 1³/₄″ thicknesses and have prepared hinge mortises and lock holes.*

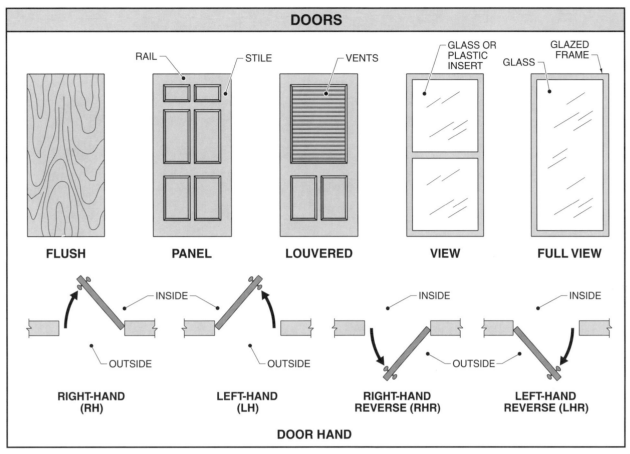

DOORS

| FLUSH | PANEL | LOUVERED | VIEW | FULL VIEW |

RAIL — — STILE — VENTS GLASS OR PLASTIC INSERT GLAZED FRAME GLASS

DOOR HAND

RIGHT-HAND (RH)   LEFT-HAND (LH)   RIGHT-HAND REVERSE (RHR)   LEFT-HAND REVERSE (LHR)

INSIDE — OUTSIDE

Figure 11-2. Common residential and commercial doors include flush, panel, louvered, view, and full view.

Metal Doors. Hollow metal doors are composed of a metal perimeter channel frame filled with a foam or fiber core and covered by a metal sheet on each side. Minimum metal thicknesses are specified for metal door frames and faces depending on the door application. See Figure 11-5. For example, commercial metal doors require 16 gauge or 18 gauge metal. Institutional use metal doors require 12 gauge or 14 gauge metal. Standard finish for metal doors is a primer applied at the factory prior to delivery at the job site. Metal doors may also receive a baked enamel finish, an applied finish such as vinyl cladding, a textured, embossed finish such as stainless steel or aluminum, or a polished metal finish.

Wood Frames and Doors

As with metal frames and doors, traffic and climatic conditions determine the design of wood frames and doors used in building construction. Wood frames and doors are commonly used in residential and light commercial applications. They may be prefinished or finished at the job site. Wood frames and doors may be painted or finished with coatings that expose the wood grain.

Metal frames should be constructed to conform to Steel Door Institute Standard SDI-100 and may be formed from 18 gauge, 16 gauge, or 14 gauge cold-rolled, hot-rolled, or galvanized steel.

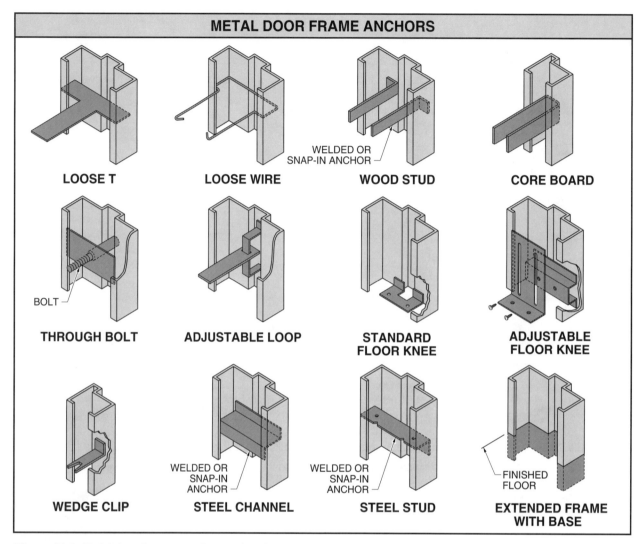

Figure 11-3. Metal door frames are fastened to the surrounding supporting structure with a variety of anchors.

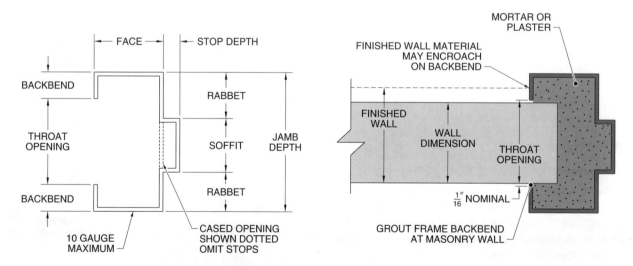

Figure 11-4. Variables in metal door frame design include jamb depth, rabbet, soffit, face, stop depth, and backbend.

STEEL DOOR MINIMUM METAL THICKNESSES		
Commercial Doors		
Item	**Gauge Number**	**Equivalent Thickness***
Door Frames Surface applied hardware reinforcement	16 16	.0598 .0598
Doors – hollow steel construction Panels and stile	18	.0478

Doors – composite construction
  Perimeter channel
  Surface sheets
	18
	22
		.0478
		.0299

Reinforcement
  Surface applied hardware
  Lock and strike
  Hinge
  Flush bolt
	16
	16
	10
	16
		.0598
		.0598
		.1345
		.0598

Glass molding	20	.0359
Glass muntins	22	.0299
Interior Doors		
Item	**Gauge Number**	**Equivalent Thickness***
Door frames, 1³/₈″ thick	18	.0478
Door frames, 1³/₄″ thick	16	.0598
Stiles and panels	20	.0359

Reinforcement
  Lock and strike
  Hinge
  Closer
	16
	11
	14
		.0598
		.1196
		.0747

* in in.

Figure 11-5. Minimum metal thicknesses are specified for commercial and interior metal doors and frames.

Wood Frames. Wood frames are most common in residential and light commercial buildings. Wood frames may contain built-up or split jambs. See Figure 11-6. Built-up jambs are made from ¾″ thick lumber. The lumber is cut to the same width as the wall thickness including finish materials. Stops and casings are applied to the wood jamb. Split jambs are made from molded wood members designed to interlock. The two interlocking jamb pieces may be slid together or apart to adjust the jamb depth. Wood frames are also used for specialty doors such as overhead doors and pocket doors. Fire-rated wood frames are also available.

Wood Doors. The general categories of wood doors include solid core, hollow core, and panel. See Figure 11-7. Solid core wood doors are made of a wood frame reinforced by a solid particleboard or staggered block

core. The frame and core are covered with veneer to create a solid wood door. Wood doors are classified as 5 ply and 7 ply. A 5 ply door is glued with Type I glue and has a crossband and face veneer on the stiles and rails. A 7 ply door is glued with Type II glue and has 3 ply manufactured hardwood skins on both faces of the stiles and rails. Hollow core doors are made of a wood frame with a mesh core. The wood frame and mesh core are covered with veneer materials to form a light-traffic door. Panel doors are made of solid pine or built-up rails and stiles that frame flush or raised panels.

Specialty Doors

Specialty doors include fire, access, coiling, folding and sliding, overhead, and entrance doors. Estimators obtain information concerning the various doors from floor plans, elevation drawings, and door schedules. Some special function doors such as revolving, coiling, and overhead doors are commonly subcontracted items for estimating purposes.

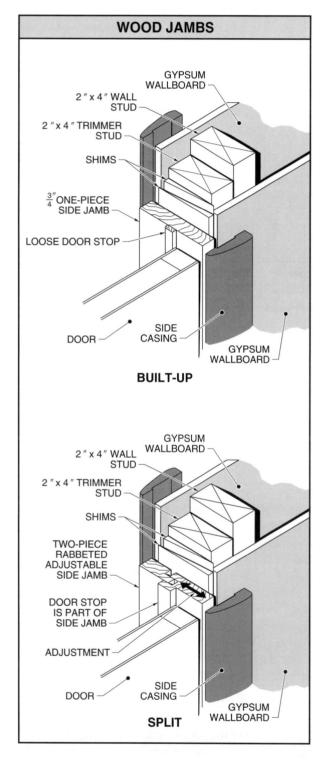

Figure 11-6. Wood jambs are made from flat finished lumber or interlocking molded members.

Fire Doors. A *fire door* is a fire-resistant door and assembly, including the frame and hardware, commonly equipped with an automatic door closer. A *door closer* is a device that closes a door and controls the speed and closing action of the door. Fire doors are specially built and installed to inhibit the transmission of fire from one building area to another. The five classifications of door openings for fire door installation are Class A, B, C, D, and E. The five classifications are based on the time that a door opening withstands penetration by fire. See Appendix. Class A openings (3 hr) are in walls separating buildings or dividing one building into separate fire areas. Class B openings (1 hr – 1½ hr) are in enclosures of vertical egress in a structure such as stairs and elevators. Class C openings (¾ hr) are in hallways and room partitions. Class D openings (1½ hr) are in exterior walls subject to fire exposure from outside the structure. Class E openings (¾ hr) are in exterior walls subject to moderate to light fire exposure.

Fire doors include composite wood and hollow metal. Composite wood fire doors have a minimum thickness of 1¾″ and are used for Class B and Class C openings. Hollow metal fire doors are formed of 20 gauge or heavier steel. All fire doors and frames must display a label listing their fire rating. See Figure 11-8.

Access Doors. An *access door* is a door used to enclose an area that houses concealed equipment. Access doors include any special doors such as sidewalk, floor, and other small doors for access to mechanical and electrical equipment. Sidewalk doors are available in sizes from 2′-0″ to 3′-6″ in 6″ increments. Sidewalk doors are made of steel or aluminum ¼″ checker or diamond plate and can withstand 300 lb per square foot of live load. Floor doors are made in a manner similar to sidewalk doors with an allowance for floor covering that can match the surrounding floor area. Access doors are commonly made of metal and are pre-mounted in a frame. Many different sizes and designs of access doors are available depending on the application.

▲ *Wood and metal doors can be purchased as prehung units from the manufacturer. The door is delivered at the job site installed in the frame with hardware attached. These prehung units save installation time in the field.*

WOOD DOORS

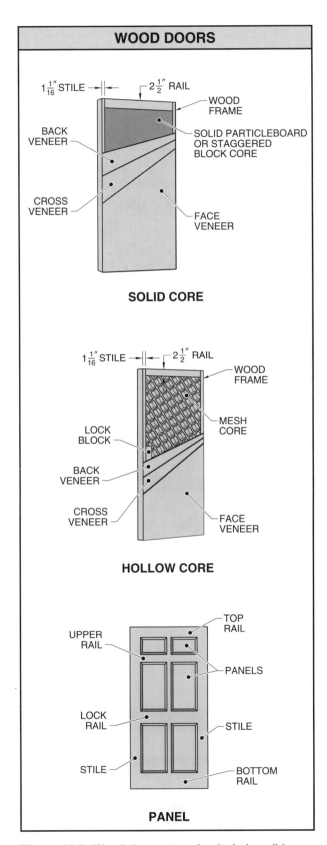

Figure 11-7. Wood door categories include solid core, hollow core, and panel.

FIRE DOOR LABELS

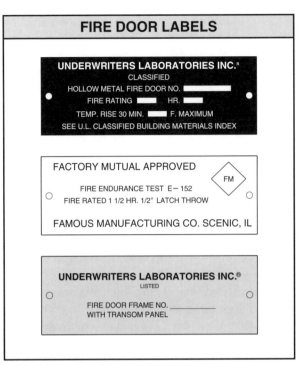

Figure 11-8. All fire doors and frames must display a label listing fire rating information.

Coiling Doors. A *coiling door* is a door that moves vertically to coil around a steel rod. Interlocking strips of galvanized steel, stainless steel, aluminum, or grillwork are fabricated into coiling doors. See Figure 11-9. A track mounted to either side of the opening holds the coil in place. Coiling doors are available in many standard sizes varying in height up to 30′ and in width up to 33′.

Figure 11-9. A coiling door moves vertically to coil around a steel rod.

Folding and Sliding Doors. A *folding door* is a door formed with panel sections joined with hinges along their vertical edges and supported by rollers in a horizontal upper track. A *sliding door* is a horizontal-moving door suspended on rollers that travel in a track that is fixed at the top of the opening or rollers mounted in the bottom of the door. Wood, metal, and reinforced fabrics are used for folding and sliding doors. Folding door tracks are suspended from the ceiling or roof. A series of rollers in the top of the folding door supports its movement. Folding door sizes vary from small sizes for residential installations to large sizes for meeting room dividers.

Sliding doors travel in a track that is fixed at the top of the opening or at the floor. Rollers fixed at the top of the opening or in the bottom of the door panel provide horizontal movement. Sliding doors are commonly constructed of wood or metal frames with large tempered glass lights. Common sliding door heights include 6'-8", 8'-0", and 10'-0". Widths vary depending on the number of sliding panels.

> ▲ *Information related to the different types of doors, frames, and assemblies is given in different sections of the specifications. All of these sections must be checked to determine the amount of material and labor costs required for each type of door.*

Overhead Doors. An *overhead door* is a door which, when opened, is suspended in a horizontal track above the opening. Overhead doors consist of a series of horizontal panels joined with hinges, allowing for folding of the sections as the door is raised and lowered. Overhead doors include fiberglass, steel, and wood panel doors. Panel and section dimensions and designs are set at the manufacturer to provide door dimensions in accordance with the design requirements. Approximate heights range up to 20' and widths up to 30'. Overhead doors are held in place and moved vertically by rollers attached to the inside of the door. The rollers move in a track mounted to the structural members on both sides of the door opening. See Figure 11-10. Springs facilitate raising and lowering the door. Overhead doors may use motors to raise and lower the door.

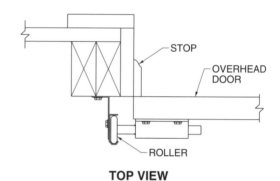

TOP VIEW

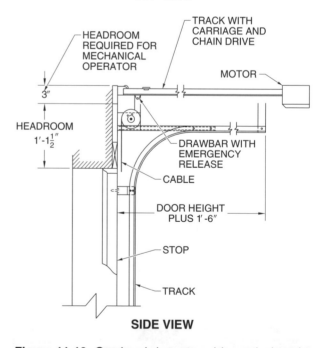

SIDE VIEW

Figure 11-10. Overhead doors travel in vertical tracks mounted to the structural members on both sides of the door opening.

For overhead doors, estimators determine the size of the opening, the type, style, finish, and size of the door, the type of door hanger, the door installation requirements, and method of operation.

Entrance Doors. An *entrance door* is the primary means of egress of most pedestrian traffic for a commercial building such as retail, office, health care, and manufacturing facilities. Entrance doors used in commercial construction include metal framed and all-glass entrances, automatic doors, and revolving doors. Metal framed entrance door systems may be fabricated from aluminum, bronze, stainless steel, or other metal finishes. The metal frames commonly support tempered glass and use heavy-duty hardware due to the high traffic requirements. Standard sizes are available depending on the manufacturer. Frameless glass doors are fitted with metal channels at the top and bottom with pivot rods. The pivot rods are fitted into recessed hinges. Tempered glass is mandatory in these applications to withstand traffic and wind loads. Standard revolving door dimensions vary from 6′-6″ in diameter to 7′-6″ in diameter in 2″ increments. Revolving door standard heights are 6′-8″ and 7′-0″. The circular enclosure for the door unit is commonly constructed of metal and glass with approximately 3¾″ allowance for the rubber sweeps at the outside edge of the revolving door sections.

Door Hardware

Door hardware materials include hinges, locksets, closers, and miscellaneous hardware such as panic exit devices, weatherstripping, thresholds, etc. The primary source of hardware information is found on the door schedule in the specifications. Architects commonly use manufacturer identification codes to specify hardware requirements at each door. Standard hardware finish codes are used to specify hardware for building construction. See Figure 11-11.

Hinges. A *hinge* is pivoting hardware that joins two surfaces or objects and allows them to swing around the pivot. Hinges allow for pivoting at the point of the door and jamb attachment. Hinges include mortised (butt hinge), concealed, ball bearing, gate, spring, strap, T, olive knuckle, paumelle, and many other designs. See Figure 11-12. Many different metals, finishes, fire ratings, closing options, and door and jamb configurations are taken into account when selecting the proper hinge for an application.

▲ *Door hardware can add considerable cost to each door. Door hardware requirements are determined most efficiently by reviewing the door schedule and specifications for the project.*

DOOR HARDWARE FINISH SYMBOLS	
Symbol	**Finish**
USP	Primed for painting
US 3	Bright brass
US 4	Dull brass
US 10	Dull bronze
US 10B	Dull bronze, oxidized, and oil rubbed
US 14	Nickel plated, bright
US 26	Chromium plated, bright
US 26D	Chromium plated, dull
US 27	Satin aluminum, lacquered
US 28	Satin aluminum, anodized
US 32	Stainless steel, polished
US 32D	Stainless steel, dull

Figure 11-11. Standard hardware finish codes are used to specify hardware for building construction.

Entrance doors require tempered glass and heavy-duty hardware to withstand traffic and wind loads.

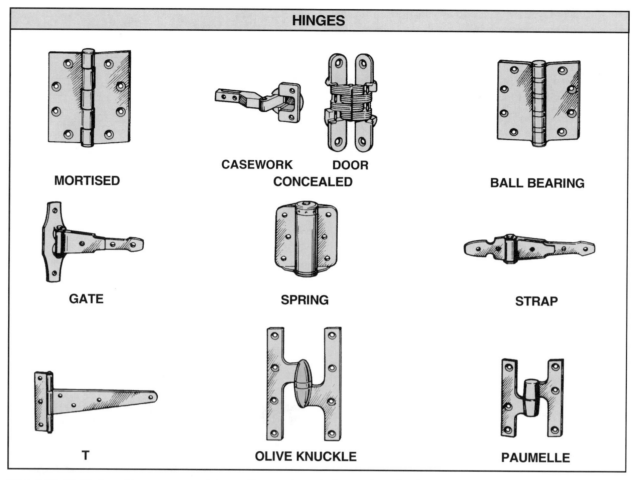

HINGES

MORTISED

CASEWORK   DOOR
CONCEALED

BALL BEARING

GATE

SPRING

STRAP

T

OLIVE KNUCKLE

PAUMELLE

Figure 11-12. Various hinges are used depending on door design and application.

Locksets. A *lockset* is the complete assembly of a bolt, knobs, escutcheon, and all mechanical components for securing a door and providing means for opening. An *escutcheon* is a protective cover placed around a door knob. A lockset provides security, locking, and manual ability to open and close a door. Locksets for swinging doors include cabinet, cylinder, deadbolt, grip, and mortise. See Figure 11-13. As with other door hardware, many different designs are available based on finish, security requirements, and amount of traffic. Deadbolt locksets are commonly used for coiling and overhead doors, with a deadbolt projecting into the door track. Sliding doors and folding doors normally incorporate a concealed lockset.

Closers. A door closer is a device that closes a door and controls the speed and closing action of the door. Door closer hardware automatically closes a door using a hydraulic device and pivots. The two primary categories of closers are hidden and surface-mounted. Hidden door closers may be mounted into the top or

bottom of the door or top or bottom of the frame depending on the closer design. See Figure 11-14. Other electrically-controlled automatic door closers are powered by various detectors. These power-activated doors are controlled by electric or pneumatic sensing mates, photoelectric cells, motion detectors, or other methods. Power-assisted opening and closing equipment is also required in some applications for access by individuals with disabilities.

Miscellaneous Hardware. Additional door hardware includes panic exit devices, handles, push plates, and door stops and holders. Panic exit devices include rim, mortise, exposed vertical rod, and concealed vertical rod. Rim panic exit devices are mounted on the surface of the door without any fitting into the door itself. Mortise panic exit devices allow for the bolt to project from the edge of the door in a manner similar to a cylinder or mortise lockset. Rod panic exit devices use a mechanism with latches at the top and bottom of the door fastened to vertical rods. See Figure 11-15.

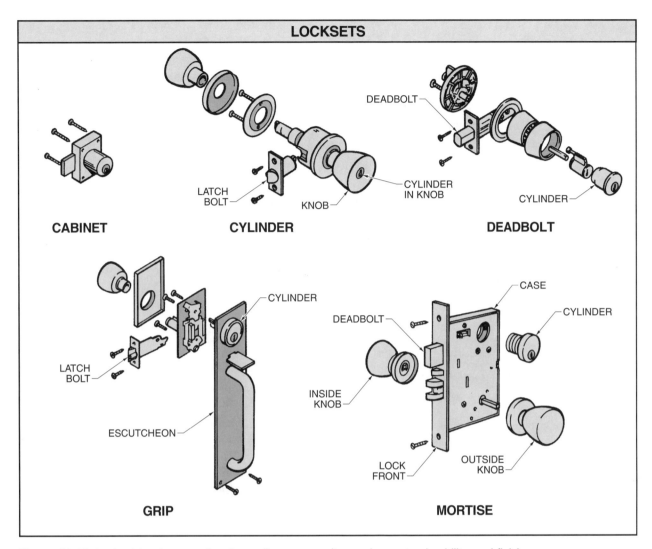

Figure 11-13. Lockset hardware varies depending on security requirements, durability, and finish.

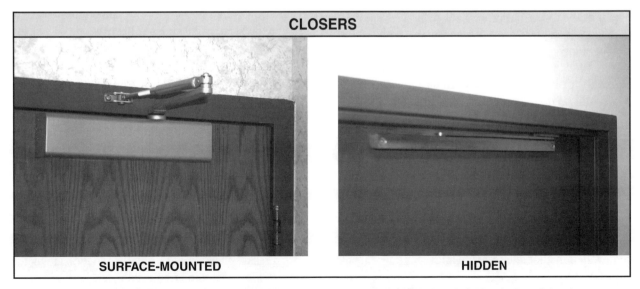

Figure 11-14. A door closer is a device that closes a door and controls the speed and closing action of the door.

Figure 11-15. Panic bars allow for quick exit from public areas in cases of emergency.

DOOR AND FRAME CONSTRUCTION METHODS

Door frames are installed as the first component of door construction. Sizes and types of frames are determined from door schedule specifications. After frames are set plumb and level, the required hardware is installed. Doors are then set and hung in place. Latching hardware is installed to complete the door construction.

Frame Construction

Door frame construction begins with checking the rough openings to ensure the proper size opening is provided for the door and frame. Allowances are made in the wall opening construction for the door, any open space required for opening of the door, frame materials, and required trim and finish. Rough opening allowances vary depending on the frame.

Swinging Door Frames. Metal and wood frames for swinging doors are set in place in different methods depending on the wall construction method used. For concrete walls, frames and jambs may be set into formwork prior to concrete placement. Care is taken to ensure the frame is set at the proper location and set plumb and level and securely braced to prevent movement during concrete placement. Blockouts may be installed in concrete forms to allow for later installation of door jambs. A *blockout* is a frame set in a concrete form to create a void in the finished con-

crete structure. For masonry walls, door frames are set in place and braced prior to laying masonry. Masonry members are set around the frames with proper anchors set in mortar joints. For steel frame and wood frame walls, door frames are set after framing is in place.

Folding, Coiling, and Overhead Door Frames. Frames for folding, coiling, and overhead doors are sized based on allowances for door size and supporting hardware. Folding doors may not require any frame material depending on their size and surrounding finish. Coiling and overhead door frames provide additional support at the header for the weight of the door and opening hardware. The side jambs for coiling and overhead doors are set in place to allow for a tight seal at the door sides and stops.

Door Construction

Doors may be prehung or hung on site. Prehung doors are set into jambs at the mill. Hinges are fastened to the door and jamb and temporary bracing is attached to hold the door in place during shipping. The prehung unit is set into the wall opening and adjustments are made to the jamb position with the door in place. Doors hung on site are set into preset jambs. Doors are fitted into the jambs with hanging hardware attached. Stops and finish materials are fastened to the doors and jambs to complete the door hanging.

Hardware. Standard allowances are made in metal doors and jambs for hinge and lockset and hardware installation. Metal doors are delivered prefabricated to receive the proper size and number of hinges. Lockset holes may also be predrilled or drilled at the job site. Holes for other hardware such as closers and panic bars are drilled and tapped at the job site. Wood doors may be premilled for hinges and locksets or drilled and fitted at the job site.

DOOR QUANTITY TAKEOFF

Items considered in takeoff for doors include frames, doors, and hardware. Estimators should prepare a schedule or work code indicating the location of each door and frame by floor or building area. As with other estimating tasks, plans should be marked to indicate that a door opening, including frame, door, and hardware, has been included in the quantity takeoff total.

Door, Frame, and Hardware Takeoff

The number of each type of door and frame is individually counted and totaled on each floor plan. Drawing symbols and architectural notes describe each door and frame. See Figure 11-16. The door schedule is cross-referenced to the floor plans to further describe the totals and types of doors. Elevation drawings may be cross-referenced with floor plans during quantity takeoff to double check the proper types and sizes of doors.

When estimating using a ledger sheet, a chart is established for each area of the structure as determined by the estimator. Columns are established for frames, doors, and hardware. Totals are entered in each row to accumulate a final total. Spreadsheets allow information entry in the same format as ledger sheets but the totals are added automatically. Estimating software allows for assembly takeoff of frames, doors, and hardware as a single unit, with the appropriate entries made by entering one item assembly with labor costs calculated based on a per unit method. See Figure 11-17. Items may also be entered individually concerning frames, doors, and hardware.

During pricing, labor costs are determined based on the average time taken to install a door frame, hang a door, and install the necessary hardware. These labor costs are based on standard industry sources or company historical data. A general contractor normally subcontracts specialty doors such as overhead, revolving, and automatic doors.

Door Takeoff. Items considered by an estimator during door takeoff include door type, possible panel matching requirements, height, width, thickness, hand (for swinging doors), transoms, hinge type and placement, lockset, door material, fire rating, and door inserts such as louvers or lights. The door schedule in the specifications is the primary source of this information. Floor plans are the primary source of quantity information.

Frame Takeoff. During quantity takeoff, frame information considered includes door type, height, width, thickness, hand (for swinging doors), transoms, hinge type and placement, lockset, frame material, and fire rating. The door schedule in the specifications is the primary source of this information. Floor plans are the primary source of quantity information.

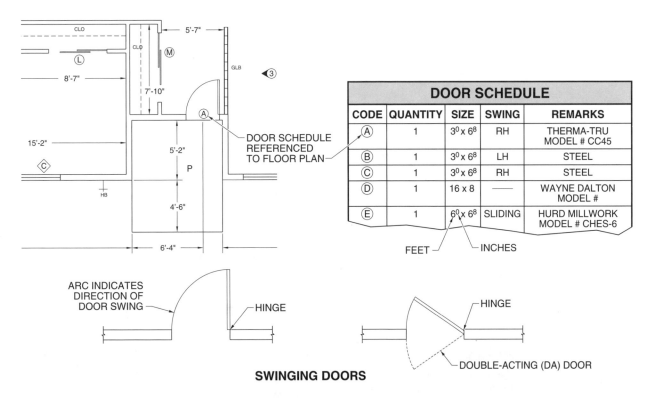

Figure 11-16. Estimators use floor plans and door schedules to perform accurate door, frame, and hardware takeoff.

ITEM
ASSEMBLY

Figure 11-17. Estimating software allows for assembly takeoff of frames, doors, and hardware as a single item with labor costs calculated based on a per unit method.

Specialty Door Takeoff. Specifications, floor plans, elevation drawings, and detail drawings provide specialty door information. See Figure 11-18. When using estimating software, specialty doors are taken off as a one-time item and not commonly entered into the database where they occur.

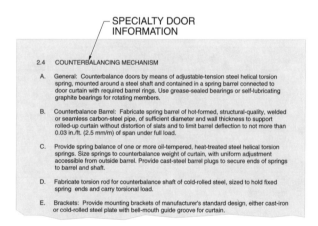

SPECIALTY DOOR
INFORMATION

Figure 11-18. Specialty door information is detailed in the specifications, floor plans, elevation drawings, and detail drawings.

Hardware Takeoff. Takeoff of door hardware items is commonly specialized by manufacturer information. Quantities for hardware are related to the number of each door and frame. Items such as hinges, locksets, panic bars, thresholds, and closers are totaled accord-

ing to the number of individual units needed for various doors. Specifications are the primary source of door hardware detail information.

WINDOWS

Materials for window construction include various metals, woods, plastics, and glass. Metals used for window frames include aluminum, steel, stainless steel, bronze, and other decorative metals. Wood windows may be painted or clad with metal or plastic. *Cladding* is a layer of material greater than .04″ thick used to improve corrosion resistance or other properties. Glass sheets are available in many different designs and types. The type of glass installed takes into account wind loads, thermal transmission, privacy, appearance, safety, and security. The majority of information concerning windows and installed glass is given in the specifications. Detail drawings may refer to a specific type of window or glass for an identified area. Exterior elevations and floor plans provide dimensional information for the sizes of windows and glass panels.

Most windows are available in steel, aluminum, or wood sash. A *sash* is the fixed or movable framework of a window that contains the glass. Many window manufacturers produce windows of varying quality depending on the weight of the sash, glass type, and other factors. Most building codes have restrictions on windows concerning minimum light, ventilation, wind load resistance, and egress requirements.

Elements considered during window design include air and water resistance, ventilation requirements, insulation, light considerations, visual and acoustical separation, safety, access, ease of operation, security, and maintenance. Architects and designers review these elements when choosing the window to be installed in each opening. Common windows include fixed, single- or double-hung, sliding, casement, hopper, awning, bay, bow, jalousie, and skylight. See Figure 11-19. The two general categories of windows are fixed and movable. Window units may be comprised of fixed and/or movable windows.

> *Aluminum window frames must be protected during construction. Corrosion may be caused when exposed aluminum surfaces are splashed with plaster, mortar, or concrete masonry cleaning solutions. The cost of aluminum window frame protection is included in the cost of the window estimate.*

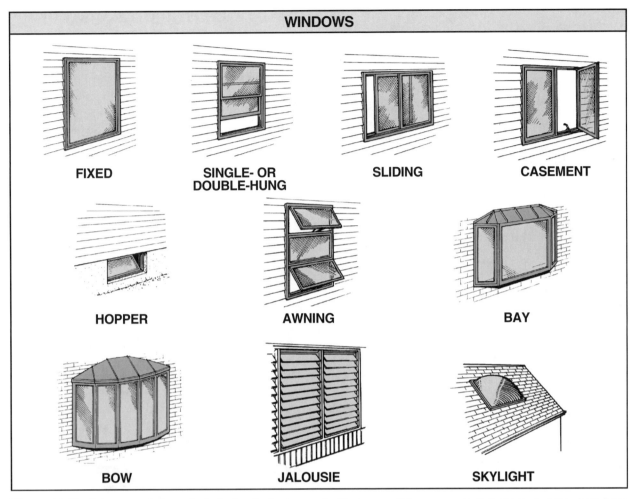

WINDOWS

FIXED

SINGLE- OR DOUBLE-HUNG

SLIDING

CASEMENT

HOPPER

AWNING

BAY

BOW

JALOUSIE

SKYLIGHT

Figure 11-19. Common windows include fixed, single- or double-hung, sliding, casement, hopper, awning, bay, bow, jalousie, and skylight.

Fixed Windows

Fixed windows are made of one piece of glass set in a frame with wood, metal, or plastic trim holding the glass in place. See Figure 11-20. Fixed glass windows are common in commercial buildings for security and safety reasons. Glass for fixed windows may be clear, reflective, or tinted. Fixed glass panels are commonly applied in curtain wall systems. Bay and bow windows are made with fixed glass panels and possibly movable panels.

Movable Windows

Movable windows are windows that may be opened and closed. Single-hung windows have one fixed window panel (usually the top window) and one vertical sliding window panel. Springs or weights are attached to the windows to make opening and closing easier. Double-hung windows consist of two window panels that slide vertically. The upper or lower half of the window may be opened. Sliding windows move in a horizontal direction. Rollers are fastened to the bottom of the sliding window frame to help them move easily. Casement windows are hinged at one side and open similar to a swinging door. Casement windows open and close with a crank and gear system built into the jamb and attached to the window frame. Hopper windows consist of a single sash hinged at the bottom of the frame. Awning windows are made of a sash hinged at the top of the frame. Jalousie windows are made of narrow pieces of glass mounted horizontally in a frame. The glass pieces are controlled by a crank and gear system. A *skylight* is a roof opening covered with glass or plastic designed to let in light. Skylights may have single, double, or triple layers of glass or plastic for insulating purposes.

Andersen Windows, Inc.

Figure 11-20. Fixed windows in commercial buildings provide security and safety.

Glazing

Glazing is the glass material installed in an opening or the job of setting and installing glass into an opening. Glazing includes sheet (window), float, and plate glass. *Sheet glass* is clear or opaque glass manufactured in continuous, long, flat pieces that are cut to desired sizes and shapes. Sheet glass is used in residential and industrial applications. *Float glass* is high-quality glass used in architectural and specialty applications. *Plate glass* is polished glass manufactured in large sheets. Plate glass is thicker and of higher quality than sheet glass. Plate glass may be bent to various curves when required. Additional varieties of glass include acoustical, cathedral, heat-absorbing tinted, heat strengthened, insulating, obscure, reflective, safety, spandrel, security, tempered, and wire glass. Mastics, caulks, sealants, rubber or plastic seals, and metal or wood stops secure glass into frames.

Plastics. High-strength plastics may be installed as glazing materials where high resistance to shattering and cracking is required and glass materials may not be sufficient to meet the architectural needs. Acrylic plastic and polycarbonate sheets are shatter- and crack-resistant thermoplastics. Plastic glazing surfaces may be susceptible to scratching. Acrylic materials have higher weather resistance than polycarbonates. Plastic glazing may also be bent into an arc or radius. Plastics are available in sizes up to 10′ × 14′ and in a wide range of thicknesses. Common applications for plastic glazing include skylights, domes, protective shields, and bullet-resistant applications.

WINDOW CONSTRUCTION METHODS

Window frames and glazing are commonly set in place after construction of structural members. Opening sizes for various windows are determined from architectural and manufacturer specifications. Openings are slightly oversized to allow for leveling and plumbing of window frames and glazing. Sealants and trim are applied after frames and glazing are set in place. Although most windows can be installed from the inside of the structure, scaffolding may be necessary in some applications.

Small, prefabricated window units are glazed at the mill and shipped to the job site ready for installation. For commercial buildings with large metal frames and glass panels, windows are glazed on site near the completion of construction to avoid breakage.

Frame Construction

Frames may be delivered to the job site with glass installed or may be glazed after set in place. Small window frames are commonly manufactured to be set as a single unit and fastened in place by leveling and securing the frame to the surrounding structural members. Frames for curtain walls and large sheet glass are set in place and leveled to receive glass or plastic. See Figure 11-21.

Andersen Windows, Inc.

Figure 11-21. Curtain wall frames may be set in place and leveled to receive glass or plastic.

> ▲ *Glass is estimated by the square foot with sizes taken from the working drawings. The different types of glazing required for a project must be kept separate because different frames may require different types of glass.*

Suction cups are attached to the face of large glass panels to lift them into place. Sealants, mullions, stops, and finish trim members are applied to secure the glass in the frame. Sealants may be wet, such as oil-based glazing compound, two-part rubber base compound, one-part elastic compound, polybutene, polyvinyl chloride, or butyl materials, or dry such as neoprene or polyvinyl chloride gaskets.

Skylights. Skylights may be manufactured with domes set into frames or they may be made of glass panels set in metal frames. Manufactured skylights are set onto preframed openings in roof systems. Skylights are set onto the roof or mounted slightly above the roof to prevent leakage. See Figure 11-22.

Andersen Windows, Inc.

Figure 11-22. Skylights are set onto the roof or mounted slightly above the roof to prevent leakage.

Curtain Walls. A *curtain wall* is a non-bearing prefabricated panel suspended on or fastened to structural members. Insulated metal panels and glass are set in a common frame and attached to structural members as a single unit. Exterior walls made up of glass panels set in metal frames may also be referred to as curtain walls.

Curtain walls include custom, commercial, and industrial. See Figure 11-23. Custom curtain walls are specifically designed for use on one building project. Commercial curtain walls are made of standardized parts and materials. Industrial curtain walls use ribbed or preformed metal sheets in standard sizes along with standard-size metal sashes.

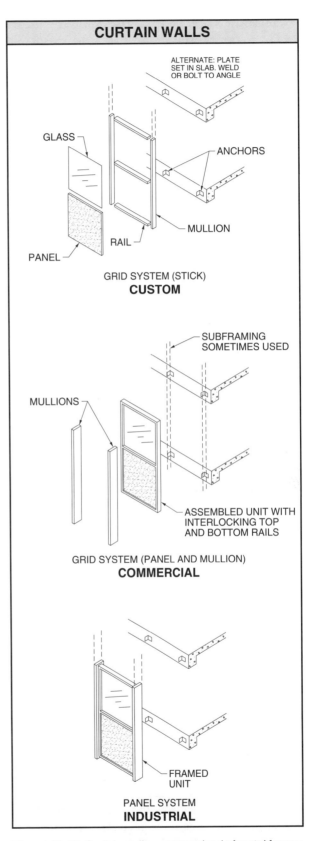

Figure 11-23. Curtain walls are comprised of metal frames holding glass and other panels in place.

Curtain walls take into account the expansion and contraction properties of all the materials in the curtain wall panels. Curtain walls are fastened to structural members with welded clips or inserts. Curtain wall panel locations are shown on elevation views. Curtain wall manufacturers may provide shop drawings that identify each panel and location by code letters and numbers.

A *mullion* is a vertical dividing member between two window units. Mullions several stories in length can be installed to cap the joints between curtain wall panels. Mullions and curtain wall frames are made of aluminum or other metals treated to be rust-resistant. Methods for sealing joints between curtain wall panels are shown on detail drawings.

WINDOW QUANTITY TAKEOFF

Information necessary for window quantity takeoff includes the grade, thickness, and type of glass to be used at each location, the type of each window, and the number and size of the glass panes. Estimation of windows requires knowledge of the various window symbols used on floor plans and elevation drawings.

Estimators should prepare a schedule or work code indicating the location of each window by floor or building area. Manufacturer identification codes may be required for window information quantity takeoff. As with other estimating tasks, plans should be marked to indicate that a window has been included in the takeoff totals.

Manufactured Unit Takeoff

The number of each window and frame is individually counted and totaled on each floor plan. A window schedule may be provided that is related to floor plans and elevation drawings to further describe the totals and types of windows required. See Figure 11-24. Manufacturer identification codes that denote the frame and glass type may be listed in the schedule.

When using a ledger sheet, a chart is established for each area of the structure as determined by the estimator. Columns are established for each type of window. Totals are entered in each row to accumulate a final total. When using a spreadsheet, columns are established for each type of window similar to ledger sheets. The spreadsheet calculates the totals automatically. When using estimating software, an estimator may enter standard windows into the item database for repeated use. See Figure 11-25.

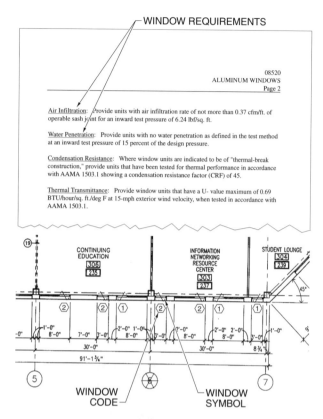

Figure 11-24. Cross-referencing window locations on floor plans with the specifications provides complete window takeoff information.

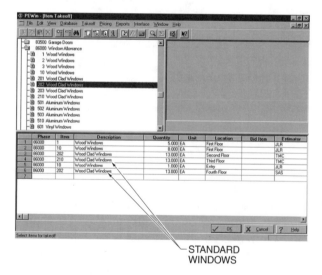

Figure 11-25. An estimator may enter standard windows into the item database for repeated use when using estimating software.

Labor costs are calculated based on a per unit method for manufactured window frame installation. Curtain wall installation labor is determined based on variables such as the height of the installation, glass type and size, and trim members. These labor costs are based on standard industry sources or company historical data. A general contractor commonly subcontracts curtain walls and large fixed glass panel installations.

Sizing. Window size is calculated from exterior elevation drawings, wall section drawings, and floor plans. The window provider normally takes on-site measurements to verify these dimensions. See Figure 11-26.

The size of small windows may be given on elevation drawings by the size of the glass. Sizes are given in inches as the width followed by the height. The two dimensions are separated by a slash line. For example, a window with a light 32″ wide and 28″ high is noted on the elevation drawings as 32/28. This dimension is written in the window on the drawing.

Curtain Wall Takeoff

Quantity takeoff for curtain walls includes the frame and sash materials, anchoring systems, and facing materials including glass, sheet metal, enameled metal, plastic, precast panels, or other architectural panels.

Providing scaffolding or lifts may be a consideration in curtain wall installation depending on job site conditions and the curtain walls being installed. Heavy curtain wall panels may require a crane or other lifting equipment to lift and hold panels in place during installation.

Curtain Wall Frame Takeoff. Estimators total the number of frame components for each section of curtain wall. The frame components include vertical and horizontal mullions, hardware required at joints where frame members meet sash members, and flashing. These members may be available in standard sizes from the manufacturer as specified by the architect or may need to be fabricated for the specific job. The quantity of each component is entered into a ledger sheet, spreadsheet, or estimating software cell.

Andersen Windows, Inc.

Curtain wall installation requirements should be checked to determine if the window is built into the wall or if the windows are slipped into openings after the building has been constructed.

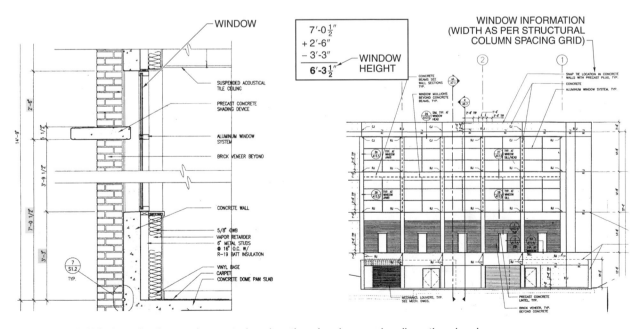

Figure 11-26. Window size is noted on exterior elevation drawings and wall section drawings.

Anchoring calculations include the number and types of anchors and the structural materials to which the curtain wall members are fastened. Anchors are designed to provide for some adjustment vertically and horizontally. Some shimming of anchors may be required depending on the quality of the structural surfaces. Estimators also total the linear feet of trim and sealant members required, such as stops, frames, and gaskets.

Curtain Wall Panel Takeoff. The size and type of each curtain wall panel member is entered into a ledger sheet, spreadsheet, or estimating software cell. Each panel should be identified as to the material (glass or metal), size, and installation location. Manufacturers of curtain wall systems use many standard panel sizes. Some standard glass sizes are noted in standard or officially published lists such as that published by the National Glass Distributors Association. Plate glass and mirrors are estimated by the square foot. The allowance for breakage is usually between 3% and 6%.

ESTIMATING SOFTWARE – CREATING AN ASSEMBLY ON THE FLY

An assembly is created in Precision estimating software by taking off several items and saving these items as an assembly. See Figure 11-27.

PRECISION ESTIMATING SOFTWARE – CREATING AN ASSEMBLY ON THE FLY

1. Open the Item Takeoff window by clicking the Item Takeoff button (   ).

2. Take off items 8110.01 10 Metal Doors, 8110.01 20 Metal Frames, 8710.01 160 Door Closers, 8710.01 170 Weatherstripping, and 8710.01 190 Kickplates.

3. Enter 7 in the Quantity field for Metal Doors and press ENTER. Highlight the quantity field again and while keeping the mouse button pressed, drag the cursor down to the last item taken off (Kickplates). Once these cells have been highlighted, right-click anywhere in the highlighted block of cells and select Fill Down from the shortcut menu.

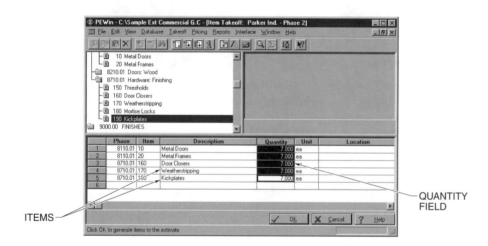

4. Click the Save as Assembly button (   ).

5. Enter 10 as the assembly number in the Assembly field. Each assembly must have a unique code. Enter Metal Door Assembly in the Description field.

6. Click on One-time (   ) to save the assembly only with this estimate. The new assembly is not saved to the database.

7. Click OK to save the assembly.

8. Click Close to return to the Item Takeoff window.

9. Click OK and Close to send this assembly to the spreadsheet.

Note: Click on the Assembly Takeoff button to see the estimate name added to the Assembly list. Double-click on Parker Ind. – Phase 2 to see the new assembly.

Figure 11-27. An assembly can be created by taking off several items and saving these items as an assembly.

Estimating

_____ **1.** Door and window materials and methods are included in Division _____ of the CSI MasterFormat™.

_____ **2.** A(n) _____ is a pane of glass or translucent material in a door or window.

_____ **3.** A(n) _____ is a horizontal member of a door, window frame, or cabinet face frame.

_____ **4.** A(n) _____ is a vertical member of a door, window frame, or cabinet face frame.

_____ **5.** A(n) _____ door is a fire-resistant door and assembly, including the frame and hardware, commonly equipped with an automatic door closer.

_____ **6.** A(n) door _____ is a device that closes a door and controls the speed and closing action of the door.

_____ **7.** A(n) _____ door is a door used to enclose an area that houses concealed equipment.

       A. access   C. barrier
       B. exit   D. entrance

_____ **8.** A(n) _____ door is a door that moves vertically to coil around a steel rod.

_____ **9.** A(n) _____ door is a door formed with panel sections joined with hinges along their vertical edges and supported by rollers in a horizontal upper track.

_____ **10.** A(n) _____ door is a horizontal-moving door suspended on rollers that travel in a track that is fixed at the top of the opening or rollers mounted in the bottom of the door.

_____ **11.** A(n) _____ door is a door which, when opened, is suspended in a horizontal track above the opening.

T   F **12.** A hinge is pivoting hardware that joins two surfaces or objects and allows them to swing around the pivot.

_____ **13.** A(n) _____ is the complete assembly of a bolt, knobs, escutcheon, and all mechanical components for securing a door and providing means for opening.

_____ **14.** A(n) _____ is a protective cover placed around a door knob.

_____ **15.** A(n) _____ is a frame set in a concrete form to create a void in the finished concrete structure.

       A. restricter   C. backplate
       B. blockout   D. neither A, B, nor C

_____ **16.** A(n) _____ is the fixed or movable framework of a window that contains the glass.

_____ **17.** A(n) _____ is a roof opening covered with glass or plastic designed to let in light.

_____ 18. _____ is the glass material installed in an opening or the job of setting and installing glass into an opening.

_____ 19. _____ glass is clear or opaque glass manufactured in continuous, long, flat pieces that are cut to desired sizes and shapes.

_____ 20. _____ glass is high-quality glass used in architectural and specialty applications.

_____ 21. _____ glass is polished glass manufactured in large sheets.

_____ 22. A(n) _____ wall is a non-bearing prefabricated panel suspended on or fastened to structural members.

T   F   **23.** Common swinging door heights are 6'-10″ and 7'-0″.

T   F   **24.** A Class A door has a fire rating of 3 hr.

T   F   **25.** Doors may be prehung or hung on the site.

Door Hand

_____ **1.** Right-hand

_____ **2.** Right-hand reverse

_____ **3.** Left-hand

_____ **4.** Left-hand reverse

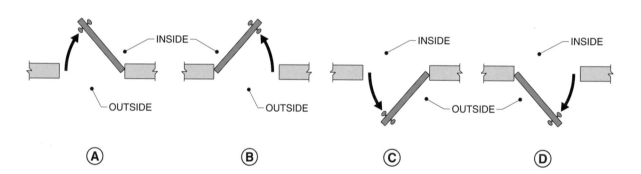

Doors

_____ **1.** Panel

_____ **2.** Flush

_____ **3.** Louvered

_____ **4.** View

_____ **5.** Full view

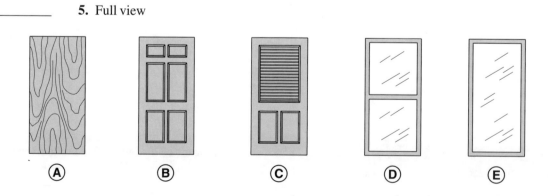

Windows

_____ 1. Awning

_____ 2. Sliding

_____ 3. Bay

_____ 4. Fixed

_____ 5. Bow

_____ 6. Hopper

_____ 7. Casement

_____ 8. Jalousie

_____ 9. Skylight

_____ 10. Single- or double-hung

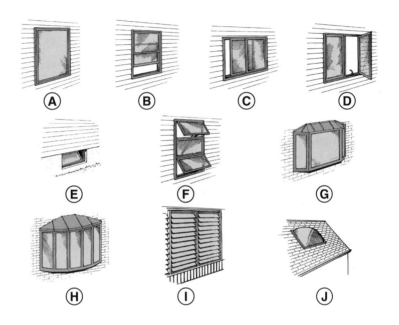

Hinges

_____ 1. Gate

_____ 2. Spring

_____ 3. T

_____ 4. Paumelle

_____ 5. Concealed

_____ 6. Mortised

_____ 7. Strap

_____ 8. Ball bearing

_____ 9. Olive knuckle

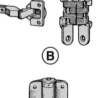

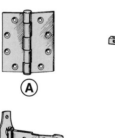

Metal Door Frame Anchors

_____   **1.** Steel channel

_____   **2.** Steel stud

_____   **3.** Loose wire

_____   **4.** Loose T

_____   **5.** Adjustable loop

_____   **6.** Adjustable floor knee

_____   **7.** Wood stud

_____   **8.** Through bolt

_____   **9.** Core

_____   **10.** Wedge clip

_____   **11.** Standard floor knee

_____   **12.** Extended frame with base

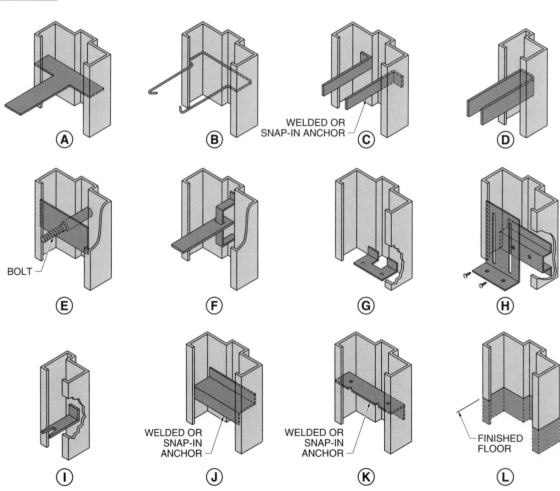

Activity 11-1 – Ledger Sheet Activity

Refer to the Door Schedule and Quantity Sheet No. 11-1. Take off the doors based on the door schedule. List each door by type and the quantity required.

DOOR SCHEDULE			
Number	**Type**	**Size**	**Material**
101	Flush solid core	3'-0" × 7'-0"	Oak
102	Panel	3'-0" × 7'-0"	STL
103	Flush	3'-0" × 7'-0"	STL
104	Flush solid core	3'-0" × 6'-8"	Oak
105	Flush	4'-0" × 7'-0"	STL
106	Flush solid core	3'-0" × 7'-0"	Oak
107	Panel	3'-0" × 6'-8"	STL
108	Panel	3'-0" × 6'-8"	STL
109	Panel	3'-0" × 7'-0"	STL
110	Panel	3'-0" × 7'-0"	STL
111	Flush	4'-0" × 7'-0"	STL
112	Flush solid core	3'-0" × 6'-8"	Oak
113	Flush solid core	3'-0" × 6'-8"	Oak
114	Flush solid core	3'-0" × 7'-0"	Oak
115	Panel	3'-0" × 6'-8"	STL

QUANTITY SHEET

Sheet No. __11-1__

Project: _____
Estimator: _____

Date: _____
Checked: _____

No.	Description	Dimensions				Unit		Unit		Unit		Unit
		W	H									
	Doors											

Activity 11-2 – Spreadsheet Activity

Refer to the cost data and Estimate Summary Sheet No. 11-2 on the CD-ROM. Take off four 3'-0" × 7'-0" commercial steel doors, frames, and hardware. The hardware includes hinges, locksets, pull handles, and push bars. Each door requires 1½ pair of hinge material. Take off twelve 3'-0" × 3'-0" aluminum casement windows. Determine the total material and labor cost for the four doors and 12 windows.

COST DATA			
Material	Unit	Material Unit Cost*	Labor Unit Cost*
Steel doors, 3'-0" × 7'-0"	ea	179.30	24.20
Frames, steel door 3'-0" × 7'-0"	ea	205.70	84.15
Hinges	pair	19.31	—
Locksets	ea	70.95	24.20
Pull handles and push bars	ea	115.50	21.84
Aluminum casement windows, 3'-0" × 3'-0"	ea	247.50	53.90

* in $

Activity 11-3 – Estimating Software Activity

Create a new estimate and name it Activity 11-3. Take off items **8110.01 10 Metal Doors, 8110.01 20 Metal Frames, 8710.01 160 Door Closers**, and **8710.01 170 Weatherstripping** for 50 metal doors and frames, 40 door closers, and weatherstripping for 50 doors. Print a standard estimate report.

Finish Materials

Interior walls and ceilings are finished with a variety of plaster and gypsum surfaces and are calculated by determining the square feet of walls and ceilings, with allowances made for door and window openings. Tile and terrazzo provide a durable and long-lasting finish for interior areas. Ceiling finishes include furred and suspended ceilings. For furred ceilings, furring is attached directly to structural members with ceiling tiles, drywall, or lath and plaster attached to the furring members. For suspended ceilings, wires, tees, cross tees, tiles, and other devices are hung from structural members to create a finish ceiling system. Commercial floor finishes include masonry, hardwood, resilient flooring, carpet, and specialized finishes such as asphalt composition floors. Wall covering materials include paints, primers, stains, sealers, and wallpaper.

PLASTER AND GYPSUM MATERIALS AND METHODS

Interior and exterior walls and ceilings are finished with a variety of plaster and gypsum surfaces. Plaster is comprised of various cementious materials and aggregates that form a plastic material when combined with water. This material hardens into a solid wall and ceiling finish. Plaster may receive different surface finishes. Plaster is held in place with lath. Gypsum is used as a component of the plaster mix or as a lath material. Gypsum is also commonly used as a wall or ceiling finish material when formed into sheets.

Plaster Finishes

Interior and exterior building surfaces may be coated with various plaster finishes. The surface of the plaster is troweled smooth, swirled, or left with a rough finish as specified by the architect in the specifications or shown on elevation drawings. See Figure 12-1. Plaster wall finish materials include portland cement plaster on metal lath or gypsum lath and thin coat

finishes applied directly to gypsum drywall. Control joints and trims are installed to create a complete finish system.

Figure 12-1. Plaster provides a long-wearing surface finish which may be smooth, swirled, or left with a rough finish as specified by the architect.

Lath. *Lath* is the backing fastened to structural members onto which plaster is applied. Lath may be metal or gypsum. Metal lath is tied, screwed, or clipped onto structural supporting members such as metal studs or furring. Metal lath is available in sheets of diamond mesh and rib expanded metal. See Figure 12-2. Metal lath sheets are commonly 27″ wide and 8′-0″ in length. Metal lath sheets are made of painted or galvanized steel and may also be manufactured with a paper backing material for vapor resistance.

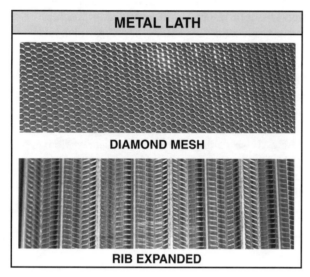

Figure 12-2. Lath is available in sheets of diamond mesh and rib expanded metal.

Gypsum lath is made of an air-entrained gypsum core faced with a paper material specially designed to act as a base for plaster application. Gypsum lath includes plain, perforated, and insulating lath. Plain gypsum lath is ³/₈″ or ¹/₂″ thick, 16″ wide, and 48″ long. Perforated gypsum lath is punched with holes to provide additional plaster adhesion. Insulating gypsum lath is coated with foil on the side applied away from the plaster. Gypsum lath sheets are screwed or clipped to structural members or metal furring channels.

Plaster. *Plaster* is a mixture of portland cement, water, and sand used as an interior and exterior wall finish. Plaster is applied in coats to the final desired thickness. The total thickness varies from ¹/₈″ to 2″ depending on the application. Plaster may be applied in two or three coats. A base (scratch) coat of plaster is applied to the lath and left with a rough finish. A second (brown) coat is applied over the surface of the base coat. A third coat is applied to add strength to the plaster fin-

ish. *Keene's cement* is a hard, dense finish plaster. Keene's cement is used for applications where a very smooth finish is required.

Plaster Accessories. Grounds, corner beads, and control joints are accessories used with lath and plaster to complete the plaster job. See Figure 12-3. A *ground* is a wood or metal shape used as an edge for plaster material and a gauge for plaster thickness. A *corner bead* is a light-gauge, L-shaped galvanized metal device used to cover outside corner joints in plaster walls. Corner beads are available in a wide variety of shapes and sizes depending on their application. A *control joint* is a thin strip of perforated metal applied on lath and plaster surfaces to relieve stress resulting from expansion and contraction in large ceiling and wall surfaces. As with corner beads, control joints are made from light-gauge metal and are available in many styles and shapes depending on the application.

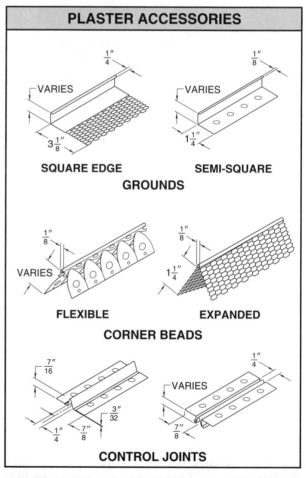

Figure 12-3. Grounds, corner beads, and control joints are accessories used with lath and plaster to complete the plaster job.

Drywall

Due to improvements in manufacturing and transportation systems, gypsum board (drywall) has become the most common material used for interior wall and ceiling finish. Drywall sheets are fastened to framing or structural members with nails or screws. See Figure 12-4. Mastic and a vapor barrier may be incorporated into the drywall installation between the back of the panel and the structural members. The joints between the sheets are sealed with joint compound, paper tape, and various metal trim members.

Hilti, Inc.

Figure 12-4. Drywall sheets are applied to structural supporting members as a finish material for walls and ceilings.

Drywall Sheets. Drywall sheets are made of a core of gypsum coated with layers of paper. Drywall sheets are manufactured in a variety of standard sizes. Sheets are available in 4′ or 5′ widths and lengths of 8′, 9′, 10′, 12′, and 14′. Thicknesses vary in ⅛″ increments beginning at ¼″. Special types of drywall are designed for fireproofing applications. Special paper finishes are applied for waterproof grades. Drywall sheets may have tapered, beveled, square, tongue-and-groove, rounded, or rounded-tapered edges.

Drywall Accessories. Drywall accessories include various fasteners, corner beads, control joints, and finishing materials. Drywall sheets are attached to wood or metal framing members and furring with nails or self-tapping screws. If screws are used for an application, the screws are spaced at a maximum of 12″ OC on ceilings and a maximum of 16″ OC on walls. If wall studs are spaced 24″ OC, the maximum vertical screw spacing is 12″. Drywall sheets may be glued to framing members with mastic. Metal corner beads, control

joints, and end caps may be applied as noted on detail drawings or specifications. Joints between sheets are finished with joint compound and paper tape. See Figure 12-5. Several coats of joint compound are applied to allow for sanding to a smooth finish.

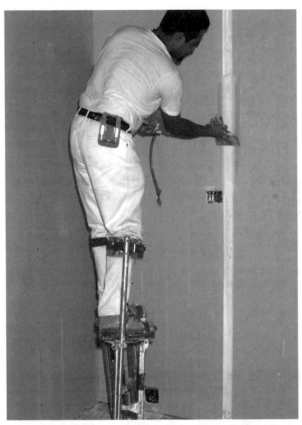

Figure 12-5. Fastener holes and joints between drywall sheets are filled with joint compound and sanded to a smooth finish.

PLASTER AND GYPSUM QUANTITY TAKEOFF

Estimators use the square feet of walls and ceilings to begin calculation for interior wall and ceiling finish material quantities. Wall length and height are used to determine interior wall finish areas, with allowances made for door and window openings. The square feet of flat ceilings for plaster and drywall ceiling finishes may be determined in the same manner as the square feet of flooring. Sloping ceiling areas are calculated in the same manner as roof sheathing for a sloping roof, with the sloped surface providing actual ceiling dimensions. Plaster and drywall quantity takeoff is commonly performed by specialty contractors and is included as a bid submitted to general contractors.

Plaster Finish Takeoff

Components in the quantity takeoff for plaster finish areas include lath, plaster, and accessories such as trim members and control joints. Lath and plaster material quantities are based on the number of square feet to be covered. Control joints are installed at locations shown on the prints or detailed in the specifications. Control joints are normally calculated by the piece or the number of linear feet. Costs are normally based on the cost per square foot from standard industry sources or company historical data. When using a ledger sheet or spreadsheet, the number of square feet required is calculated for each room or area and entered into the appropriate row and column. Each area contains additional ledger sheet columns for lath, plaster, accessories, and labor. When using estimating software, item or assembly takeoff methods are used to calculate lath and plaster materials. See Figure 12-6.

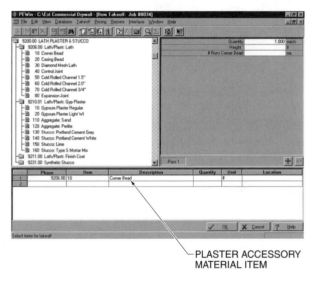

PLASTER ACCESSORY
MATERIAL ITEM

Figure 12-6. Estimating software item or assembly takeoff may be used to calculate lath and plaster materials.

Plaster. Plaster quantities are calculated by the square foot or by the square yard regardless of the number of coats used. Estimators make allowances for wall openings by deducting the number of square feet of openings from the total wall area. Manufacturer reference information is used to transfer the number of square yards of required coverage to the number of tons and sacks of plaster material and the amounts of aggregate required. See Figure 12-7.

Lath and Accessories. The amount of lath for plaster walls is taken from the plaster calculations in square feet or in square yards. The type of lath is indicated on wall section details or provided in the specifications. The sheet size used for wall lath depends on the height of the ceiling and wall lengths. Plaster accessories are calculated by the linear feet of each type of material required such as corner beads and control joints.

A general rule for determining the number of nails necessary for metal lath is to use 9 lb of 3d nails per 1000 sq ft of lath. Nailing requirements vary according to the type and weight of the lath. These requirements may be specified in local building codes or in manufacturer specifications.

Drywall Takeoff

Drywall thickness and face finish are shown on detail drawings and room finish schedules. Takeoff is based on square feet of coverage. Estimators reference the plan drawings to determine the number of layers of gypsum material required and the direction of application. In some areas, two or more layers of drywall may be specified. For common commercial wall applications, drywall sheets may be applied with the long dimension of the sheet placed vertically. Residential applications commonly use horizontal installation. The orientation of the sheets may affect the length of the sheets ordered. The square feet of coverage for each drywall sheet depends on the length of the sheets installed. Drywall sheet lengths are matched to room sizes to avoid an excessive number of joints between sheets. Joint compound materials are also based on square foot calculations. Trim members are taken off by the linear foot or by the piece.

As with lath and plaster finishes, ledger sheet and spreadsheet takeoff include columns for the square foot quantity in each area, the type and number of drywall sheets required, the linear feet of trim members, joint finishing materials, and labor. Estimating software may perform drywall finish material calculations by the item or assembly methods. Drywall finish materials may be included in a typical wall assembly for interior framed walls. See Figure 12-8.

> ▲ *Plaster furring and framing member spacing depends on the strength of the furring and the lath used and should be indicated on the plans. Manufacturers, trade associations, and building codes publish tables showing maximum spacing to meet fire and safety standards.*

GYPSUM BASECOAT PLASTER COVERAGE						
Plaster	Mix	Ratio*	Approximate Compressive Strength Dry**	Approximate Coverage Per Ton of Gypsum Basecoat		
				Gypsum Lath†	Metal Lath†	Unit Masonry†
Red Top Gypsum and Two-Purpose Plasters	sand	2.0	875	180	114	140
	sand	2.5	750	206	131	160
	sand	3.0	650	232	148	181
	perlite	2.0	700	176	112	137
	perlite	3.0	525	224	143	174
	vermiculite	2.0	465	171	109	133
	vermiculite	3.0	290	215	137	168
Structo-Base	sand	2.0	2800 min	154	99	120
	sand	2.5	1900 min	185	118	144
	sand	3.0	1400 min	214	136	167
Structo-Lite	regular	—	700	140	89	109
Red Top Wood Fiber	neat	—	1750	85	54	66
	sand	1.0	1400	135	86	105

* aggregate (vol)/basecoat (wt) in cu ft/100 lb
** in lb/sq in.
† in sq yd/ton

Figure 12-7. Estimators determine the total square yards of coverage for plaster material and use manufacturer information to determine plaster quantities.

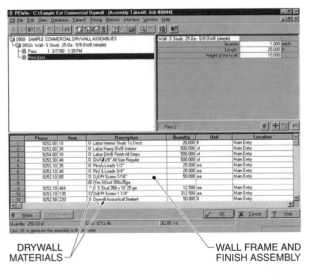

DRYWALL MATERIALS — WALL FRAME AND FINISH ASSEMBLY

Figure 12-8. Drywall is commonly included in wall frame and finish assemblies in estimating software systems.

Drywall Sheets. The number of drywall sheets is determined by calculating the square feet of wall or ceiling surface. The square feet of drywall for flat ceilings is determined in the same manner as for subflooring. The total square feet of floor area is calculated to obtain the square feet of ceiling area. Eight percent is added for waste. Slanted (sloped or cathedral) ceilings are estimated using the same method as for roof sheathing. Rafter length is multiplied by the length of the room to calculate the square feet of drywall required.

The square feet of drywall required for wall finish installation is estimated by totaling the linear feet of wall and multiplying by the wall height. For example, a 24'-6" × 13'-0" room has 75'-0" lf of wall ([24.5' × 2] + [13' × 2] = 75'). Multiplying 75 lf by 8' of wall height equals 600 sq ft (75' × 8' = 600 sq ft). Eight percent is added for waste to obtain 648 sq ft (600 sq ft × 1.08 = 648 sq ft) of drywall needed to cover the walls. Deductions are made for door and window openings.

The drywall for the ceiling is calculated the same as for subfloor sheathing. For example, a 24'-6" × 13'-0" room has 318.5 sq ft of ceiling (24.5' × 13' = 318.5 sq ft). This value is multiplied by a waste factor of 8% to obtain 343.98 sq ft (318.5 × 1.08 = 343.98 sq ft). The total number of square feet of drywall for walls and ceiling is 991.98 sq ft (648 sq ft + 343.98 sq ft = 991.98 sq ft).

Drywall Accessories. Corner bead quantities are calculated by the linear foot. The number of corners requiring beading are determined and the total length of bead is calculated for each corner. For example, to determine the number of feet of corner bead required around a rectangular plastered opening, the floor-to-ceiling height is multiplied by four. Curved surfaces requiring beading are estimated for length of beading by using geometric formulas for linear curves.

For drywall sheets, allow 10 lb of nails or screws per 1000 sq ft of board. The number of fasteners per sheet varies with the sheet size. Fastener requirements may be noted in the local building code or manufacturer specifications. Joint compound and paper tape for seam concealing are estimated based on formulas linked to the number of square feet of wall and ceiling area. Manufacturer information is used to perform these calculations. See Figure 12-9.

Porter-Cable Corp.

Proper sanding of joint compound ensures that the surface of the drywall is free from ridges or other blemishes.

TILE AND TERRAZZO MATERIALS AND METHODS

Tile is a thin building material made of cement, fired clay, glass, plastic, or stone. *Terrazzo* is a mixture of cement and water with colored stone, marble chips, or other decorative aggregate embedded in the surface. Tile and terrazzo provide very durable and long-lasting finishes for interior areas. Tile may be applied to walls and ceilings, and may be used as a ceiling finish material in shower areas. Terrazzo is a floor surfacing material that is placed wet and is finished and ground to a smooth, long-wearing finish. Information concerning these materials is found on floor plans, detail drawings, interior elevation drawings, and specifications.

Tile

Tile may be ceramic or quarry tile. Ceramic and quarry tile are available in many sizes, shapes, thicknesses, colors, and finishes. Ceramic and quarry tile may be applied into a cement base or set with mastic. See Figure 12-10. Where a cement base is specified, a bed of portland cement is set with metal lath reinforcing material. A thin layer of mortar is applied on top of the reinforcing material. The tile is set directly into the thin layer of mortar. For mastic application, a base of gypsum board or other surface finish is coated with adhesive using a grooved trowel. The tile is set into the mastic.

JOINT COMPOUND SPECIFICATIONS		
Product	**Container Size**	**Approximate Coverage***
Sheetrock Ready-Mixed Joint Compound	12 lb pail 42 lb pail 61.7 lb pail 48 lb carton 50 lb carton 61.7 lb carton	138 lb
Sheetrock Lightweight All-Purpose Joint Compound (Plus 3)	1 gal. pail 4.5 gal. pail 4.5 gal. carton 3.5 gal. carton	9.4 gal.
Sheetrock Powder Joint Compound	25 lb bag	83 lb
Sheetrock Lightweight All-Purpose Joint Compound (AP Lite)	20 lb bag	67 lb
Sheetrock Setting-Type Joint Compound (Durabond)	25 lb bag	72 lb
Sheetrock Setting-Type Joint Compound (Easy Sand)	18 lb bag	52 lb

* per 1000 cu ft

Figure 12-9. The quantity of joint compound required for drywall finishing is based on the number of pounds or gallons of material per 1000 sq ft.

Figure 12-10. Ceramic and quarry tile may be applied into a cement base or set with mastic.

Tile installation is normally started in the middle of the work area and progresses out to the perimeter. Cutting is done with special saws or cutters designed to produce clean cuts or breaks. After installation, the spaces between the tile are filled with grout. *Grout* is a thin, fluid mortar made of a mixture of portland cement, fine aggregate, lime, and water. The excess grout is removed while wet and the surface of the tile is polished after the grout has hardened.

Ceramic. Ceramic tile is manufactured from clay powder formed using a process that involves firing the clay while wet or a process in which clay powder is compressed. Various manufacturers produce many different shapes and sizes of ceramic tile. The most common shapes are square, rectangular, octagonal, hexagonal, and Spanish. Ceramic tile has a nominal thickness of $5/16''$. Surface finishes may be bright or matte.

Many ceramic tile trim members are available for installation at corners, caps, intersections, and bases. See Figure 12-11. Many different ceramic tile accessories are available, such as soap, towel, and toilet paper holders. Ceramic tile accessories may be recessed or surface-mounted.

Quarry. Quarry tile is formed of unglazed ceramic tile made of clay or shale. Quarry tile is primarily used as a floor finish material. The most common shapes are square, rectangular, hexagonal, and Spanish.

Bullnose, cove, and windowsill trim members are also available. Nominal thicknesses for quarry tile are $3/8''$, $1/2''$, and $3/4''$.

Adhesives and Grouts. Tile adhesives include portland cement or premixed mastics. For portland cement set applications, materials required include sand, cement, and metal lath reinforcement. These materials are set in several layers to hold the tile in place. Premixed adhesives are available in 1 gal. or 5 gal. quantities. Premixed adhesives are applied directly to the wall, floor, or ceiling surface with a notched trowel as specified by the manufacturer.

Grout is applied in a semiliquid state to fill the joints between tiles. Grout is available in many colors and is purchased in powder form that is mixed with water. Grout is applied to fill all cracks and is cleaned smooth after reaching a final set. The type of adhesive method and grout color are provided in the specifications or on detail drawings.

Terrazzo

Stone chips are mixed with a cement base to form terrazzo. Terrazzo is placed in a manner similar to concrete floor slabs, formed, and screeded to the desired shape. See Figure 12-12. Terrazzo placement systems include setting on a sand cushion, bonding to an underbed, monolithic placement, thin set, and structural. After the material is set, it is ground to a smooth finish, exposing the stone chips and concrete. Stone chips used in terrazzo include marble, onyx, granite, quartz, and silica.

Plans, specifications, and room finish schedules indicate the required tile flooring to be installed.

CERAMIC TILE TRIM MEMBERS

Member	Sizes* Height	Sizes* Length	Member	Sizes* Height	Sizes* Length
BULLNOSE / **SURFACE BULLNOSE**	$1\frac{3}{8}$	$1\frac{3}{8}$	**BEAD**	$\frac{3}{4}$	6
	2	6			
	$4\frac{1}{4}$	$4\frac{1}{4}$			
	$4\frac{1}{4}$	6			
	$4\frac{1}{4}$	$8\frac{1}{2}$			
	6	$4\frac{1}{4}$			
	6	6			
	$8\frac{1}{2}$	$4\frac{1}{4}$			
COVE / **THIN LIP COVE**	$1\frac{3}{8}$	$1\frac{3}{8}$	**COUNTER TRIM**	$2\frac{1}{4}$	6
	2	6		$2\frac{1}{4}$	$4\frac{1}{4}$
	$4\frac{1}{4}$	$4\frac{1}{4}$			
	$4\frac{1}{4}$	6			
	$4\frac{1}{4}$	$8\frac{1}{2}$			
	6	$4\frac{1}{4}$	**CURB TILE**	$2\frac{1}{2}$	6
	6	6		5	6
	$8\frac{1}{2}$	$4\frac{1}{4}$		6	6
BASES	$3\frac{1}{4}$	6	**SURFACE CURB**	$4\frac{1}{4}$	$4\frac{1}{4}$
	$4\frac{1}{4}$	$4\frac{1}{4}$		$5\frac{1}{4}$	$4\frac{1}{4}$
	$4\frac{1}{4}$	6			
	6	6			

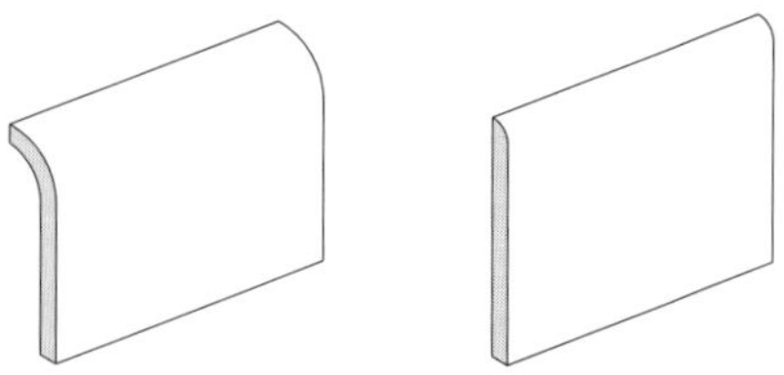

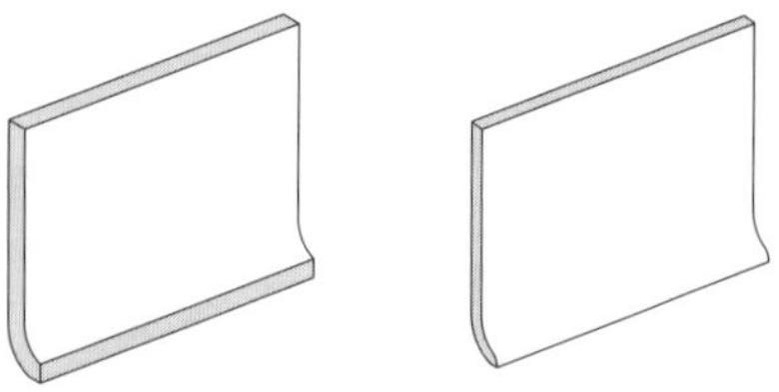

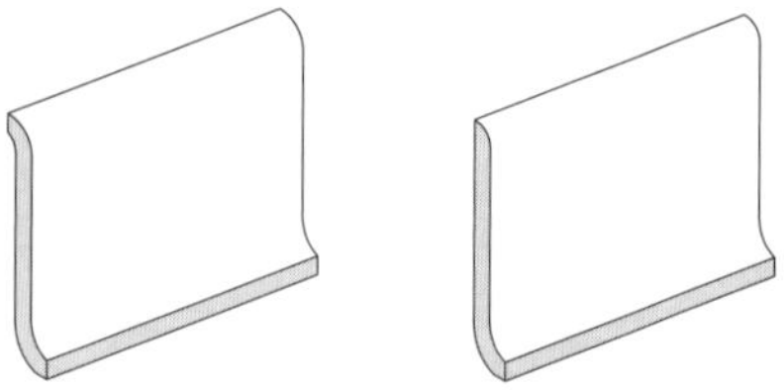

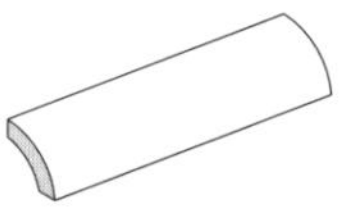

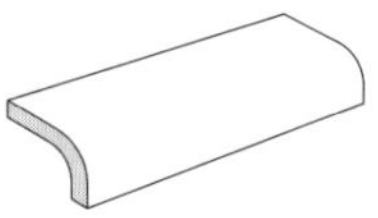

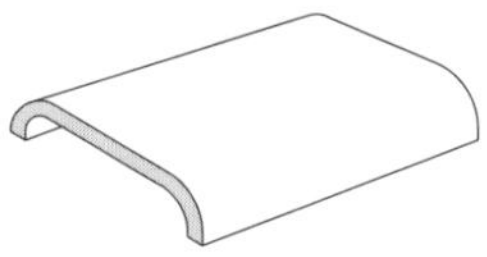

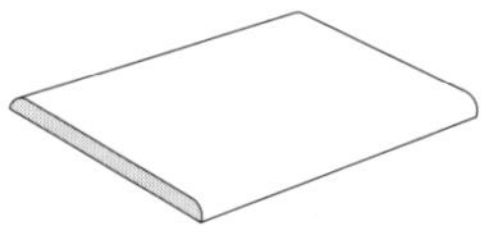

* in in.

Figure 12-11. Ceramic tile trim members are installed at corners, caps, intersections, and bases.

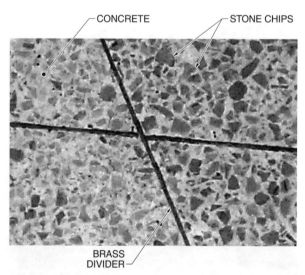

Figure 12-12. Terrazzo is formed by stone chips mixed with a concrete base.

The four basic types of terrazzo include standard, Venetian, palladiana, and rustic. Standard terrazzo is formed with small stone chips. Venetian terrazzo is formed with large stone chips with some small chips integrated to fill voids. Palladiana terrazzo is composed of random pieces of marble, $\frac{3}{8}''$ to $1''$ thick and up to $15''$ across, set as slabs with smaller chips filling openings. Rustic terrazzo is made by pressing the chips into the surface of the concrete base and eliminating the heavy grinding common with other applications. Brass dividers are installed to separate terrazzo floor areas and achieve the desired patterns on the floor surface. Some precast terrazzo members such as stair treads, window sills, cove bases, and other materials may be integrated in the installation.

Binders. Terrazzo binders are the materials used to bond the terrazzo chips together. Portland cement or chemical materials are used in combination with terrazzo chips. Colors are added to the binder material to enhance the finished appearance. White or gray portland cement is used in combination with limestone or synthetic mineral pigments. Chemical binders include epoxy, polyester, polyacrylate, and latex.

TILE AND TERRAZZO QUANTITY TAKEOFF

The square feet of coverage area is used to determine tile and terrazzo material quantities. See Figure 12-13. Estimators calculate the wall, floor, or ceiling areas to be covered and deduct areas of door, window, and floor openings. As with other materials, the ledger sheet and spreadsheet contain rows and columns for various materials and labor. Estimating software uses item or assembly takeoff that includes all components required for construction. Tile and terrazzo work are commonly performed by specialty contractors and are included as a bid submitted to general contractors.

Ceramic and Quarry Tile Takeoff

Ceramic and quarry tiles are calculated in square feet of coverage. Moldings needed to finish edges are taken off in linear feet of the perimeter of the tile area. Trim pieces such as bullnoses, caps, and base strips are also calculated by the linear foot. Estimators must check each tile application to determine the type, size, tile finish, adhesive method, accessories, and grout. Quantities of adhesive and grout are determined according to manufacturer recommendations for the number of square feet of coverage per gallon. See Figure 12-14.

Terrazzo Takeoff

The estimator must determine the required subflooring because terrazzo may be applied in several different ways. The amount of underbed fill must be calculated if the terrazzo is to be bonded to an underbed. The underbed is calculated in cubic yards, as if it were a poured concrete slab. The amount of terrazzo topping is then calculated separately in cubic yards as a poured concrete slab with the variables being the square feet of floor area and the depth of the terrazzo finish.

When terrazzo is separated from the concrete slab, a thin bed of dry sand is laid with tar paper laid on top. The underbed is placed and the terrazzo topping is applied. Cubic yards are calculated for these materials, with the tar paper calculated as the area of the floor. When terrazzo is

applied on wood floors, tar paper and an overlay of wire mesh is installed. Brass dividers and precast accessories are taken off in linear feet. The depth of the brass strips is considered when taking off brass dividers.

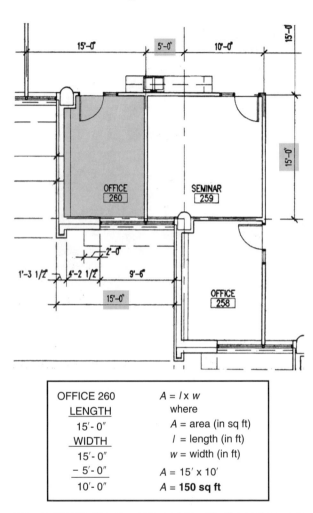

OFFICE 260	$A = l \times w$
LENGTH	where
15'- 0"	A = area (in sq ft)
WIDTH	l = length (in ft)
15'- 0"	w = width (in ft)
– 5'- 0"	$A = 15' \times 10'$
10'- 0"	$A = \textbf{150 sq ft}$

Figure 12-13. Room width and length dimensions are used to determine the number of square feet of tile or terrazzo floor coverage required.

METAL WALL FRAMING MATERIALS AND METHODS

Light-gauge metal wall framing use has been increasing in many commercial applications. Metal runners and studs are installed in a manner similar to wood plates and studs. See Figure 12-15. Layout methods are similar to wood framing, and finish materials are fastened with self-tapping screws. Metal framing systems are resistant to insects, decay, and fire and allow for installation of piping and wiring without excessive drilling.

PRODUCT DESCRIPTION AND USE: Color Tile 990W Ceramic Wall Tile Adhesive is designed for interior installation of ceramic and mosaic wall tile. Use over drywall, concrete, and plaster walls. Three hour open time provides easier installation. Adhesive is moisture resistant and conforms to ANSI Standard A.135.1 Type II. Do not use for setting of fixtures.
PREPARATION: No wall tile installation is better than the surface over which it is installed. Satisfactory results depend on proper preparatory work. For complete detailed instructions, refer to Color Tile Instruction Booklet. (1) Apply over clean, structurally sound surfaces. Surface must be free from all wax, dirt, and grease by cleaning with Color Tile recommended cleaner. Rinse and allow to dry. Remove poorly bonded paint by sanding or stripping. Remove old adhesive with Color Tile Old Adhesive Remover. (2) Fill all cracks with Color Tile 707 Redi-Mix or Speed Patch. Allow to harden 4–6 hours. (3) GYPSUM DRYWALL: WET AREAS AROUND TUBS. Where gypsum wallboard is used, cut 1/4″ around tub areas and fill with Color Tile Caulk or Color Tile 990W Ceramic and Mosaic Wall Tile Adhesive. (4) Apply two coats of Color Tile 880P Latex Floor and Wall Primer over all areas to be tiled. Allow final coat of 880P to dry.
ADHESIVE APPLICATION: Maintain adhesive and tile at room temperature for 24 hours before and after installation. Spread Color Tile 990W Adhesive with a Color Tile 8V notched trowel held at a 60° angle. Do not spread any more adhesive in an area than can be tiled within 3 hours. Spread adhesive with notched edge of trowel. Set tiles by using a slight twisting motion. DO NOT SLIDE TILES. If skin coat has formed on adhesive, scrape off dried adhesive and apply fresh material.
CLEANING: Remove excess adhesive by cleaning with Color Tile Cement Remover. Apply with a clean, dry cloth. Use only Color Tile Cement Remover to avoid damage to tiles. Do not attempt to grout for a period of 16–24 hours or until tiles are firmly set. Drying time is dependent on temperature and humidity.
COVERAGE: Approximately 10 square feet per quart.
CAUTION: KEEP OUT OF THE REACH OF CHILDREN. DO NOT TAKE INTERNALLY. Close cover when not in use.
NOTICE: Since the use of this product is beyond the control of seller, liability or damages, either incidental or consequential, shall be limited to replacement of the product.

COVERAGE INFORMATION

Figure 12-14. Manufacturer information concerning the amount of coverage per gallon of adhesive is consulted in determining the amount of adhesive required.

Interior walls and ceilings between structural members are framed with light-gauge metal. Metal framing members, including runners and studs, are fastened to structural members. Metal furring may be applied to the face of masonry walls to allow for fastening of finish materials. Rough frame openings are created for doors and windows. Plan views, detail drawings, and architectural notes give wall dimensions, opening dimensions, framing, and finish information. See Figure 12-16.

Runners are fastened to structural members at the locations of interior walls.

STUD   RUNNER

Figure 12-15. Metal framing members create lightweight, non-load bearing walls in commercial buildings.

Runners

A *runner* is a U-shaped channel fastened to the floor and ceiling in a metal framing system. The base of the runner is fastened to the supporting structural members. Runners are shaped similar to metal studs and are designed to receive the metal studs that fit into the open face of the runner. Runners are also referred to as track. Runners are fastened to structural members with self-tapping screws or pins from powder-actuated tools. The most common runner widths are 3½″ and 5½″. Runners are secured with fasteners located 2″ from each end and spaced a maximum of 24″ OC.

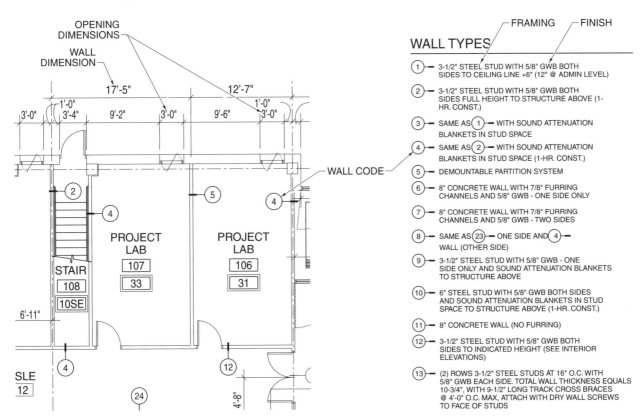

Figure 12-16. Plan views, detail drawings, and architectural notes give wall dimensions, opening dimensions, framing, and finish information for metal frame walls.

Studs

Metal studs are roll-formed from corrosion-resistant steel in several stud widths. Metal studs are manufactured with punched holes to facilitate electrical and plumbing installation. A wide range of designs are available for various applications. See Figure 12-17. Studs are twisted into place between runners at the floor and ceiling and fastened to the runners with screws or a crimping tool.

Furring

Furring is wood or metal strips fastened to a structural surface to provide a base for fastening finish material. Metal furring is manufactured in several Z- or U-shapes having various depths and widths. Metal furring is normally fastened with self-tapping screws or pins from powder-actuated tools. Finish materials are fastened to the furring.

> ▲ *Metal load bearing and non-load bearing studs differ in that load bearing studs are fabricated from heavier materials than non-load bearing studs and may have wider flanges.*

METAL WALL FRAMING QUANTITY TAKEOFF

Metal wall, door, and window locations and sizes are shown on plan views. Detail drawings, elevation drawings, and specifications provide additional information concerning the metal gauge, wall width, furring, and fastening methods. See Figure 12-18. When using a ledger sheet or spreadsheet, component quantities are entered into individual cells. When using estimating software, the required amount of runners, studs, and furring is automatically calculated based on entry of wall length and height dimensions. These items may also be included with plaster or drywall calculations in an assembly. Metal wall framing takeoff is commonly performed by specialty contractors and included as a bid submitted to general contractors.

Runner Takeoff

Runner quantities are calculated in linear feet. The total length of the metal walls is multiplied by two to determine the linear feet of runner for the top and bottom track. Additional linear feet of runner may be required for soffits or additional window, door, or opening framing. Estimators specify the linear feet, gauge, and width for runners.

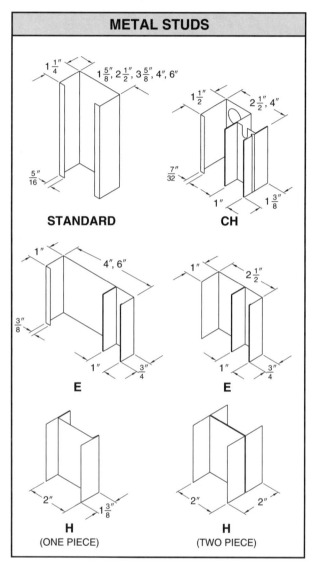

METAL STUDS

STANDARD

CH

E

E

H
(ONE PIECE)

H
(TWO PIECE)

METAL RUNNERS

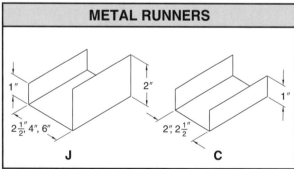

J

C

Figure 12-17. Metal studs are roll-formed from corrosion-resistant steel in several stud widths and are available in a wide range of designs for various applications.

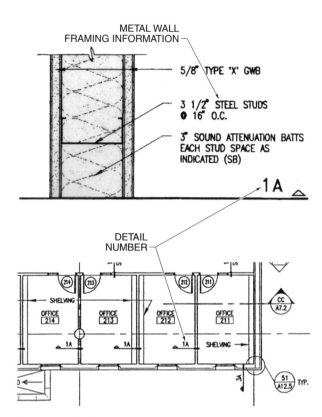

METAL WALL
FRAMING INFORMATION

5/8" TYPE 'X' GWB

3 1/2" STEEL STUDS
@ 16" O.C.

3" SOUND ATTENUATION BATTS
EACH STUD SPACE AS
INDICATED (SB)

1A

DETAIL
NUMBER

Figure 12-18. Plan views, detail drawings, elevation drawings, and specifications are used to obtain complete information concerning metal wall framing and finish.

Stud Takeoff

The method used for estimating the number of wall studs depends on the stud spacing. For 16″ OC studs, the ratio of wall length to studs is .75. For 24″ OC studs, the ratio of wall length to studs is .5. To calculate the number of studs, multiply the wall length by .5. For example, with studs spaced 16″ OC for an 85′-6″ wall, 85.5′ is multiplied by .75 for an estimate of 64.125 studs (85.5′ × .75 = 64.125 studs). Adding 5% for waste equals a total of 67.33 studs (64.125 × 1.05 = 67.33 studs). This is rounded up to 68 studs. Other elements considered when determining the number of studs needed are additional studs required at intersecting walls and studs that may be omitted at large door or window openings. One stud may be added to the calculated amount to end wall framing at a corner. Metal stud framing may require heavy-gauge metal studs placed at either side of a window or door opening. Estimators should check the specifications and detail drawings to determine if additional heavy-gauge studs are required for window or door openings. Estimators specify the number of metal studs, the length required based on the wall height, the metal gauge, and the width.

Furring Takeoff

Metal furring channel may be taken off in the same manner as metal runners or metal studs. For most metal furring, the number of linear feet is calculated based on the OC spacing determined from plan information and the length and width of the area to be furred. For example, a masonry wall 50′ long and 18′ high may require furring channel placed 24″ vertically OC for the fastening of drywall. The total linear feet of furring is calculated by multiplying the wall height (18′) by the number of rows of furring (50′ ÷ 2′ = 25 rows). One additional row is added for the end of the wall. In this example, 450 lf of furring is required (25 × 18′ = 450 lf). An addition is made for waste depending on the type of material used.

CEILING MATERIALS AND METHODS

Ceiling applications include furred and suspended ceilings. For furred ceilings, furring is attached directly to structural members, such as steel bar joists or concrete beams and slabs. The furring may be wood strips or metal members. Prefinished ceiling tiles, drywall, or lath and plaster may be attached to the furring members. For suspended ceilings, a system of wires, tees, cross tees, tiles, and other devices are hung from structural members to create a finish ceiling system.

Suspended Ceilings

A *suspended ceiling* is a ceiling hung on wires from structural members. Suspended ceilings are hung at a predetermined distance below structural members to create a finished ceiling. Hanger wires are fastened to structural members above the finished area. The hanger wires support metal gridwork. Metal gridwork is designed to support lay-in tiles, concealed grid tiles, gypsum drywall, or lath and plaster. The height of the finished ceiling surface above the floor is given on reflected ceiling plan notes, interior elevation drawings, or the room finish schedule in the specifications.

Metal Framing Systems. The metal framing system for suspended ceilings consists of interlocking main runners and cross tees. The height of the metal gridwork is determined after hanger wires are in place. Suspended ceiling main and cross tees are leveled with a rotating laser that projects a level beam of light around the entire space. L-channels are fastened to the walls around the perimeter of the area to support ceiling finish materials. See Figure 12-19. Light fixtures are set

into the gridwork and additional hanger wires are installed as necessary to support the weight of the fixtures. Metal runners may be designed to allow for prefinished metal channels to be clipped onto the underside of the grid. Gridwork may also be used to attach drywall with self-tapping screws.

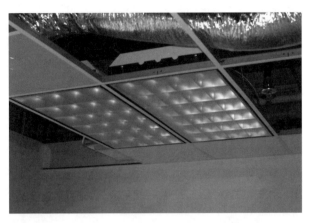

Figure 12-19. Suspended ceiling systems create a finished ceiling by means of metal gridwork and decorative panels.

Panels. Suspended ceiling panels are made of fiberboard and finished with decorative surfaces. The panels are commonly available in 1′ × 1′ or 2′ × 2′ squares or 2′ × 4′ rectangles. Lay-in tiles are set into the gridwork that remains exposed. Concealed grid tiles are supported at each edge by splines that tie the tile together and hide the gridwork. A *spline* is a thin strip of wood inserted into grooves in edges of adjoining members to align and reinforce the joint.

In drywall suspended ceilings, the drywall sheets are applied and finished in a manner similar to other drywall installations. For plaster ceilings, metal, expanded wire, or gypsum lath may be attached to the suspended gridwork. Other finish panels or strips for suspended ceilings include metal pans, wood strips, and other decorative finishes.

CEILING QUANTITY TAKEOFF

Architectural plan views may include reflected ceiling plans. A *reflected ceiling plan* is a plan view with the view point located beneath the object. Each ceiling finish is shown using ceiling symbols. Symbols also show locations for air diffusers, exhaust fans and intakes, and ceiling-mounted lighting fixtures. See Figure 12-20. Additional information concerning air diffusers and intakes is shown on mechanical prints. Ceiling-mounted light fixtures are described in the electrical prints. Fire sprinkler locations may also be shown on reflected ceiling plans.

The quantity of materials required to finish a ceiling is calculated by determining the total area in square feet. Any large openings are deducted in the takeoff. Ledger sheet, spreadsheet, and estimating software are commonly structured to show ceiling quantity takeoff information organized by room or area. Ceiling quantity takeoff is commonly performed by specialty contractors and is included as a bid submitted to general contractors.

Suspended Ceiling Takeoff

Material quantities to be taken off for suspended ceilings include hanger wire, clips or fasteners for the wire to the structural ceiling, main tees, cross tees, L-channel, splines, and ceiling finish materials. See Figure 12-21. Variables determined for suspended ceiling takeoff include the distance from the structural ceiling to the finish ceiling, the height of the finish ceiling above the floor, the grid system and spacing, and the type of finish. Estimators must include suspended ceiling installation requirements such as rolling scaffolds for low ceilings or special scaffolding for high ceilings.

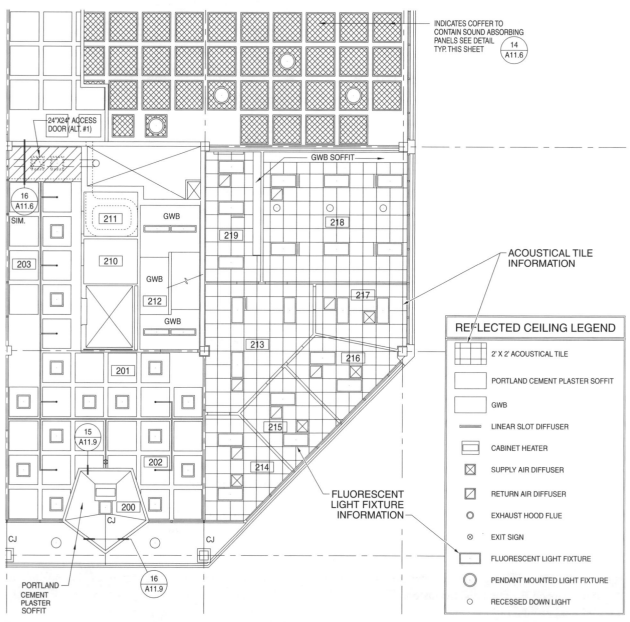

Figure 12-20. Reflected ceiling plans show ceiling finishes using ceiling symbols and a reflected ceiling legend.

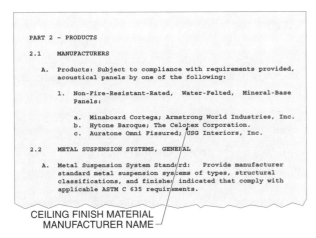

```
PART 2 - PRODUCTS

2.1   MANUFACTURERS

  A.  Products: Subject to compliance with requirements provided,
      acoustical panels by one of the following:

      1.  Non-Fire-Resistant-Rated, Water-Felted, Mineral-Base
          Panels:

          a.  Minaboard Cortega; Armstrong World Industries, Inc.
          b.  Hytone Baroque; The Celotex Corporation.
          c.  Auratone Omni Fissured; USG Interiors, Inc.

2.2   METAL SUSPENSION SYSTEMS, GENERAL

  A.  Metal Suspension System Standard:   Provide manufacturer
      standard metal suspension systems of types, structural
      classifications, and finished indicated that comply with
      applicable ASTM C 635 requirements.
```

CEILING FINISH MATERIAL
MANUFACTURER NAME

Figure 12-21. Specifications may give manufacturer names for specific ceiling finish materials.

Metal Framing Systems. The amount of hanger wire required is determined by calculating the distance between the structural ceiling and the finish ceiling, consulting manufacturer recommendations concerning hanger wire spacing, and calculating the linear feet of wire required. Hanger wire is ordered from suppliers in rolls of various lengths. The linear feet of main tees is calculated from reflected ceiling plans. Main tees are 12′ long. The spacing of the tees provides the number of runs. The length of each run is multiplied by the number of runs to determine the linear feet of main tees required. The number of cross tees required is based on main tee spacing and panel size. Panels 2′ × 2′ require a main tee and cross tee every 2′. Panels 2′ × 4′ require a cross tee every 4′. Wall L-channel length is calculated as the linear feet of the perimeter of the suspended ceiling area.

Panels. Suspended ceiling panels and finishes are taken off by the square feet of area to be covered. Deductions are made in the square feet required for light fixtures, skylights, and other large ceiling openings that eliminate individual panels. Deductions are not made for small ceiling openings cut into panels such as small air diffusers or small light fixtures. Estimators determine the number of square feet of ceiling finish material, the size of each panel (1 sq ft, 4 sq ft, or 8 sq ft) and the finish specified.

> ▲ *Suspended ceiling systems permit access to pipes and electrical fixtures above the finished ceiling, are easier, quicker, and less costly to install than drywall, reduce noise, and increase heating and cooling system efficiency.*

FLOORING MATERIALS AND METHODS

Commercial floor finishes include masonry, hardwood, resilient flooring, carpet, and specialized finishes such as asphalt composition floors, interlocking rubber, and various sports surfaces. Masonry floor surfaces include brick, marble, and stone. Wood floors are made of various types of wood that are cut and finished in different patterns. Resilient flooring is a thin material laid onto a strong subsurface. Carpet provides a soft, sound-absorptive finish.

Masonry

Masonry floor finishes are built of brick or stone. High-traffic areas and exterior areas may use masonry as a finish floor. See Figure 12-22. Brick or stone flooring materials may be set in an open pattern to allow water to drain through, set with mastic, or set in a mortar bed with grout similar to ceramic or quarry tile. Sealant and filler strips may also be applied between masonry flooring materials to allow for expansion and contraction. When using a masonry floor, care should be taken to ensure the material used provides adequate traction.

Figure 12-22. Brick is installed over structural flooring such as concrete to create a long-wearing finish floor.

Brick and Stone. Solid brick for flooring is commonly set on a concrete base. Patterns and joint details are shown on detail drawings and floor plans. Herringbone and basket weave are the two most common brick flooring patterns. Stone provides a long-wearing floor surface. Marble and precast stone are the most common stone floor materials. Stone is also used for stair treads on metal or concrete supporting members.

Wood

Wood flooring is commonly made from hardwood materials. Hardwood flooring materials include oak, maple, beech, and birch. Various grading systems are used with different wood species. Fastener size and spacing also varies among the grading systems and species. The finest grades of wood flooring minimize or exclude defects such as knots, checks, streaks, and open grain. Light-use wood flooring is $3/8''$ to $5/8''$ thick. Normal-use wood flooring is $25/32''$ thick. Heavy-use wood flooring is $1^1/32''$ to $1^{21}/32''$ thick.

Wood flooring may be fastened with nails, screws, or mastic depending on the floor design and manufacturer requirements. Wood flooring may be fastened directly to the structural floor below or placed on furring strips. See Figure 12-23.

Figure 12-23. Wood flooring consists of hardwood pieces fastened in place and finished with varnish or other protective finish.

Prefinished wood flooring is stained and varnished prior to installation. Unfinished wood flooring requires filling, sanding, staining, and varnishing upon completion of installation. Unfinished wood floors require significantly more on-site labor than prefinished wood floors.

Parquet and Block. Wood parquet flooring is made of individual square or hexagonal panels that interlock when set in place. The most common thickness of parquet flooring panels is $5/16''$, with some panels available in thicknesses up to $3/4''$. Square and hexagon dimensions vary by manufacturer, with the most common size being $12''$ × $12''$. These panels are available in many designs and may be unfinished or prefinished. Parquet flooring panels are most commonly set in place with mastic.

Block flooring is made of individual wood blocks set on end with the end grain of the blocks exposed. Blocks vary in size from $3'' \times 6''$ to $4'' \times 8''$ and in thickness from $2''$ to $4''$. Block flooring is fastened to the subfloor material with mastic. Block flooring is commonly used in heavy industrial applications where it is glued onto a concrete slab.

Strip. Wood strip flooring consists of individual interlocking pieces of hardwood fastened to concrete with mastic or nailed, screwed, or stapled to wood decking. Wood strips are available in various widths, wood species, lengths, and finishes. Work begins at one end of the area to be covered and proceeds across in one direction, with each strip interlocking into the previously placed piece.

Resilient Flooring

Resilient flooring includes vinyl sheets, vinyl tile, rubber tile, linoleum, and asphalt tile. Resilient flooring may be placed directly onto a concrete or wood floor. For wood floors, a high-grade plywood underlayment or hardboard is installed to provide a sturdy support for the resilient surface. A moisture barrier may be set in place prior to installation of resilient flooring as required by the architect.

Resilient flooring is held in place with mastic applied with a notched trowel. Sheets or individual pieces are cut, set in place, and pressed into the mastic with weighted rollers. Fillers may be used at the seams of resilient flooring to chemically fill cracks and seams and provide a high-quality finish.

Vinyl Sheets. Vinyl sheets consist of a vinyl wear surface bonded to a backing. The vinyl wear surface may contain PVC resins, fillers, plasticizers, pigments, or decorative vinyl chips. The vinyl resins and plasticizers together are referred to as the binder. Vinyl sheets are commonly manufactured in rolls 6′ wide. Widths, lengths, and patterns vary with the manufacturer. Grades of vinyl sheets vary from thin sheets used for residential applications to thick sheets used in commercial applications. Classifications of vinyl sheets include Type I and Type II. In Type I, the binder constitutes at least 90% of the wear layer. In Type II, the binder constitutes at least 34% of the wear layer. The flooring types are divided into grades based on minimum wear layer thickness. See Figure 12-24. Vinyl sheet backings are classified as Class A (fibrous non-asbestos formulated backing), Class B (non-foamed plastic backing), and Class C (foamed plastic backing). The physical properties considered for vinyl sheet flooring include overall thickness, wear layer thickness, residual indentation, flexibility, and resistance to chemicals, heat, and light.

VINYL SHEET GRADES		
Type	Grade	Minimum Wear Layer Thickness*
I	1	.020
	2	.014
	3	.010
II	1	.050
	2	.030
	3	.020

* in in.

Figure 12-24. Vinyl sheet grades are based on minimum wear layer thickness.

Tile. Resilient vinyl tile is made in 9″, 12″, 18″, or 24″ squares of various thicknesses. Similar to vinyl sheets, many different finishes of resilient vinyl tile are available. Classifications of resilient vinyl tile include Composition 1 (non-asbestos formulated), Composition 2 (containing asbestos), Class 1 (solid color tile), Class 2 (through-pattern tile), and Class 3 (surface-pattern tile). The physical properties considered in product selection include size, thickness, impact deflection ability, and resistance to solvents. Square resilient vinyl tiles are fastened to the floor subsurface with mastic. The primary installation concern is fitting of the seams between pieces and keeping the installation lines straight. Installation begins from the middle of the area to be covered and proceeds outward to the perimeter.

Base Moldings. A *base molding* is the molding at the bottom of a wall at the intersection with the floor. Rubber and vinyl base moldings used to provide a border between walls and floors are classified as Type I Rubber, Type II Vinyl Plastic, Class 1 – Vinyl Chloride, and Class 2 – Vinyl Acetate. The physical properties considered for rubber and vinyl base moldings include thickness, height, length, flexibility, and resistance to heat, light, detergents, and alkalis. Base moldings are applied using mastic after other floor finishes are in place. The mastic is applied to the back of the base molding and the base molding is pressed into place against the surface of the wall.

Carpet

Carpet is a floor covering composed of natural or synthetic fibers woven through a backing material. Variables in carpet characteristics include the traffic load requirements, color, texture, pattern, gauge (pitch), type and height of pile and face yarns, weave (construction style), and installation method. Pitch is based on the number of yarn ends in 27″ of carpet width. A *tuft* is the cut or uncut loop comprising the face of a carpet. If a carpet is tufted, the gauge is the spacing of the needles across the width of the tufting machine in fractions of an inch. The number of tufts per square inch indicates the carpet gauge multiplied by the number of lengthwise tufts per inch.

Pile is the upright yarns that are the wearing surface of a carpet. Pile height is the measurement from the surface of the backing material to the top of the loop of tuft. *Carpet face weight* is the total weight of pile yarns measured in ounces per square yard. A *denier* is the unit of fineness of yarn. The denier is used to indicate the size of a single carpet yarn. *Ply* is the number of single ends of yarn twisted to create a large yarn. *Backing* is the material that supports carpet pile yarns. Carpets may contain a primary and secondary backing. Backing materials include jute, polypropylene, kraftcord, cotton, polyester, or rayon.

Commercial-grade carpets are unrolled and glued to the supporting floor or stretched across padding and fastened around the perimeter of the area. See Figure 12-25. Seams are sewn or connected with heat tape. Specifications contain the schedule for carpet placement in the room finish schedule. Additional carpet information in the specifications includes the manufacturer design, weight, backing, and installation instructions.

Figure 12-25. Commercial-grade carpets are unrolled and glued to the supporting floor or stretched across padding and fastened around the perimeter of the area.

Various designs are created by using individual carpet squares of various colors and patterns.

Carpet sizes vary in width from 4'-6" up to 15'. Lengths vary according to the manufacturer and type of carpet. Foam rubber or sponge rubber may be placed under a carpet as a pad to provide additional cushioning.

Weaves. *Weave* is the pattern or method used to join surface and backing carpet yarns. Carpet weaves (construction) include tufted, axminster, fusion bonded, knitted, velvet, and wilton. Tufted carpet is formed by inserting face yarn through manufactured backing, with yarns held in place by a latex coating on the backing. Tufted carpet is the most common commercial carpet. Axminster carpet has a heavy-ribbed backing that allows almost all yarn to appear at the carpet surface and allows carpet to be rolled lengthwise only. Fusion bonded carpet embeds pile yarns and backing in a vinyl paste, resulting in a very dense pile cut that nearly eliminates backing deterioration. Knitted carpet is manufactured by weaving the face and back simultaneously. Velvet carpet is produced by looping yarn over wires in the looming process. Wilton carpet is similar to velvet carpet, with the difference being that additional types and colors of yarns may be incorporated into the weave.

FLOORING QUANTITY TAKEOFF

Each room and area on the architectural floor plan is commonly numbered or coded. The room code referenced in the specifications or general notes in the finish schedule describes the finish flooring to be installed. See Figure 12-26. For large areas, the architect may note the floor finish on the architectural plans. Specific detail drawings may include a note about the floor finish material at a particular location. The amount of material required to finish a floor is determined by the total area to be covered in square feet or square yards, depending on the material. Any large openings such as stairwells are deducted during flooring quantity takeoff.

The plans, schedules, and specifications for each area are studied to determine the materials of the existing or proposed subsurface. The subsurface material, such as concrete or wood, affects the amount of work applied to the surface. In addition to the flooring material, the estimator must include any base moldings and adhesives.

ROOM FINISH SCHEDULE

ROOM NO.	ROOM NAME	FLOOR	BASE	WALLS NORTH	EAST	SOUTH	WEST	WALLS FNS'H.	HT.	REMARKS
	HOUSING UNIT Ra									
H101	SALLYPORT	CS	–	PA	PA	PA	PA	PA	11'-4"	
H102	SALLYPORT	CS	–	PA	PA	PA	PA	PA	11'-4"	
H103	CONTROL	VCT	V	PA	PA	PA	PA	ACP-1	9'-0"	Paint ceiling black
H104	TOILET	CT	CT	CT	CT	PA	CT	PA	8'-0"	
H105	TOILET	CT	CT	PA	CT	CT	CT	PA	8'-0"	
H106	CONFERENCE	CS	–	PA	PA	PA	PA	ACP-1	8'-0"	
H107	COUNSELOR	VCT	V	PA	PA	PA	PA	ACP-1	8'-0"	
H108	CASE WORKER	VCT	V	PA	PA	PA	PA	ACP-1	8'-0"	
H109	ELECT SECURITY	CS	–	–	–	–	–	–	–	
H110	RECORD STORAGE	CS	–	PA	PA	PA	PA	ACP-1	8'-0"	
H111	UNIT MANAGER	CS	–	PA	PA	PA	PA	ACP-1	8'-0"	
H112	COUNSELOR	VCT	V	MET	PA	PA	SGFT	ACP-1	8'-0"	
H113	STORAGE	CS	–	–	–	–	–	–	–	
A101	HC CELL	CS	–	–	–	–	–	–	–	
A102	CELL	CS	–	PA	PA	PA	PA	EXP	8'-0"	
A102	CELL	CS	–	PA	PA	PA	PA	EXP	8'-0"	

Figure 12-26. Floor finish schedules contain room code information that relates to the specifications which indicate the required floor finish.

When using a ledger sheet and spreadsheet, each area is identified in the first column. Dimensions and areas (in square feet or square yards) are added in subsequent columns, depending on the floor finish material. Material quantities are entered into following columns and rows including finish material, mastics, fasteners, and trim. When using estimating software, item takeoff or assembly takeoff may be used. The number of square feet or square yards is determined from plan views and the proper items entered into the spreadsheet from the database. See Figure 12-27. Flooring material takeoff is commonly performed by specialty contractors and is included as a bid submitted to general contractors.

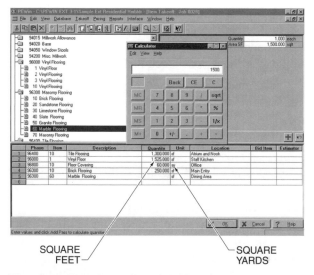

SQUARE FEET   SQUARE YARDS

Figure 12-27. Length and width dimensions are entered into estimating software to determine the number of square feet or square yards of flooring material.

Masonry Takeoff

The area to be covered with masonry flooring is taken off in square feet to be covered. Estimators review plan views, specifications, and detail drawings to calculate the number of square feet, the type of brick or stone, and the fastening method. The subsurface material may affect labor costs, as does accessibility of the area to which the materials must be delivered and placed.

Brick and Stone. The size and style of brick placement determines the number of bricks required. A common 4″ × 8″ brick covers 32 sq in. (.22 sq ft). The total square feet of coverage area is divided by .22 to determine the number of bricks required. For stone, the supplying quarry is consulted to obtain information concerning the coverage per ton for the stone specified. Plans may also indicate the depth of the mortar bed for setting brick or stone and mortar requirements between masonry units. The cubic yards of mortar is calculated based on mortar bed depth and the length and width of the area to be covered.

Wood Takeoff

Wood flooring is taken off by the square foot. Wood flooring of oak, maple, or other decorative woods is shown on architectural plan views. Wood may be prefinished or sanded, stained, and finished after installation. Enlarged floor plans are provided where specific wood finishes and patterns are designed. In applications without special pattern requirements, hardwood flooring information is contained in the room finish schedule.

Parquet and Block. Parquet and block flooring is taken off by the square foot or by the piece for the area to be covered. The area of coverage for each square of parquet or piece of block is divided into the overall square feet of the area to be covered. Mastic quantities are based on the square feet of coverage per gallon as per manufacturer specifications.

The type of subflooring used affects the wood flooring estimate. For parquet or block flooring applied to a concrete slab, a mastic fill and dampproofing may be required. These items are calculated in relation to the area covered.

Strip. The amount of material required for interlocking hardwood plank floors is taken off in square feet of floor to be covered. Dampproofing may be required between the rough floor and the finished floor. Fasteners or mastic quantities are based on manufacturer recommendations per square foot of floor coverage.

Resilient Flooring Takeoff

The quantity of resilient flooring required is taken off by the square foot or by the square yard. Felt backing, if required, is ordered to the same size as the resilient flooring. The amount of adhesive required is calculated by the gallon. One gallon of adhesive is used for each 100 sq ft of resilient flooring or the manufacturer specifications are checked concerning coverage area. Base and metal moldings and threshold materials are taken off by the linear foot.

Vinyl Sheets. Wall-to-wall dimensions are taken to the widest part of the area to be covered. Cutting is required around closets, floor drains, and other obstructions. In some instances, excess pieces of resilient sheet flooring can be used in closets and other small areas. Roll or seamless flooring is taken off by the square yard. The width, roll length, and pattern of the material determine the waste factor. The supplier or manufacturer can provide waste information for the specific material required.

Tile. Resilient floor tiles, whether vinyl, asphalt, cork, or rubber, are taken off by the square foot or by the piece in relation to the square feet of floor area. The area of a single tile is divided into the total square feet of the area to be covered to determine the number of tiles needed. The amount of mastic required is also based on the total square feet of floor area and manufacturer coverage information.

Carpet Takeoff

Carpet is taken off by the square yard, with allowances made for carpet roll width, length, and pattern. For example, a room 24'-6" wide by 13'-0" long has an area of 318.5 sq ft (24.5' × 13' = 318.5 sq ft). Dividing 318.5 sq ft by 9 equals 35.38̄ sq yd (318.5 sq ft ÷ 9 = 35.38̄ sq yd) of carpeting to cover the room. A minimum of 36 sq yd of carpet is ordered to cover the floor. Widths and lengths of carpet vary according to the style and manufacturer, and may affect the final quantity required. In pricing, carpet is priced by the square yard. Carpet tiles are estimated by the same methods used for tile.

Rolls. Estimators obtain information from the carpet supplier concerning roll width and length for various carpet installations. Carpet width and length affect the number of rolls required and the waste factor. For example, in an area 10'-0" × 11'-0", the estimator calculates carpeting 12'-0" wide because standard widths of carpeting include 12'-0" and 15'-0". The estimate includes 10 lf of carpeting 12' wide. The amount of carpeting in the estimate is 120 sq ft (13.3 sq yd) whereas the actual area to be covered requires 110 sq ft (12.22 sq yd). Due to this material consideration, carpet rolls usually have a built-in waste factor.

WALL COVERING MATERIALS AND METHODS

Information concerning application of the proper wall covering materials is commonly provided in Division 9 of specifications written in accordance with the CSI MasterFormat™. See Figure 12-28. Wall covering materials include paints, primers, stains, sealers, and wallpaper. Items considered include coverage properties and rates, surface preparation, application equipment and procedures, and curing requirements.

Paint Properties

Paint is a mixture of minute solid particles (pigment) suspended in a liquid medium (vehicle). The pigment provides hiding power and color. The vehicle combines the volatile solvent (thinner) and the binder, which bonds the pigment particles into a cohesive paint film.

Volatile organic compounds (VOCs) are the portion of a paint mixture that easily evaporates after application. VOCs are associated with the paint solvent. The *solvent* is the volatile portion of the vehicle, such as turpentine and mineral spirits, that evaporates during the drying process. Volatile organic compounds evaporate from various paints and are often limited on government and public work projects.

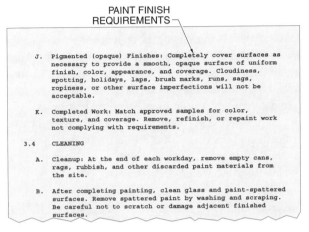

PAINT FINISH
REQUIREMENTS

J.   Pigmented (opaque) Finishes: Completely cover surfaces as necessary to provide a smooth, opaque surface of uniform finish, color, appearance, and coverage. Cloudiness, spotting, holidays, laps, brush marks, runs, sags, ropiness, or other surface imperfections will not be acceptable.

K.   Completed Work: Match approved samples for color, texture, and coverage. Remove, refinish, or repaint work not complying with requirements.

3.4   CLEANING

A.   Cleanup: At the end of each workday, remove empty cans, rags, rubbish, and other discarded paint materials from the site.

B.   After completing painting, clean glass and paint-spattered surfaces. Remove spattered paint by washing and scraping. Be careful not to scratch or damage adjacent finished surfaces.

Figure 12-28. Specifications provide information concerning paint finish materials.

Dry film thickness is the thickness of the resultant mass after the VOCs have evaporated. A paint coating must be applied to the correct dry film thickness to ensure proper performance. Paint dry film thicknesses are measured in mils. One mil is equal to $1/1000$ of an inch (.001″). Paints and other coatings are designed to be applied to a specific dry film thickness to perform in the manner required by the architect.

The proper dry film thickness of a coating depends on the method of curing, type of coating, function, substrate, etc. Some coatings that rely solely on evaporation of VOCs for curing produce relatively thin coatings. Examples include alkyd and waterborne acrylic primers and finishes. For the substrate, surface irregularities are a major consideration. For example, a block filler for use over concrete masonry units must provide a sufficient film thickness to fill the porosities of the substrate. The average dry film thickness (DFT) of block fillers is 10 mils. However, the coating may be 30 mils thick in the porosities and must cure properly at that film thickness. Dry film thickness may be critical to the function of some coatings. For example, primers with at least 75% metallic zinc in the dried film are applied at 2 mils to 3 mils minimum to ensure that sufficient metallic zinc is available to protect the steel substrate.

> ▲ *Paint failures are usually the result of improper surface preparation. For best results, surfaces should be clean, dry, and free from foreign matter.*

Liquid Wall Finish Materials

A wide range of materials and methods are used for application of the various liquid wall finish materials. See Figure 12-29. Liquid wall finish materials include primers, paints, stains, and sealers. Each of these materials requires specific application and curing methods for proper performance. Review of the different liquid wall finish materials is required to ensure proper quantity takeoff.

Figure 12-29. Paint and other liquid finishes are rolled, brushed, or sprayed onto surfaces to be finished.

Primers. A *primer* is a coating applied to new and old surfaces to improve the adhesion and final quality of the finish coating. Primers include latex, alkyd, stain-covering, zinc, and epoxy. A *latex primer* is a water-based primer that dries by the evaporation of water. Latex primers are used over gypsum drywall, plaster, and concrete. An *alkyd primer* is a thermoplastic or synthetic resin-based primer. Alkyd primers are used on new wood and require mineral spirits or paint thinner for cleanup. Stain-covering primers are formulated to cover water stains or knots in bare wood that bleed through ordinary primers and mar the finished paint surface. Stain-covering primers are available in water-based, oil-based, and shellac-based mixtures. Zinc and epoxy primers are used in environments where harsh conditions exist or in industrial applications. Zinc and epoxy primers are also used for exterior steel applications.

Paints. Paints include oil, latex, water-reducible, alkyds, epoxy, urethane-modified alkyds, and aluminum paint. The paint selected depends on the surface characteristics and the desired finish.

Oil paint contains pigments that are commonly suspended in linseed oil, a dryer, and mineral spirit thinner. Linseed oil serves as the binder for the pigment, the dryer controls drying time, and the thinner controls the flowability of the paint. As the thinner evaporates, the mixture of pigments and oil gradually dries to an elastic skin as the oil absorbs oxygen from the air and cures. Oil paints are used for interior and exterior applications.

Latex paint is made of a dispersion of fine particles of synthetic resin and pigment in water. Latex paint is quick-drying and thinned with water. Primer is not required for interior applications except over bare metal or wood or over highly alkaline surfaces. Exterior latex paint can be applied directly to previously painted surfaces. On new wood, latex paint should be applied over a primer.

Exterior building surfaces may be painted to add variety and enhance the overall appearance of a structure.

Water-reducible paints include latex paints and products based on new synthetic polymers. Most latex paints dry by solvent evaporation. Synthetic polymer paints dry by a combination of solvent evaporation and chemical cross-linking. *Chemical cross-linking* is a curing process in which the curing is dependent on molecular action between the coating and a hardener (curing compound) as opposed to evaporation. Chemical cross-linking requires the blending of two materials (two-component coatings) and a digestion time before the coating is applied. The blending of specific materials results in chemical cross-linking and outstanding performance features. These products are also known as water-based or water-borne paints.

Alkyd finishes contain synthetic thermosetting resins as a binding agent. Alkyd finishes are produced in flat, semi-gloss (low-luster), and high-gloss sheens. Flat finishes produce a rich, softly reflective surface. Alkyd flat finishes are often applied to painted walls and ceilings, metal, fully-cured plaster, wallboard, and woodwork without a primer. When required, the primer should be of a similar material. An alkali-resistant primer should be used for high-alkaline surfaces. Semi-gloss (low-luster) alkyd finishes contain some sheen to contrast with flat-finished wall surfaces. High-gloss alkyd enamels are often used for high service and wash applications.

Epoxy paint consists of a two-part formulation that is mixed immediately prior to application. Epoxy paints are very hard and are used for protecting materials such as steel, aluminum, and fiberglass. Epoxy paint dries to a high gloss, creating a smooth tile-like finish.

Urethane-modified alkyds are one-component coatings with high abrasion resistance for use on wood floors, furniture, paneling, and cabinets. Urethane-modified alkyds are used for interior and exterior applications requiring chemical and stain resistance and color and gloss retention. Urethane-modified alkyds are applied to primed steel, aluminum, and masonry that is subjected to high acidity and alkalinity. Urethane-modified alkyds have high resistance to salt, steam, grease, oils, coolants, solvents, and general maintenance machinery fluids. A gloss or semi-gloss finish provides corrosion and abrasion resistance and maintains gloss and color for long periods of time.

Aluminum paint is an all-purpose coating formulated with varnish as the vehicle for aluminum flake pigment. As the paint dries, the aluminum flakes rise to the coating surface, providing a reflective coating that is resistant to weathering. Aluminum paint is used on exterior chain link fencing, interior wood, metal, or masonry. When formulated with an asphalt base, aluminum paint has high adhesion and water resistance when applied to asphalt composition materials.

Stains. A *stain* is a pigmented wood finish used to seal wood while allowing the wood grain to be exposed. Stain may be clear or contain pigments to tint the wood by penetrating into the wood grain. Stain may be oil-based, latex-based, or supplied in a gel form. Stain is commonly sealed with varnish or shellac.

Sealers. A *sealer* is a coating applied to a surface to close off the pores and prevent penetration of liquids to the surface. Common wood sealers include varnish, enamel, and shellac. *Varnish* is a clear or tinted liquid consisting of resin dissolved in alcohol or oil that dries to a clear, protective finish. Varnish dries and hardens through evaporation of the solvent, oxidation of the oil, or both. Varnish is used for interior and exterior applications where a hard, glossy, moisture-resistant finish is required. Varnish is also commonly applied as an interior wood floor finish. *Enamel* is a paint with a large amount of varnish. Enamel has the same qualities as varnish and can be used for interior and exterior applications. A primer is required for high-quality interior work. *Shellac* is a coating made from special resins dissolved in alcohol. Shellac is used as a finish for wood floors, trim, and furniture. Shellac is used as a surface-preparation coat to obtain an even stain tone on porous or soft wood such as pine. Shellac is also used to change the tone of an already shellacked surface by tinting it with alcohol-soluble aniline dye. Instead of restaining some work, pigmented shellac (shellac enamel) is used as a sealant over stained finishes to create a uniform surface finish.

Wall Covering Materials

Wall covering materials include wallpaper, fabric wall covering, acoustical wall covering, and wood veneer wall covering. *Wallpaper* is an interior patterned-paper or vinyl sheet wall covering. Wallpaper is made with a wide variety of surface finishes. See Figure 12-30. Standard American wallpaper products are available in widths of 27″, 36″, and 54″. European wallpaper products are most commonly available in 20″ widths. Wallpaper may be prepasted or unpasted. Prepasted wallpaper is supplied with the paste applied to the back. Vinyl and vinyl-coated wallpaper provide a water-resistant surface. Vinyl and vinyl-coated wallpaper is available in a variety of textures. Foil and mylar

wall coverings have a highly reflective, thin metal coating. Paintable wall coverings have a neutral color and are designed to be painted after application. Grasscloths are textured wall coverings woven from natural fibers. Flocked wall coverings have raised fiber patterns. Embossed wall coverings are wallpapers stamped to create a relief pattern.

Figure 12-30. Wall covering materials such as wallpaper are fitted and secured in place by various adhesives.

WALL COVERING QUANTITY TAKEOFF

Wall covering quantities are calculated in square feet, linear feet, or by the item. Paint coverage is based on the number of square feet covered per quart or gallon. Estimators must calculate the square feet of area covered by a gallon of the proper type of paint, primer, stain, or sealer. For items taken off in linear feet, such as trim members, the number of linear feet of material covered per quart or gallon is determined. Items taken off individually include fixtures such as bollards, door frames, and fire cabinets.

In addition to paint materials, other products included in the paint estimate include putty, fillers, scaffolding, protection of surrounding areas, application equipment such as sprayers and compressors, and personal protective equipment for craftworkers during paint application. Estimators must also determine from the specifications whether one or more coats are required and the approximate coverage for each coat. See Figure 12-31.

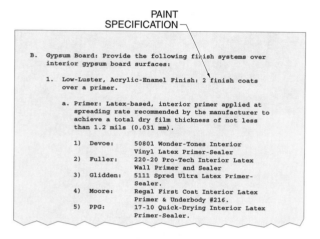

PAINT
SPECIFICATION

B. Gypsum Board: Provide the following finish systems over
   interior gypsum board surfaces:

   1. Low-Luster, Acrylic-Enamel Finish: 2 finish coats
      over a primer.

      a. Primer: Latex-based, interior primer applied at
         spreading rate recommended by the manufacturer to
         achieve a total dry film thickness of not less
         than 1.2 mils (0.031 mm).

         1) Devoe: 50801 Wonder-Tones Interior
                      Vinyl Latex Primer-Sealer
         2) Fuller: 220-20 Pro-Tech Interior Latex
                      Wall Primer and Sealer
         3) Glidden: 5111 Spred Ultra Latex Primer-
                      Sealer.
         4) Moore: Regal First Coat Interior Latex
                      Primer & Underbody #216.
         5) PPG: 17-10 Quick-Drying Interior Latex
                      Primer-Sealer.

Figure 12-31. Paint specifications indicate the number of coats of finish for various materials.

Ledger sheets and spreadsheets are structured by each room, area, wall, floor, or ceiling, depending on the complexity of the paint finish schedule. The number of square feet of coverage for each area and each type of coating are entered into the proper row and column and totaled. Labor costs and waste percentages vary depending on the material, number of coats, application, and location. Door and window areas are commonly included in the square footage for wall calculations for paint labor pricing. Company historical data and standard industry information are used in costing these items. Painting is commonly performed by specialty contractors and is included as a bid submitted to general contractors.

Paint Coverage Takeoff

Solids by volume is an indicator of the square feet of coverage for a gallon of paint at given dry film thickness. A 100% solids by volume coating material without solvents covers 1604 sq ft at 1 mil (.001″) dry film thickness. A 50% solids by volume coating material covers 802 sq ft at 1 mil dry film thickness. Estimators refer to coating manufacturer specifications for each material concerning actual solids by volume and coverage rates. See Figure 12-32. The solids by volume of a coating is approximately proportional to the volatile organic compounds (VOCs), with the exception of water-based coatings.

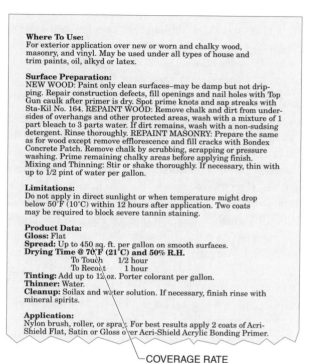

Where To Use:
For exterior application over new or worn and chalky wood, masonry, and vinyl. May be used under all types of house and trim paints, oil, alkyd or latex.

Surface Preparation:
NEW WOOD: Paint only clean surfaces–may be damp but not dripping. Repair construction defects, fill openings and nail holes with Top Gun caulk after primer is dry. Spot prime knots and sap streaks with Sta-Kil No. 164. REPAINT WOOD: Remove chalk and dirt from undersides of overhangs and other protected areas, wash with a mixture of 1 part bleach to 3 parts water. If dirt remains, wash with a non-sudsing detergent. Rinse thoroughly. REPAINT MASONRY: Prepare the same as for wood except remove efflorescence and fill cracks with Bondex Concrete Patch. Remove chalk by scrubbing, scrapping or pressure washing. Prime remaining chalky areas before applying finish. Mixing and Thinning: Stir or shake thoroughly. If necessary, thin with up to 1/2 pint of water per gallon.

Limitations:
Do not apply in direct sunlight or when temperature might drop below 50°F (10°C) within 12 hours after application. Two coats may be required to block severe tannin staining.

Product Data:
Gloss: Flat
Spread: Up to 450 sq. ft. per gallon on smooth surfaces.
Drying Time @ 70°F (21°C) and 50% R.H.
        To Touch   1/2 hour
        To Recoat   1 hour
Tinting: Add up to 12 oz. Porter colorant per gallon.
Thinner: Water.
Cleanup: Soilax and water solution. If necessary, finish rinse with mineral spirits.

Application:
Nylon brush, roller, or spray. For best results apply 2 coats of Acri-Shield Flat, Satin or Gloss over Acri-Shield Acrylic Bonding Primer.

COVERAGE RATE

Figure 12-32. Paint manufacturers provide information concerning paint coverage rates.

The actual coverage rate of any coating is less than the theoretical coverage rate calculated from the solids by volume. This is due to waste during mixing, coating residue left in containers and applicators, inefficiency variations in the methods of application (spray, brush, or roller), environmental conditions such as wind, inconsistent film thickness, and surface porosity and texture. Material spread rates vary depending on the surface being coated. Estimators must include waste percentages to compensate for these losses based on industry standards or company historical data.

Primers, Paints, Stains, and Sealers. The quantity of primer, paint, stain, or sealer is determined by calculating the square feet of area to be covered and dividing by the square feet of coverage per quart or gallon of material. For walls, deduct the areas of windows and doors if they are not to be painted. Labor rates for various applications vary for wall, ceiling, door, window, or other trim work. Labor rates for walls are commonly based on overall wall area without door or window deductions. These are commonly based on company historical data or standard industry information.

Standard latex primer for gypsum drywall typically covers between 270 sq ft and 300 sq ft per gallon. Latex finish coatings cover approximately 300 sq ft per gallon. For interior door frames and window openings, approximately 1 qt of paint is required per coat for 4 openings. The areas of doors are calculated by multiplying door length by width and then multiplying by two due to the two door sides. An additional two or three square feet may be added to compensate for panel bevels and irregular surfaces. Stain and sealer for doors covers about 500 sq ft per gallon. Estimators generally add 1 qt each of stain and sealer for every 5 stained doors. Doors may also be estimated individually.

Approximately 3 qt of paint per coat is required for 20 windows or 20 openings. Ten openings require approximately $1/2$ gal. of paint per coat. Approximately $1/3$ the amount of paint required for wood window frames is taken off for metal window frames. For painting shutters, the area of both sides of the shutters is determined to calculate the square feet of area. When enamel paint is applied to window frames and other openings, 1 qt of primer is allowed for every four openings. The enamel provides about the same coverage as ordinary paint but sometimes it covers slightly less. For painted stairs, the areas of treads, risers, and stringers are calculated by multiplying the length by the width.

Wall Covering Material Takeoff

The amount of wall covering materials required may be taken off as the number of rolls, linear yards, or square yards required to cover an area. The number of rolls is determined by the total wall area (in square feet) divided by the area of coverage for each roll (in square feet). Deductions may be made for large door and window openings. Small openings are cut around and are not deducted from overall wallpaper coverage. Wood veneer materials for wall covering are taken off and bid according to the square feet of area.

> ▲ *Wall covering materials should be opened and inspected after delivery to ensure all rolls are from the same dye lot. Wall covering containing defects or pattern mismatches should be replaced by the manufacturer.*

A common waste factor for wall coverings is approximately 15%. A small repeating wallpaper pattern results in less waste than a large repeating pattern. A pattern with a horizontal match wastes less paper than a pattern with a drop or alternate match.

Wallpaper is commonly available in single, double, and triple rolls. The proper square feet of coverage is determined from the product label. The per roll coverage value is divided into the total square feet of area to be covered. The answer is rounded up to the nearest whole number to determine the required number of rolls. Borders are ordered by the linear yard. Paste is calculated by the pound. The amount of paste needed depends on the weight of the wallpaper.

ESTIMATING SOFTWARE – ASSEMBLY TAKEOFF ITEM SUBSTITUTION

Precision estimating software enables items to be substituted in an assembly. Assembly takeoff item substitution is performed by following a standard procedure. See Figure 12-33.

Grasscloth is made from vegetable fibers that are laminated to a paper backing and add a natural texture to the applied area.

ESTIMATING SOFTWARE – ASSEMBLY TAKEOFF ITEM SUBSTITUTION

1. Take off assembly 0932s Wall – S Studs 25 Ga – $^5/_8$ GWB (simple) by selecting assembly 0932s in the Assembly Takeoff window. *Note:* Any changes made to the assembly for this estimate will not affect the original assembly.

2. In the Takeoff grid, highlight any field in the row containing item 9253.30 40 GWB $^5/_8''$ All Size Regular.

3. Right-click in the highlighted area.

4. Choose Substitute Item from the shortcut menu. The Item List opens at the highlighted item.

5. Double-click item 9253.30 30 GWB $^1/_2''$ All Size Regular to place it in the Takeoff grid and substitute it for the original item 9253.30 40 GWB $^5/_8''$ All Size Regular.

6. Take off the wall by entering 50 for Length and 8 for Height of the Wall.

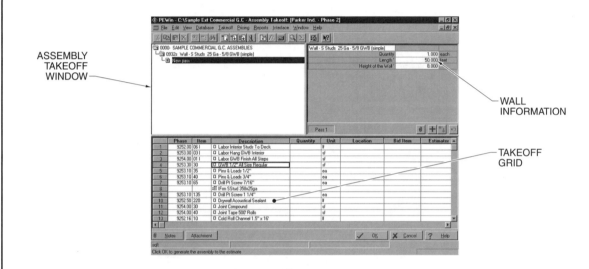

7. Hit ENTER or click Add Pass to generate item quantities to the Takeoff grid. Click OK and Close to send these items and their quantities to the spreadsheet.

Figure 12-33. Items in an assembly can be substituted without affecting the database.

Review Questions
Finish Materials

Name _____ Date _____

Chapter 12

Estimating

_____ **1.** _____ is the backing fastened to structural members onto which plaster is applied.

_____ **2.** _____ is a mixture of portland cement, water, and sand used as an interior and exterior wall finish.

T   F **3.** Tile is a thick building material made of cement, fired clay, glass, plastic, or stone.

_____ **4.** A(n) _____ is a U-shaped channel fastened to the floor and ceiling in a metal framing system.

_____ **5.** A(n) _____ ceiling is a ceiling hung on wires from structural members.

_____ **6.** _____ is a floor covering composed of natural or synthetic fibers woven through a backing material.

_____ **7.** _____ thickness is the thickness of the resultant mass after the VOCs have evaporated.

_____ **8.** A(n) _____ primer is a water-based primer that dries by the evaporation of water.

T   F **9.** Enamel is a clear or tinted liquid consisting of resin dissolved in alcohol or oil that dries to a clear, protective finish.

_____ **10.** A(n) _____ is a wood or metal shape used as an edge for plaster material and a gauge for plaster thickness.

_____ **11.** _____ is a mixture of cement and water with colored stone, marble chips, or other decorative aggregate embedded in the surface.

_____ **12.** _____ is wood or metal strips fastened to a structural surface to provide a base for fastening finish material.

_____ **13.** A(n) _____ molding is the molding at the bottom of a wall at the intersection with the floor.

T   F **14.** Weave is the pattern or method used to join surface and backing carpet yarns.

T   F **15.** A stain is a coating applied to new and old surfaces to improve the adhesion and final quality of the finish coating.

_____ **16.** A(n) _____ is a coating applied to a surface to close off the pores and prevent penetration of liquids to the surface.

_____ **17.** A(n) _____ is a thin strip of perforated metal applied on lath and plaster surfaces to relieve stress resulting from expansion and contraction in large ceiling and wall surfaces.

T   F **18.** Pile is the upright yarns that are the wearing surface of a carpet.

_____ **19.** A(n) _____ primer is a thermoplastic or synthetic resin-based primer.

_____ **20.** Chemical _____ is a curing process in which the curing is dependent on molecular action between the coating and a hardener (curing compound) as opposed to evaporation.

T   F   **21.** Shellac is a coating made from special resins dissolved in alcohol.

_____   **22.** A(n) _____ is a thin strip of wood inserted into grooves in edges of adjoining members to align and reinforce the joint.

T   F   **23.** A denier is the unit of fineness of yarn.

_____   **24.** A(n) _____ is a pigmented wood finish used to seal wood while allowing the wood grain to be exposed.

_____   **25.** _____ is an interior patterned-paper or vinyl sheet wall covering.

Ceramic Tile Trim Members

_____   **1.** Bead

_____   **2.** Bases

_____   **3.** Curb tile

_____   **4.** Bullnose

_____   **5.** Thin lip cove

_____   **6.** Surface curb

_____   **7.** Counter trim

_____   **8.** Cove

_____   **9.** Surface bullnose

Metal Studs

_____   **1.** CH

_____   **2.** H (two piece)

_____   **3.** H (one piece)

_____   **4.** Standard

_____   **5.** E

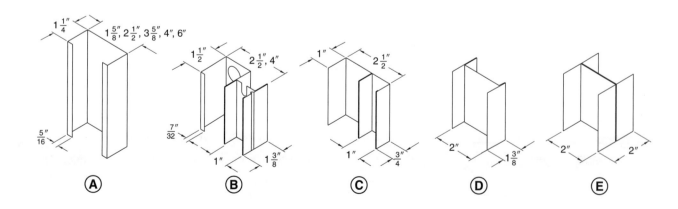

Activities
Finish Materials

Name _____ Date _____ **Chapter**

Activity 12-1 – Ledger Sheet Activity

Refer to Print 12-1 and Quantity Sheet No. 12-1. Take off the square feet of field tile VCT-1 and VCT-2, and border tile VCT-3 for corridor 314.

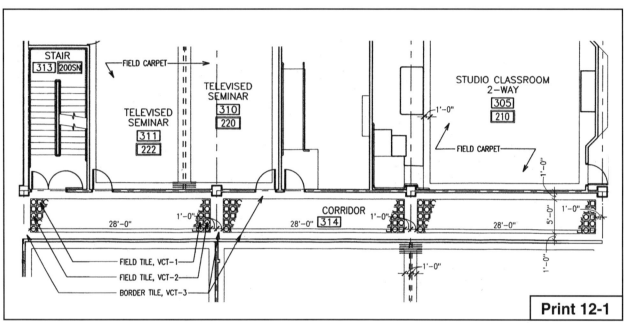

Print 12-1

QUANTITY SHEET

Sheet No. ___12-1___

Project: _____   Date: _____
Estimator: _____   Checked: _____

No.	Description	Dimensions					Unit		Unit		Unit		Unit
		L	W										
	Field tile VCT-1												
	Field tile VCT-2												
	Border tile VCT-3												

Activity 12-2 – Spreadsheet Activity

Refer to Print 12-2 and Quantity Sheet No. 12-2 on the CD-ROM. Take off the number of 2′ × 2′ acoustical tiles and linear feet of wall runner for rooms 131, 132, and 133. All three rooms have a suspended ceiling 8′-0″ above the floor consisting of 2′ × 2′ acoustical tile. Room 132 is finished with 6″ × 6″ ceramic tile from floor to ceiling on all four walls. Take off the number of tiles. Do not make deductions for the door opening. Rooms 131 and 133 are finished with two coats of paint. Take off the gallons of paint for the walls in rooms 131 and 133. Assume 400 sq ft of coverage per gallon.

Activity 12-3 – Estimating Software Activity

Refer to Print 12-3 on the CD-ROM. Create a new estimate and name it Activity 12-3. Take off **Assembly 0932s Wall – S Studs 25 Ga - 5/8 GWB (simple)** for wall type 4 for the Televised Seminar, Control Room, and Studio Classroom on print 12-3. Do not make deductions for door openings or angular walls. The wall height is 8′-0″. Print a standard estimate report.

Mechanical

Mechanical systems include piping, fixtures, and equipment for plumbing, fire protection, heat generation, refrigeration, and air distribution systems. Mechanical system materials include piping, accessories, fittings, valves, and fixtures that make up a complete system. Mechanical prints show piping, connection, valve, and fixture information. Mechanical prints may also contain a schedule detailing the equipment in each occupied space. Heating, ventilating, and air conditioning systems control the temperature and humidity of areas to be conditioned. Mechanical prints show the heating, cooling, and ventilating systems. Plan and elevation views and detail drawings provide installation and fabrication information.

MECHANICAL SYSTEM MATERIALS AND METHODS

Mechanical system specifications in Division 15 of the CSI MasterFormat™ include building services, process, and fire protection piping, plumbing fixtures and equipment, heat-generation, refrigeration, and heating, ventilating, and air conditioning (HVAC) equipment. Also included are HVAC instrumentation and controls and mechanical system testing, adjusting, and balancing. Mechanical and plumbing prints and specifications provide piping, connection, valve, and fixture information. Plumbing prints show piping and fittings with symbols and abbreviations. See Figure 13-1. See Appendix.

Portions of a floor plan may be drawn to a large scale and provided on separate detail or shop drawings when plumbing and mechanical systems are complex and require detailed descriptions. The large scale prints allow for clear viewing of plumbing and mechanical systems in a specific portion of the structure. Mechanical prints may also contain a schedule listing details of the equipment in each occupied space. See Figure 13-2.

General contractors often obtain bids for plumbing, mechanical, and fire protection systems from subcontractors who specialize in the construction of these systems. Estimators for mechanical systems are commonly specialized in their knowledge of piping systems, installation and labor requirements, equipment requirements, available fixtures and fittings, and other technical aspects of these systems.

Mechanical System Materials

Mechanical system materials include all piping, accessories, fittings, valves, and fixtures that make up a complete system. For drainage and waste piping, details of the system, including roof and floor drains, are given with architectural notes and numbers that refer to a schedule contained in the general notes on mechanical plumbing prints or the written specifications. Diameters for roof and floor drains are given on plan views and isometric drawings. For water supply systems and compressed air and gas systems, similar schedules are included along with isometric plumbing drawings showing piping, valves, and fixtures. See Figure 13-3.

PLUMBING/PIPING SYMBOLS

Symbol	Name	Symbol	Name
	ELBOW UP		CIRCULATING PUMP (POINTS IN DIRECTION OF FLOW)
	ELBOW DOWN		
	TEE UP		PRESSURE GAUGE
	TEE DOWN		VALVE (AS INDICATED OR SPECIFIED)
	CONCENTRIC REDUCER/INCREASER		
	ECCENTRIC REDUCER/INCREASER		GLOBE VALVE
	UNION		VALVE IN RISER
	RISE/DROP IN PIPE		ANGLE VALVE
	VENT THRU ROOF		CHECK VALVE
	PIPE SLEEVE		PRESSURE AND TEMPERATURE RELIEF VALVE
	EXPANSION JOINT		
	FLEXIBLE CONNECTION		PRESSURE-REDUCING VALVE (POINTS TOWARD LOW PRESSURE)
	PIPE ANCHOR		
	PIPE GUIDE		GAS VALVE
	CAP		SOLENOID VALVE
	ROOF DRAIN		VALVE W/TAMPER SWITCH
	ROOF DRAIN (ABOVE)		HOSE BIBB/WALL HYDRANT
	CLEAN-OUT (EXPOSED)		HOSE BIBB (ELEVATION)
	CLEAN-OUT (FLUSH TO FLOOR OR GRADE)		DOWNSPOUT NOZZLE
			FIRE DEPARTMENT CONNECTION
	FLOOR DRAIN		CIRCUIT SETTER

PLUMBING/PIPING ABBREVIATIONS

AFF	ABOVE FINISH FLOOR	GTV	GAS TANK VENT
AFG	ABOVE FINISH GRADE	HW	HOT WATER
AH	AIR HANDLING UNIT	HWC	HOT WATER CIRC
AR	ACID RESISTANT	IE	INVERT ELEVATION
ARV	ACID RESISTANT VENT	LAB	LABORATORY
ARVTR	ACID RESISTANT VENT THRU ROOF	MFG	MANUFACTURER
ARW	ACID RESISTANT WASTE	NC	NORMALLY CLOSED
AT	ATTENUATOR	NO	NORMALLY OPEN
BLR	BOILER	OSA	OUTSIDE AIR
CLG	CEILING	PRV	PRESSURE-REDUCING VALVE
CO	CLEAN-OUT	RA	RETURN AIR
COIW	CLEAN-OUT IN WALL	RC	ROOF COWL
COTF	CLEAN-OUT TO FLOOR	RD	ROOF DRAIN
COTG	CLEAN-OUT TO GRADE	SV	SUMP PUMP VENT
CW	COLD WATER	TYP	TYPICAL
DIV	DIVISION	UH	UNIT HEATER
DWG	DRAWING	UON	UNLESS OTHERWISE NOTED
EF	EXHAUST FAN	VTR	VENT THRU ROOF
ET	EXPANSION TANK	WHA	WATER HAMMER ARRESTER
EXH	EXHAUST	W/	WITH

Figure 13-1. Plumbing prints show piping and fittings with symbols and abbreviations.

Division 15 of the CSI MasterFormat™ includes pipe, fittings, valves, and pumping equipment for liquids and dry products that are part of an industrial process.

PLUMBING FIXTURE SCHEDULE

ITEM	FIXTURE	MANUFACTURER	MODEL #	MOUNTING	TYPE	MATERIAL
P1	WATER CLOSET	AMERICAN STANDARD	2174.138	FLOOR	SIPHON-ACTION	VITREOUS CHINA
P2	"HANDICAPPED" WATER CLOSET	AMERICAN STANDARD	2218.143	FLOOR	SIPHON-ACTION	VITREOUS CHINA
P3	LAVATORY	AMERICAN STANDARD	0476.028	COUNTER	SELF-PRIMING	VITREOUS CHINA
P4	"HANDICAPPED" LAVATORY	AMERICAN STANDARD	0476.029	COUNTER	SELF-PRIMING	VITREOUS CHINA
P5	SHOWER	LASCO	1866-C	WALL/FLOOR	MODULE	FIBERGLASS
P6	"HANDICAPPED" SHOWER	LASCO	3636-BPS	WALL/FLOOR	MODULE	FIBERGLASS
P7	SINK	ELKAY	LRAD 1722	COUNTER	SELF-PRIMING	STAINLESS STEEL
P8	KITCHEN SINK	DAYTON	GE-23321	COUNTER	SELF-PRIMING	STAINLESS STEEL
P9	LAUNDRY SINK	MUSTEE	MODEL NO. 10	COUNTER	SELF-PRIMING	FIBERGLASS
P10	SERVICE SINK	STERN-WILLIAMS	HL-1300	FLOOR	SQUARE	TERRAZO
P11	"HANDICAPPED" LAVATORY	AMERICAN STANDARD	0373.080	WALL	BACKSPLASH	VITREOUS CHINA
P12	BATH TUB	LASCO	2008-80	FLOOR	–	FIBERGLASS
P13	WASHING MACHINE SUPPLY BOX	GUY GRAY	B-200	WALL	CORNER	STEEL WITH ENAMEL FINISH

NOTES:
① OFFSET TRAP BACK AGAINST WALL ② MOUNT PER ADA & WAC REQUIREMENTS ③ PROVI

⑤ PROVIDE GRAB BARS & FOLD UP SEATS ⑥ PROVIDE T-35 HOSE & WALL HOOK, T-40 STAINLESS STEEL

Figure 13-2. Schedules on mechanical prints indicate specific equipment for each building space.

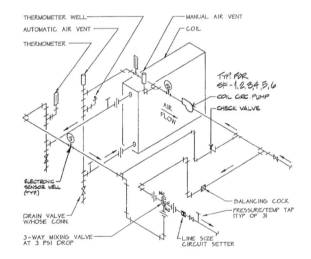

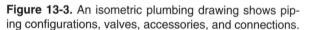

TYPICAL SF COIL PIPING DETAIL ①

Figure 13-3. An isometric plumbing drawing shows piping configurations, valves, accessories, and connections.

Piping Materials. Pipe for mechanical systems is made from copper, brass, aluminum, steel, stainless steel, cast iron, ductile iron, and polyvinyl chloride (PVC). Each pipe material is available in various diameters and grades for different applications. Estimators consult the specifications to ensure the proper pipe and grade are used in the estimate.

Copper pipe is available in Type K, Type L, Type M, Type G, Type DWV, Type ACR, and medical gas.

Type K copper pipe is color-coded green and is used for underground water supply, fire protection, and HVAC applications. Type L copper pipe is color-coded blue and is used for interior water supply, fire protection, liquefied petroleum (LP) gas, and HVAC applications. Type M copper pipe is color-coded red and is used in domestic water, service and distribution, fire protection, solar, fuel oil, HVAC, and snow melting applications. Type G copper pipe is color-coded yellow and is used for natural gas and LP gas applications. Type DWV copper pipe is color-coded yellow and is used for drain, waste, and vent applications. Type ACR copper pipe is color-coded blue and is used for air conditioning and refrigeration applications. Medical gas copper pipe is color-coded green or blue and is used for medical gas applications. Nominal diameters vary from ¼″ to 6″ with a variety of wall thicknesses based on type and diameter.

Brass pipe is available in standard and extra-strong weights. Nominal diameters for brass pipe vary from ¹/₈″ to 6″ with a variety of wall thicknesses based on grade and diameter. Aluminum pipe is a lightweight pipe used for water piping. Nominal diameters vary from ¼″ to 3″. As a comparison, ½″ nominal brass pipe weighs .934 lb/ft and ½″ nominal aluminum pipe weighs .294 lb/ft.

Steel pipe may be seamless or welded. Steel pipe is available in a variety of ASTM specifications. For example, Type A53 is a seamless or welded steel pipe, black or hot dip galvanized, used for general service. Type A106 is a seamless carbon steel pipe for high-temperature service. Seamless steel pipe is available in nominal diameters from ¹/₈″ to 48″. Welded steel pipe is available in nominal diameters from ¹/₈″ to 24″ and is used for standard water, gas, air, or steam applications. Weights for seamless and welded steel pipe include standard, extra-strong (XS), and double extra-strong (XXS). Stainless steel pipe is installed where corrosion resistance and freedom from contamination are required. Stainless steel pipe is available in nominal diameters from ¹/₈″ to 12″.

Cast iron pipe is produced by pouring molten iron into a mold that defines the final shape. Centrifugal force may be used to ensure high-quality pipe with greater consistency. Iron is identified by color or physical properties. The most common iron from which cast iron soil pipe and fittings are produced is gray iron. Different grades of cast iron pipe are available based on diameter and design at each end of the pipe. Cast iron pipe is available in single hub, double hub, and hubless designs. See Figure 13-4.

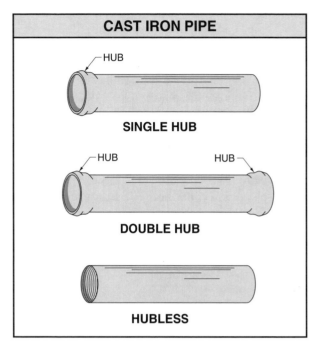

Figure 13-4. Cast iron pipe is available in single hub, double hub, and hubless designs.

Ductile iron is a form of cast iron that has magnesium added to the molten gray iron mixture. This reforms the graphite flakes in gray iron into spherical particles. Standard wall thickness of ductile iron pipe is based on the pressure class of the pipe. Pressure classes include 150 psi, 200 psi, 250 psi, 300 psi, and 350 psi. Ductile iron pipe is available in nominal diameters from 4″ to 64″.

PVC pipe is available in cementing only or cementing or threading grades. Wall thickness is greater for cementing or threading grades. PVC pipe is available in nominal diameters from ¼″ to 6″. PVC pipe is a general application pipe used for a variety of waste and water supply applications. As with all piping, local building codes should be consulted to determine proper applications.

Waste and Vent Piping. Waste pipes collect waste water from sinks, lavatories, toilets, and roof, surface, and floor drains and carry the water to sanitary sewers or treatment facilities. Materials used for waste piping include cast iron and PVC pipe. Plumbing prints indicate the size, purpose, type, and elevation of each pipe. This information is shown on plan views and isometric drawings. Detail drawings related to the plan views provide reference points for the isometric drawings. See Figure 13-5. Pipe sizes are given in inches, which indicate the inside diameter of the pipe. The outside diameter of pipe varies depending on the material used for the pipe and the schedule of the pipe. The material used and the schedule affects the pipe wall thickness and outside diameter. The purpose of the pipe and materials carried by the pipe are given with abbreviations and symbols. For example, a broken line with an abbreviation of RDL on a plan view indicates a roof drain line. Cross-referencing of the plan views with the plumbing abbreviations and symbols describes the use for each pipe. Specifications contain information concerning the pipe used for various applications. General notes on the plumbing portion of the mechanical prints may provide information about the material used for waste pipes.

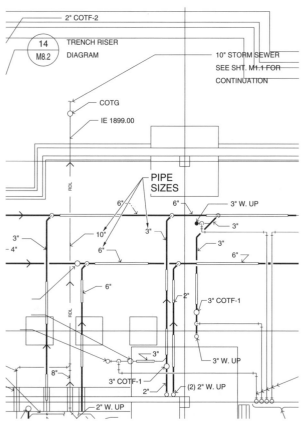

Figure 13-5. Estimators review detail drawings to ensure all piping information is determined for each location.

A *vent pipe* is a pipe that removes odors and gases from the waste piping and exhausts them away from inhabited areas. Vent pipes branch off from the liquid-carrying waste pipes and commonly terminate above the roof line. Vent pipes are shown on plan views and isometric drawings in a manner similar to waste pipes. See Figure 13-6. Plumbing prints also indicate locations for placement of cleanout accesses in the waste piping.

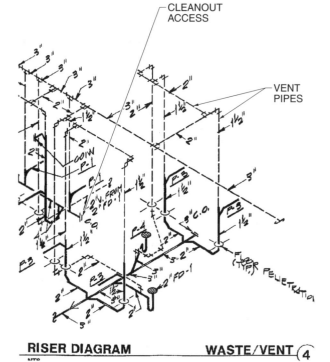

Figure 13-6. Vent pipes are shown on isometric drawings in a manner similar to waste pipes.

Supply Piping. *Supply piping* is piping that delivers water from the source to the point of use. The two most common materials for supply piping are copper and PVC. As with waste piping, many different fittings are used to join the pipe. The material and diameter of the supply piping is indicated on plumbing and mechanical prints and in the specifications.

Process Piping. *Process piping* is piping installed in industrial facilities for compressed air, vacuum, gas, or fuel. Pipes for process piping are described in the same manner as waste and water supply piping. Special abbreviations and symbols are used on plumbing plan views to describe process piping. Detail drawings are given where process piping systems are connected to tanks for fuel supply or vacuums. Plan views and isometric drawings of the piping and valve configurations are also provided.

Fittings, Accessories, and Valves. Various fittings, accessories, and valves are available to create plumbing piping systems according to mechanical prints. A *pipe fitting* is a device fastened to ends of pipes to terminate or connect individual pipes. Pipe fittings include elbows, tees, couplings, nipples, reducers, wyes, crosses, caps, plugs, unions, and traps. See Figure 13-7.

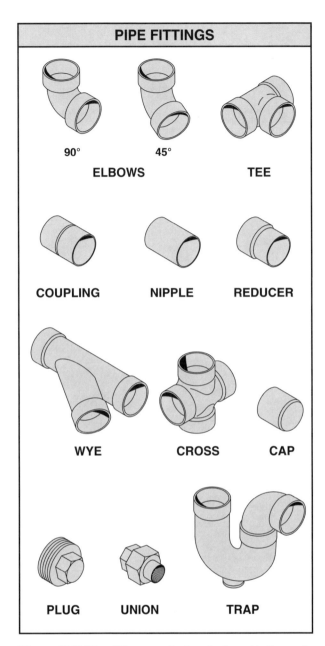

PIPE FITTINGS

90°   45°
ELBOWS   **TEE**

COUPLING   **NIPPLE**   **REDUCER**

WYE   **CROSS**   **CAP**

PLUG   **UNION**   **TRAP**

Figure 13-7. Pipe fittings are devices fastened to the ends of pipes to terminate or connect individual pipes.

Waste and vent pipes and fittings are connected by various methods depending on the pipe used. Cast iron waste pipes and fittings are connected with molten lead or fastened with couplings. PVC waste pipes are fitted and glued together with a variety of fittings. Steel pipes may be threaded or welded. Waste pipes are attached to structural members with hangers and fasteners to support them at the proper elevations. Details of these connections are not included in the plumbing portion of the mechanical prints. A general note concerning

connections to structural members may be included in the mechanical notes or the specifications.

Cast iron soil pipes and fittings are installed according to Cast Iron Soil Pipe Institute (CISPI) and American Society for Testing and Materials (ASTM) standards. Three standards for cast iron soil pipe referenced in model plumbing codes are CISPI 310-97, ASTM C1277-97, and ASTM C1173-97. CISPI 310-97 covers couplings used to connect hubless cast iron soil pipe and fittings for sanitary and storm drain, waste, and vent (DWV) piping applications. These couplings are the most commonly used couplings with cast iron pipe. ASTM C1277-97 covers shielded couplings joining hubless cast iron soil pipe and fittings. ASTM C1173-97 covers flexible transition couplings for underground piping systems.

A *valve* is a device that controls the pressure, direction, or rate of fluid flow. Different valves are used depending on the flow regulation required and the material being regulated. Valves used in mechanical systems include gate, butterfly, globe, pressure-reducing, check, and pressure-relief valves. See Figure 13-8. A *gate valve* is a valve that has an internal gate that slides over the opening through which fluid flows. A *butterfly valve* is a valve that controls fluid flow by a square, rectangular, or round disc mounted on a shaft that seats against a resilient housing. A 90° rotation of the shaft moves the valve from fully open to fully closed. Gate valves and butterfly valves are used as stop valves for basic ON and OFF service. A *globe valve* is a valve that controls the flow of liquid by means of a circular disc. Globe valves are used where throttling of flow is required. A *pressure-reducing valve* is a valve that limits the maximum pressure at its outlet, regardless of the inlet pressure. Globe valves and pressure-reducing valves are used to reduce the pressure of liquid, air, or steam in a piping system. A *check valve* is a valve that allows flow in only one direction. Check valves are designed to prevent the reversal of flow. A *pressure-relief valve* is a valve that sets a maximum operating pressure level for a circuit to protect the circuit from overpressure. Pressure-relief valves open automatically when internal pipe pressure exceeds a preset limit.

Fixtures. Plumbing and mechanical system fixtures include drinking fountains, sinks, toilets, water heaters, and tubs. Cast iron tubs are made from gray iron. Other fixtures are commonly made from porcelain or porcelain-coated iron. Suppliers offer a wide variety of styles, grades, and models of plumbing fixtures.

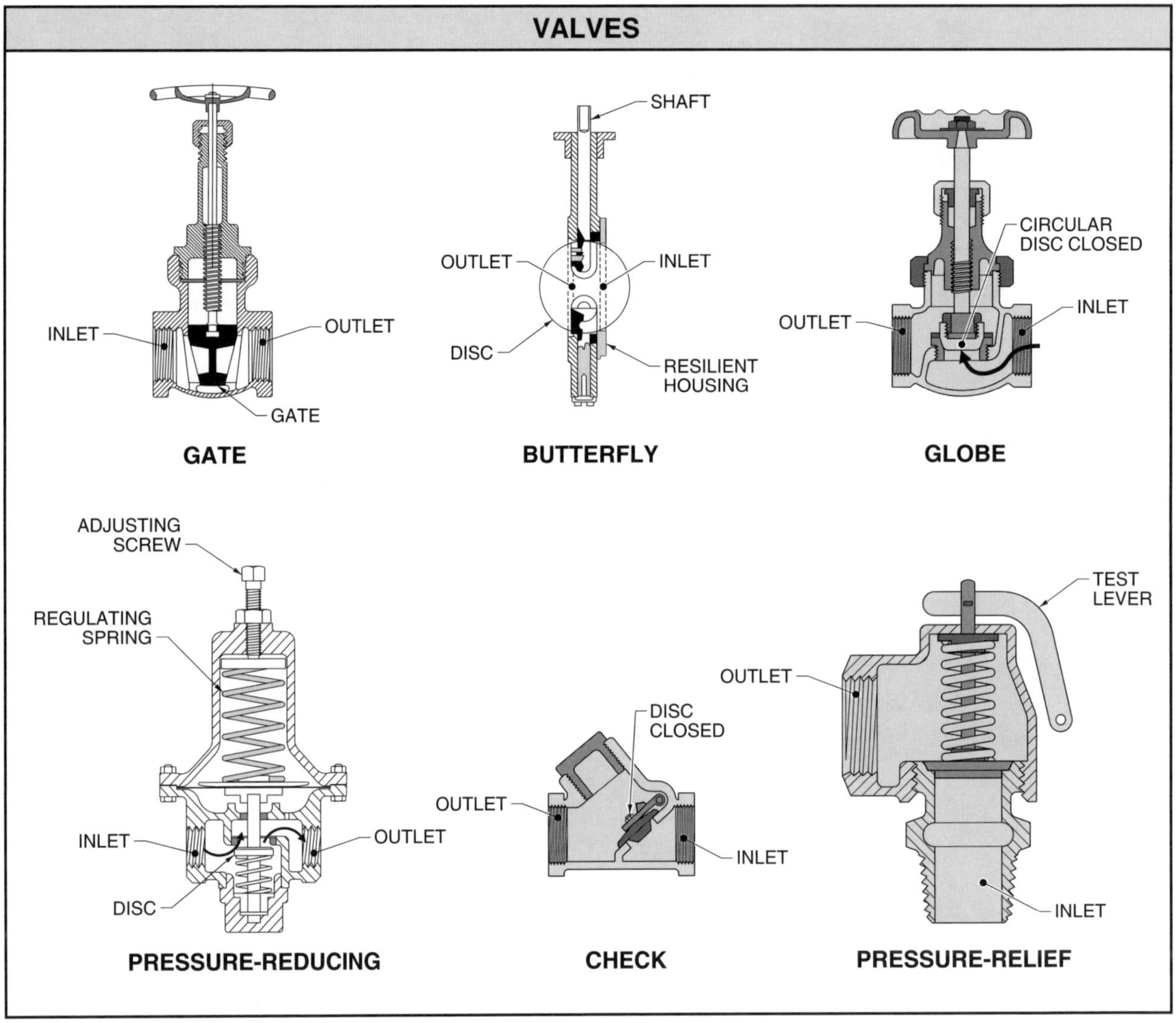

VALVES

Figure 13-8. Valves used in mechanical systems include gate, butterfly, globe, pressure-reducing, check, and pressure-relief valves.

A plumbing fixture schedule is included in the mechanical prints. Notes on plan views and isometric plumbing drawings relate to the plumbing fixture schedule. See Figure 13-9. Schedule information may include the manufacturer name and product number, sizes, fittings, methods for attachment to the structure, sizes of vent, waste, and supply piping, and general notes. Detail drawings for connections to fixtures or equipment, such as hot water tanks, are given where necessary.

Mechanical System Installation Methods

Piping for mechanical systems is installed after structural members are built to support piping systems. Integration of the piping system into the structure must continue during structural construction. For poured-in-place concrete floors, walls, and roofs, piping may be set in place prior to placement of the concrete, or chases may be provided for piping to pass through the floor, wall, or roof. In cases where this is not done prior to or during concrete placement, a core drill may be used to drill through the concrete and create a hole for pipe placement. Fixtures are installed during structural and finish construction.

Most devices used in plumbing systems, such as pipe, fittings, and valves, are identified by markings indicating the type and size of the device. The metal used for each device can normally be determined by markings on the device.

Features:

- 20″ x 17″ Counter-mounted Lavatory
- 4″ Faucet centers
- Front overflow
- Self rimming
- Supplied with template and
  color matched sealant

Nominal Dimensions:

- 20 3/8″ x 17″

Bowl Sizes:

- Height .... 5 5/8′
- Width .... 16′
- Depth .... 10′

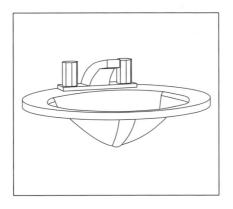
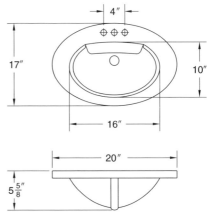

Symbol	**Fixture**	**Mfr**	**Model No.**	**Mounting**	**Type**	**Material**	**Size**	**Drain**	**Trap**	**W**	**V**	**HW**	**CW**
P1	Water closet	American Standard "Afwall"	2477.016	Wall	Siphon jet	White vitreous china	Elongated bowl	—	—	4″	2″	—	1″
P2	Urinal	American Standard "Jetbrook"	6570.022	Wall	Blowout	White vitreous china	21″ × 14½″ × 15⅛″	—	—	2″	2″	—	1″
P3	Lavatory	American Standard "Aqualyn"	0476.028	Counter	Self-rimming	White vitreous china	Oval 20″ × 17″	American Standard #7723.018	1½ × 17 GA	1½″	1½″	⅜″	⅜″

PLUMBING FIXTURE SCHEDULE

Figure 13-9. Estimators rely on manufacturer information for plumbing fixture options. A plumbing fixture schedule identifies the fixtures installed at each location noted on floor plans.

Pipe Installation. Pipes are installed prior to wall, ceiling, and floor finishes. Water supply pipes are commonly stubbed and capped as necessary where plumbing fixtures are to be attached. Copper pipes are connected by solder and fittings. PVC piping and fittings are glued. Cast iron piping is threaded and joined with fittings and sealants to prevent leaks. Piping for high-pressure applications may be joined by welding.

The ends of waste and vent pipes may be stubbed and capped where fixtures are connected after finishes are applied. Piping systems may be subjected to a pressure test in which a certain amount of pressure is built up inside the piping system for a certain period of time. Leaks are detected and fixed as required.

Care should be taken to consult local building codes and specifications concerning support requirements for various piping. For example, for multiple joints within a 4′ length, hangers are located at alternate couplings or hubs. Vertical stacks require support at each floor.

Fixture Installation. Plumbing and mechanical fixtures such as pumps, storage tanks, and filtration equipment are installed as part of the piping system installation. Manufacturer specifications concerning types of support and lubrication required are followed during installation. Finish fixtures such as toilets, drinking fountains, and lavatories are commonly installed after floor, ceiling, and wall finishes have been applied. See Figure 13-10.

MECHANICAL SYSTEM QUANTITY TAKEOFF

Mechanical system estimators list all pipes, fittings, and fixtures. After the list is created, prices are obtained from a supplier or the prices are entered by the estimator from a catalog and/or price list. A standard overhead expense is then added to material costs.

Unit estimating may be used on small jobs. Unit estimating is estimating of material and labor prices in a single step based on unit items. Unit prices are based on company historical data and previous experience. Unit plumbing pricing should include the cost of the pipes, fixtures, fittings, faucets, and labor for an average installation. Unit prices are commonly based on an average installation and average price fixtures. Additional costs are added for complicated installations or when high-quality fixtures are specified.

PLUMBING FIXTURES

Figure 13-10. Plumbers install fixtures after floor, ceiling, and wall finishes have been applied.

Pipes are installed before completion of wall, ceiling, and floor finishes, and are stubbed and capped for future fixture connection.

Labor and Material Takeoff

Labor costs are calculated based on company historical records or standard industry information. Variables include the local labor market supply and the installation conditions. Additional labor costs may be required based on trenching for pipes, hanging pipes at a high

distance, or other unusual installation requirements. Material takeoff requires that all piping, accessories, and fixtures are included. Complete and accurate plan markup and calculation is essential. The type, material, diameter, size, and grade of each pipe and fixture must be determined for the final bid.

Labor. Estimating plumbing labor can be performed by calculating labor costs based on a percentage of material costs plus detailed calculations for trenching and other items with little material costs, or by estimating labor based on a percentage cost of an average job. Labor costs may be based on the number of pipe joints plus a certain percentage of fixture values and detailed calculations for trenching.

As a general rule, the installation costs for a basic plumbing fixture may be taken as 50% of the total fixture cost. For high-quality fixtures and materials, a rate of approximately 40% is used for labor. For hot water heaters and pumps, labor is calculated at approximately 30% of material cost.

For example, a basic toilet and tank with fittings and hardware may cost $200. A labor cost of $100 (50% × $200 = $100) is added to this for a total finish installation cost of $300. Where high-quality fittings or expensive gold or high-quality metals are used, material costs for the same job may be $350. In this instance, a labor cost of $140 (40% × $350 = $140) is added for a total finish installation cost of $490.

Waste Piping. The specifications and mechanical and plumbing drawings are reviewed to determine the type, diameter, and grade of pipe required at each waste pipe location. Elevations for waste pipes are given to the inside elevation at the bottom of the flow line of the pipe. Elevations are specific to a location and provide reference points for pipe placement used in calculating pipe length. Elevations for waste piping between these elevation reference points are indicated as a percentage of slope. A slope of 1% is equal to a change in elevation of 1' over a distance of 100'. For example, a 6% slope on a 100' run of pipe is calculated by multiplying the length of the run by 6%. This equals a slope of 6' (100' × .06 = 6'). The amount of pipe needed is equal to the hypotenuse of a right triangle with one leg equal to 100' and the other leg equal to 6'. The amount of pipe needed is equal to 100.18' ($\sqrt{(100')^2 + (6')^2}$ = 100.18'). See Figure 13-11. When using a spreadsheet or estimating software, these formulas may be preset in the system to allow for entry of length and slope to automatically determine the pipe length required.

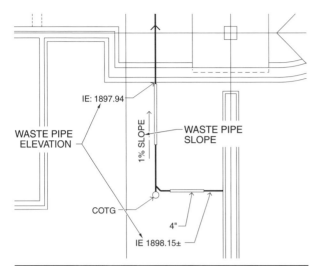

IE: 1897.94

WASTE PIPE
ELEVATION

1% SLOPE

WASTE PIPE
SLOPE

COTG

4"

IE 1898.15±

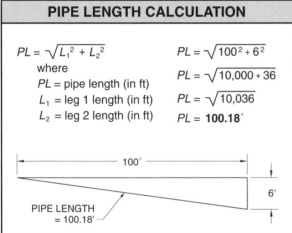

PIPE LENGTH CALCULATION

$$PL = \sqrt{L_1^2 + L_2^2}$$

where
PL = pipe length (in ft)
L_1 = leg 1 length (in ft)
L_2 = leg 2 length (in ft)

$$PL = \sqrt{100^2 + 6^2}$$
$$PL = \sqrt{10,000 + 36}$$
$$PL = \sqrt{10,036}$$
$$PL = \mathbf{100.18'}$$

100'

6'

PIPE LENGTH
= 100.18'

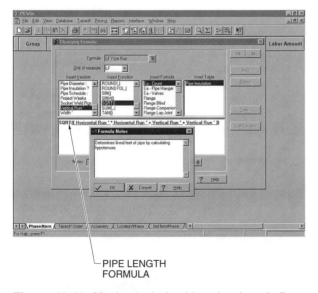

PIPE LENGTH
FORMULA

Figure 13-11. Mechanical plumbing drawings indicate waste pipe elevations and slopes. Pipe length is calculated by determining the hypotenuse of a right triangle or by formulas in estimating software.

> *Piping that is not properly installed and supported can sag at the joints, causing it to leak. Improperly installed and supported drainage piping can sag or shift from its proper slope, causing portions of the drain line to form traps, which block the pipe.*

On plan views, vertical waste pipes are indicated with open circles and an architectural note. Exact dimensions for the placement of waste pipes are not given on the mechanical prints. Gridlines and some major structural members are shown on the mechanical prints. Exact locations are determined by obtaining dimensions from architectural and structural prints. For example, where a waste pipe for a sink is to be placed through an interior wall, the location of the wall is determined from dimensions on the architectural prints.

Supply Piping. Mechanical plumbing prints indicate the sizes and locations of pipes for carrying hot and cold water. Pipe diameters are given as the inside diameter of the supply piping. Exact pipe locations are not dimensioned. Dimensions for supply piping locations are obtained from architectural prints. Pipe information is given on plan views and isometric drawings in a manner similar to waste piping diagrams. Information about the type and grade of pipe used is provided in the specifications. A special set of drawings may be included for fire suppression piping indicating the location of piping and the number of sprinkler heads and valves required. The number of feet of length of each type of pipe is entered into a ledger sheet, spreadsheet, or estimating software. See Figure 13-12.

Pipe hangers are part of the piping estimate and are adjustable to allow the proper slope of the pipe.

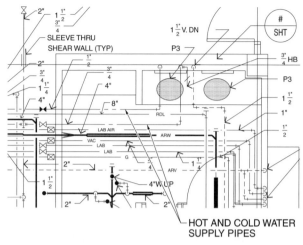

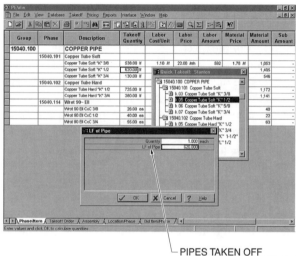

PIPES TAKEN OFF
BASED ON LINEAR FEET

HOT AND COLD WATER
SUPPLY PIPES

Figure 13-12. Hot and cold water pipes shown on mechanical plumbing prints is estimated based on linear feet.

Fittings, Accessories, and Valves. Types and locations of fittings, accessories, and valves are shown in the specifications and on plan views and isometric drawings. Standard symbols indicate the types of fittings and valves installed. Estimators may need to determine some of the types of pipe fittings and accessories required based on the length of pipe, number of joints, number of turns, pipe support system, and number of changes of elevation of the piping. Each type of fitting and accessory is entered in the ledger sheet, spreadsheet, or estimating software total. Valves are counted as individual items based on the detail drawings and information in the specifications. See Figure 13-13.

Fixtures. Each plumbing fixture is counted and entered in the total as an individual item. A plumbing fixture schedule is included in the mechanical prints. Notes

on plan views and isometric drawings relate to the plumbing fixture schedule. Schedule information may include manufacturer name and product number, sizes, fittings, methods for attachment to the structure, vent sizes, waste and supply piping, and general notes.

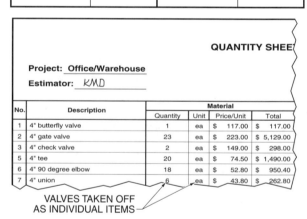

Figure 13-13. Standard symbols are used to indicate valves in piping systems which are taken off as individual items.

▲ *An estimator may be required to approximate fittings when preparing an estimate. Fittings normally account for between 25% (for simple runs) to 50% (for complex runs) of the cost of the piping.*

HVAC SYSTEM MATERIALS AND METHODS

HVAC systems control the temperature and humidity of the area to be conditioned. HVAC systems include components based on the system, fuel, and air circulation requirements. HVAC systems include hydronic and forced-air systems. Estimators determine the system used and interpret specifications and drawings concerning the materials and methods for each system.

A *hydronic system* is a system that uses water, steam, or other liquid to condition building spaces. A *forced-air system* is a system that uses warm or cool air to condition building spaces. Hydronic and forced-air systems require various automatic controls to create an efficient system. The controls regulate fuel flow and consumption, temperature levels, and humidity, and provide safety in case of system emergencies. HVAC system estimators include all controls, piping, and wiring requirements in the takeoff and estimate.

General contractors commonly obtain subcontractor bids for HVAC work. HVAC system estimators are specialized in their knowledge of heating and cooling equipment, piping, ductwork, filtration systems, installation, labor, equipment requirements, available fixtures and fittings, and other technical aspects of HVAC systems.

Hydronic Systems

Hydronic systems are used where heating or cooling equipment is located far from building spaces that require conditioning. Hydronic systems consist of a boiler, chiller, circulating pump, piping system, terminal units, and controls. See Figure 13-14. A hydronic heating system uses a boiler to heat water or create steam. A *boiler* is a closed metal container (vessel) in which water is heated to produce heated water or steam. Natural gas or electric heating elements are used to heat water or create steam in a boiler. A hydronic cooling system uses a chiller to cool water. A *chiller* is the component in a hydronic air conditioning system that cools water, which cools the air. Water or steam is then pumped through a piping system to the areas to be conditioned. A *terminal unit* is a device that transfers heat or coolness from the water or steam in a piping system to the air in building spaces. A circulating pump circulates the conditioned water through the piping system.

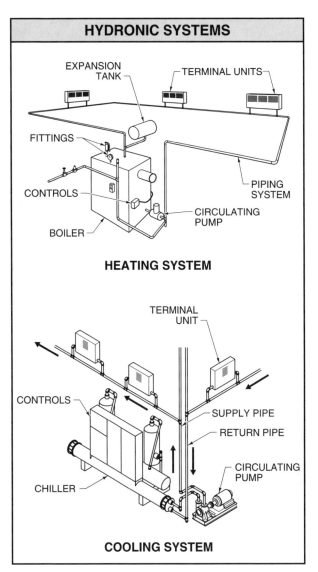

Figure 13-14. Hydronic systems consist of a boiler, chiller, circulating pump, piping system, terminal units, and controls.

The entire process is controlled by thermostats and other control devices. A *thermostat* is a temperature-actuated electric switch that controls heating and/or cooling equipment. Thermostats at each terminal unit indicate the need for heat or cooling and activate valves or fans to bring the temperature to the proper level. Water is pumped back to the chiller or boiler. Each portion of the hydronic system is shown on plan views and elevation drawings.

Boilers. Boilers may be low pressure or high pressure boilers. A *low pressure boiler* is a boiler that has a maximum allowable working pressure (MAWP) of up to 15 pounds per square inch (psi). Low pressure boilers are commonly used in warehouses, factories, schools, and residential applications. A *high pressure boiler* is a boiler that has an MAWP above 15 psi and over 6 BHP. A *boiler horsepower (BHP)* is the evaporation of 34.5 lb of water per hour at a feedwater temperature of 212°F. High pressure boilers are commonly used for industrial operations and the generation of electricity.

Boilers burn fuel or use electric resistance to heat water. In combustion boilers, the fuel is burned in a combustion chamber. In electric boilers, copper rods are electrically charged to create heat. The rods are encased in a waterproof tube that heats the surrounding water.

An *expansion tank* is a tank that allows the water in a hydronic heating system to expand without raising the water pressure to dangerous levels. Expansion tanks are installed in hydronic heating systems. The size of the expansion tank is determined by the total heating capacity of the system.

Cleaver-Brooks

Boilers and their related equipment provide hot water and/or steam for heat and industrial processes.

Chillers. A chiller uses mechanical compression refrigeration or absorption refrigeration to cool water. *Mechanical compression refrigeration* is a refrigeration process that produces a refrigeration effect with mechanical equipment. *Absorption refrigeration* is a refrigeration process that uses the absorption of one chemical by another chemical and heat transfer to produce a refrigeration effect. Chillers are used for air conditioning systems that produce large amounts of cooling. Water is piped through a chiller, pumped to building areas to be cooled, passed through terminal units through which air is circulated to distribute the cooling, and returned to the chiller to be cooled again.

Piping Systems. Hydronic system piping systems include all pipes, valves, fittings, and flues. Piping systems distribute the water from the boiler or chiller, through the circulating pump, to the terminal units, and back to the boiler or chiller. In a hydronic system, the piping system is full of water at all times. Piping systems may be one-, two-, three-, and four-pipe systems. In a one-pipe system, water passes through each terminal unit in a continuous flow back to the boiler or chiller. A one-pipe system is an economical system to install and gives good temperature control when used in small buildings. In a two-pipe system, separate piping systems are installed for supply and return. Two-pipe systems are used in medium- to large-size residential and commercial buildings.

In a three-pipe system, two supply pipes and one return pipe are used. Three-pipe systems are used when different parts of a system require heating and cooling simultaneously. In a four-pipe system, separate piping systems are installed for supply and return for heating and cooling. Four-pipe systems are expensive to install but provide excellent control of air temperature. Pipes are identified as heating hot water supply, heating hot water return, chilled water supply, and chilled water return. Chilled water supply and return pipes are used for cooling operations. Piping, valves, and fittings used are similar to those used in plumbing and process piping systems. See Figure 13-15.

A *flue* is a heat-resistant passage in a chimney that conveys smoke or other gases of combustion. Flues may be made of masonry materials, stainless steel, or sheet metal. Design engineers determine the stack diameter based on the boiler output. Piping is normally installed at the bottom of the flue or chimney to remove condensation.

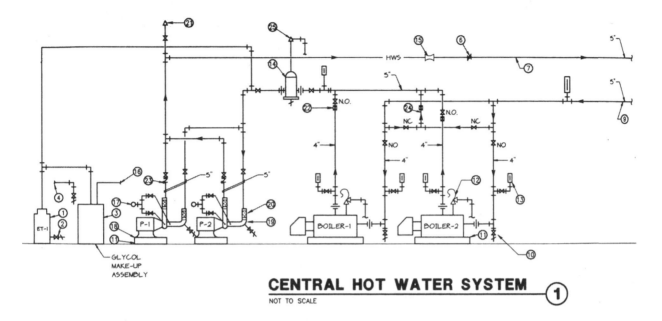

CENTRAL HOT WATER SYSTEM ①

NOT TO SCALE

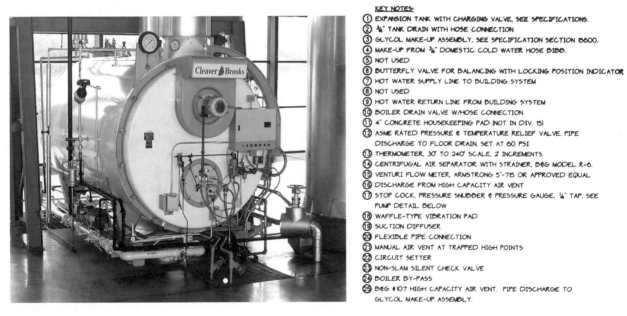

KEY NOTES:

① EXPANSION TANK WITH CHARGING VALVE, SEE SPECIFICATIONS.
② ¾" TANK DRAIN WITH HOSE CONNECTION
③ GLYCOL MAKE-UP ASSEMBLY, SEE SPECIFICATION SECTION 15600.
④ MAKE-UP FROM ¾" DOMESTIC COLD WATER HOSE BIBB.
⑤ NOT USED
⑥ BUTTERFLY VALVE FOR BALANCING WITH LOCKING POSITION INDICATOR
⑦ HOT WATER SUPPLY LINE TO BUILDING SYSTEM
⑧ NOT USED
⑨ HOT WATER RETURN LINE FROM BUILDING SYSTEM
⑩ BOILER DRAIN VALVE W/HOSE CONNECTION
⑪ 4" CONCRETE HOUSEKEEPING PAD (NOT IN DIV. 15)
⑫ ASME RATED PRESSURE & TEMPERATURE RELIEF VALVE. PIPE
   DISCHARGE TO FLOOR DRAIN. SET AT 60 PSI
⑬ THERMOMETER, 30 TO 240° SCALE, 2 INCREMENTS.
⑭ CENTRIFUGAL AIR SEPARATOR WITH STRAINER, B&G MODEL R-6.
⑮ VENTURI FLOW METER, ARMSTRONG 5"-715 OR APPROVED EQUAL
⑯ DISCHARGE FROM HIGH CAPACITY AIR VENT
⑰ STOP COCK, PRESSURE SNUBBER & PRESSURE GAUGE, ¼" TAP, SEE
   PUMP DETAIL BELOW
⑱ WAFFLE-TYPE VIBRATION PAD
⑲ SUCTION DIFFUSER
⑳ FLEXIBLE PIPE CONNECTION
㉑ MANUAL AIR VENT AT TRAPPED HIGH POINTS
㉒ CIRCUIT SETTER
㉓ NON-SLAM SILENT CHECK VALVE
㉔ BOILER BY-PASS
㉕ B&G #107 HIGH CAPACITY AIR VENT. PIPE DISCHARGE TO
   GLYCOL MAKE-UP ASSEMBLY.

Figure 13-15. Mechanical prints include identification for each pipe run in an HVAC system.

Controls. Hydronic system control devices include boiler operating, safety, and terminal unit controls. Boiler operating controls include an aquastat and a pressure control. An *aquastat* is a temperature-actuated electric switch used to limit the temperature of boiler water. An aquastat senses the temperature of boiler water and controls the burner firing to maintain proper boiler water temperature. A *pressure control* is a pressure-actuated mercury switch that controls the boiler burner(s) by starting and stopping the boiler burner(s) based on the pressure inside the boiler.

Boiler safety controls include high-temperature limit controls and safety valves. The high-temperature limit control is a temperature-actuated electric switch that senses boiler water temperature. The high-temperature limit control shuts OFF the boiler burner(s) if the temperature gets excessively high. A *safety valve* is a valve that prevents the boiler from exceeding its MAWP. The safety valve exhausts hot water and steam from the boiler when the pressure becomes excessively high. The safety valve functions as the last pressure regulation safeguard device on the boiler.

Terminal unit controls include zone valves, low-temperature limit controls, thermostats, and blowers. A *zone valve* is a valve that regulates the flow of water in a control zone or terminal unit of a building. A *control zone* is any part of a building that is controlled by one controlling device. Zone valves regulate the flow of water through the piping system and terminal units based on thermostatic readings. A *low-temperature limit control* is a temperature-actuated electric switch that energizes the damper motor and shuts the damper if the ventilation air temperature drops below a setpoint. Low-temperature limit controls allow outside air to be used in a system but shut OFF the air flow when the temperature drops below freezing to prevent system damage. A thermostat controls valves, pumps, and blower motors to regulate the temperature in each area of the building. Blowers in terminal units are controlled by thermostats and diffuse air into each area of the building.

Forced-Air Systems

A forced-air system is a system that uses warm or cool air to condition building spaces. In forced-air systems, air is heated by a heat source such as natural gas, hot water, or electric coils, or cooled by refrigeration. The air is then circulated and distributed through ductwork to condition the building spaces. Forced-air systems consist of furnaces, air conditioners, air handlers, ductwork, and controls. Humidifiers may be installed to add moisture to the air. Air filters are installed to remove dust and dirt particles. Forced-air systems are used in small structures due to the cost of running ductwork over long distances.

Furnaces. A *furnace* is a self-contained heating unit that includes a blower, burner(s), heat exchanger or electric heating elements, and controls. A furnace safely and efficiently transfers heat from the point of combustion or electrical coils to the surrounding air. In combustion furnaces, the firebox is the area in which fuel is burned. In electric furnaces, electric current flows through a high-resistance wire to create heat.

Air Conditioners. An *air conditioner* is the component in a forced-air system that cools the air. Air conditioners contain an evaporator, compressor, condenser, and expansion valve. An *evaporator* is a heat exchanger that absorbs heat from the surrounding air to evaporate refrigerant. Air is passed over the evaporator and distributed to areas to be cooled. A *compressor* is a mechanical device that compresses refrigerant. The pres-

surized refrigerant vapor is discharged into the condenser. The *condenser* is a heat exchanger that removes heat from high-pressure refrigerant vapor. The refrigerant condenses and is circulated through an expansion valve. An *expansion valve* is a device that reduces the pressure on liquid refrigerant by allowing the refrigerant to expand.

Air Handlers. An *air handler* is a device used to distribute conditioned air to building spaces. Air handlers include a fan(s), filters, movable dampers, and heating and cooling coils. Air handler drive system requirements vary depending on the horsepower required and the speed of each fan. Ceiling and exhaust fans are also included in this portion of the estimating process. Ceiling and exhaust fan information is based on the horsepower, speed, and required diameter of the fan blades. Mechanical prints provide information about air handlers and fans on schedules, which are cross-referenced to mechanical plan views and elevations. Each air handler and fan is shown by overall diameter of the blades, the size of the motor and speed, and the amount of air the fan moves. Air handlers are integrated into the ductwork system. See Figure 13-16.

Ductwork. *Ductwork* is the distribution system for forced-air heating or cooling systems. Ductwork is formed from galvanized sheet metal or plastic to distribute air throughout the heating, cooling, and ventilating system. Ductwork for heating and cooling may be internally or externally insulated to minimize heat transfer. Sheet metal ductwork is prefabricated at a shop and transported to the job site for installation. Standard sizes and types of plastic ductwork are also available. Plastic ductwork is generally used for applications where flexible ductwork is necessary. Symbols on mechanical prints indicate the material used for ductwork. See Figure 13-17. See Appendix.

In forced-air systems, return air ductwork is installed to remove air from conditioned areas and return it for reheating or recooling. In ventilation systems, ductwork is installed to exhaust air from the building to the outside or to areas to be conditioned and recycled. For food preparation areas of manufacturing operations, large quantities of air may need to be removed quickly and replaced by makeup air. *Makeup air* is air that is used to replace air that is lost to exhaust.

> ▲ *When taking off HVAC systems, the cost of the temperature control system (thermostat, control wiring, etc.) is normally included in the cost of the equipment.*

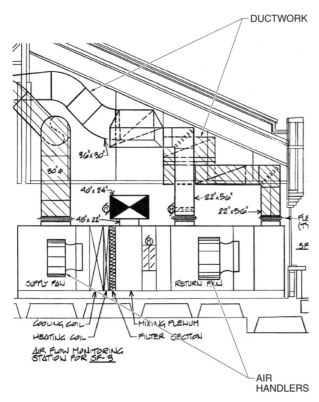

Figure 13-16. Air handlers include fans to move large volumes of air through HVAC systems.

DUCTWORK SYMBOLS			
DUCTWORK	**SYMBOL**	**DUCTWORK**	**SYMBOL**
NEW		INTERNAL INSULATION (SUPPLY)	
DUCTWORK TO BE REMOVED		STAINLESS STEEL	
RECT SUPPLY (SECTION)		SPECIALTY EXHAUST	
RECT RETURN (SECTION)		AIR TRANSFER SLEEVE	
RECT EXHAUST (SECTION)		FLEXIBLE DUCT	
RECT OSA (SECTION)		FLEXIBLE CONNECTION	
ROUND SUPPLY (SECTION)		ROUND EXHAUST (SECTION)	
ROUND RETURN (SECTION)		ROUND OSA (SECTION)	

Figure 13-17. Standard symbols are used to indicate specific ductwork required.

A *louver* is a ductwork cover containing horizontal slats that allow the passage of air but prevent rain from entering. Louvers are installed where ventilation ductwork passes through exterior walls or roofs. A *register* is a device that covers the opening of supply air ductwork. Registers contain a damper for controlling air

flow. A *damper* is a movable plate that controls and balances air flow in a forced-air system. A *grill* is a device that covers the opening of return air ductwork. Grills are sized and located to return the supply air back to the furnace. An *air filter* is a porous device that removes particles from air. Mechanical or electrostatic air filters may be installed in ductwork to remove dust and fumes from the air. Filtration of air within a structure is part of the heating, cooling, and ventilation system. Electronic filter systems use electrostatic energy that attracts particles and removes them as air flows across the energized filters.

Controls. Forced-air system controls include power controls, operating controls, and safety controls. Power controls include disconnects, fuses, and circuit breakers. A *disconnect* is a switch that disconnects electrical circuits from motors and machines. Disconnects stop the flow of electricity when manually activated. A *fuse* is an overcurrent protection device with a fusible link that melts and opens the circuit when an overload condition or short circuit occurs. A *circuit breaker* is an overcurrent protection device with a mechanism that automatically opens the circuit when an overload condition or short circuit occurs. Fuses and circuit breakers stop the flow of electricity when activated by an overcurrent condition.

Forced-air system operating controls include transformers, thermostats, and blower controls. A *transformer* is an electric device that uses electromagnetism to change voltage from one level to another. Transformers change the voltage in an electrical circuit to allow for operation of low-voltage control devices. Thermostats turn ON or OFF heating or cooling equipment based on the temperature in the area to be heated or cooled. A *blower control* is a temperature-actuated switch that controls the blower motor in a furnace.

Forced-air system safety controls include limit switches, pilot safety controls, pressure switches, and stack switches. A *limit switch* is a switch that contains a bimetal element, which senses the temperature of the surrounding air. Limit switches turn a furnace OFF in case of overheating. A *pilot safety control* is a safety control that determines if the pilot light is burning. Pilot safety controls do not allow further gas flow if the pilot light is extinguished. A *pressure switch* is an electric switch operated by the amount of pressure acting on a diaphragm or bellows element. Pressure switches are used to turn OFF air conditioning systems if refrigerant pressure becomes excessively high. A *stack switch* is a mechanical combustion safety control that contains a bimetal element that senses flue gas temperature and converts it to mechanical motion. Stack switches are safety devices for fuel flow into a furnace.

HVAC SYSTEM QUANTITY TAKEOFF

Heating, ventilating, and air conditioning needs require complex systems to provide the temperature control and air circulation required for large public areas, offices, meeting rooms, classrooms, and work areas. Mechanical prints show the heating, cooling, and ventilating systems. Plan and elevation views and detail drawings provide installation and fabrication information. Specifications, general notes, and schedules contain information concerning the manufacturer designs and equipment codes. Shop drawings may also be produced by the subcontractor or the supplier to guide in the proper installation of equipment, piping, and ductwork. Mechanical prints are interrelated with architectural, structural, and electrical prints for temperature control systems.

When calculating HVAC system labor rates, estimators use company historical data or various labor unit industry standard resources. Company historical data is generated by tracking labor units on each installation and maintaining company records concerning installation times under various conditions for various pieces of equipment. Industry standard resources include the *Mechanical Contractor's Estimating Manual*, the *National Plumbing and Heating Contractor's Manual*, the *Sheet Metal Contractor's Labor Rates*, and *Wends Sheet Metal Manual*.

Heating and Cooling Piping Takeoff

Piping to be considered by the estimator includes piping for water in a hydronic system, fuel to the boiler or furnace, piping for expansion tanks, and flues. Estimators determine from plan drawings if the hydronic piping system is a one-, two-, three-, or four-pipe system. The fuel used to fire the boiler or furnace is determined, along with the location of the utility service meter. The distance between a boiler and expansion tank determines the amount of piping required for the expansion tank. Flue requirements are determined by the architect and HVAC engineer based on a variety of building and fire protection codes.

Hydronic Systems. Locations for pumps for hot water supply and hot water return are shown on mechanical prints. Expansion tanks connected to boilers are

also shown. Schedules on mechanical prints provide pump information, including the manufacturer name and model number, flow rate in gallons per minute, and motor size and speed.

Piping connections between a boiler, pump(s), and expansion tank are given on mechanical prints. Piping information between the boiler, pump(s), and expansion tank includes the inside diameter of the pipes and valve and meter types and locations. Pipe notations indicate the purpose of the pipe and the direction of water flow. See Figure 13-18. The position and size of terminal units are indicated on working drawings. The estimator makes calculations concerning the amount of pipe and elbows, valves, tees, and other fittings based on this information. Flue details are provided by the architect, HVAC engineer, or boiler manufacturer.

Pipe is taken off and entered into a ledger sheet, spreadsheet, or estimating software based on linear feet of each pipe based on diameter and pipe material. Valves and fittings are taken off as individual items. Flue members are based on the vertical linear feet of flue required.

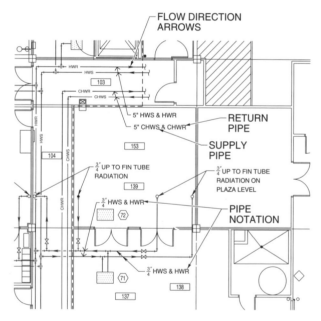

Figure 13-18. Water flow direction noted on mechanical prints determines pump and fitting installations.

▲ *Piping is sized for each piping application. Undersized piping can cause backed-up sewer pipes, low water volume at faucets and showers, and insufficient heating and cooling.*

Heating Equipment Takeoff

Heating equipment considered by the estimator includes boilers, expansion tanks, and furnaces. Specifications in CSI MasterFormat™ Division 15 are often very detailed concerning boiler gross output capacity, maximum allowable working pressure, and approved manufacturers. Boiler and furnace information, including auxiliary devices such as expansion tanks, is located in Division 15 and also on mechanical, plumbing, and HVAC plan drawings.

Boilers. Items taken into account by design engineers in the selection of a boiler include the number of passes that hot gases make through the system, the input rating, gross output rating, net unit output, and efficiency. The input rating of a boiler is equal to the input in British thermal units per hour (Btu/hr) per unit of fuel. The gross output rating of a boiler is equal to the heat output of a boiler when fired continuously. Net unit output is equal to the gross output rating multiplied by the percentage of heat loss due to initial heat-up. Efficiency factors are based on the fuel used. Coal has an efficiency rating of 65% to 75%. Natural gas and oil have efficiency ratings of 70% to 80%. Electricity has an efficiency rating of 95% to 100%.

Information about the boiler type, size, capacity, and heating load may be included on a schedule in the mechanical prints or in the specifications. Boiler locations and types may be shown on plumbing, mechanical, or HVAC plan views and elevation drawings. Specific locations are not given on these prints. Dimensions are obtained from the architectural and structural plan views. In addition to boiler location, mechanical and electrical prints show flue piping, water piping, and power and control wiring connections to boilers.

Each boiler and expansion tank is taken off as an individual item and entered into the estimating system. See Figure 13-19. Boiler manufacturers provide information concerning the necessary hookup, installation, and safety equipment requirements for various sizes and types of boiler systems. Estimators should consult suppliers to obtain current information and pricing on these items. Company historical records can provide information concerning labor costs for various boiler equipment installations.

▲ *Boilers are manufactured for specific purposes and are unique in design, maximum allowable working pressure, heat output, fuel, and fabrication location.*

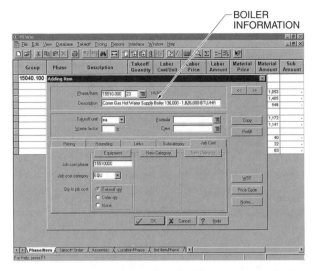

Figure 13-19. Estimating software allows the entry of boiler information into an estimate as an individual item.

Furnaces. Mechanical prints indicate the location of unit heaters and the required fuel piping and flue information. See Figure 13-20. Information concerning electric furnaces and unit heaters is contained in the electrical prints. Estimators should determine the furnace required from specification information and mechanical or HVAC drawings. As with boilers, each furnace is taken off as an individual item and entered into a ledger sheet, spreadsheet, or estimating software. Supplier information is required in selecting the proper make and model of furnace to meet the heating requirements.

Cleaver-Brooks

Estimators consult with manufacturers and suppliers concerning boiler pricing information and hookup, installation, and safety requirements.

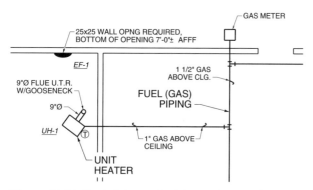

Figure 13-20. Fuel piping and unit heater location is shown on mechanical plan views.

Air Conditioning Equipment Takeoff

Mechanical prints provide plan views, detail drawings, and schedules of all components of cooling systems, including chillers, air conditioners, piping, pumps, and terminal units. Cooling tower locations are shown on plan views and site plans. Air conditioners are shown on electrical prints. In commercial applications, air conditioners may be placed on rooftops.

Cooling towers are taken off as individual items with consideration given to air and water flow requirements.

Chillers. In a manner similar to boilers, specifications note detailed chiller information. Chiller information includes approved compressors, required circuit breakers, insulation requirements, refrigerant circuit details, accessories, and approved manufacturers. Estimators use information in the specifications and on plumbing, mechanical, and HVAC drawings to take off each chiller as an individual item for entry into the estimate.

Cooling Towers. Cooling tower takeoff requires consideration of the air flow design, water basin requirements, fan and motor types, water distribution system, control system, and location of the cooling towers in relation to chillers and other mechanical equipment. As with boilers and chillers, the architect may provide a list of approved manufacturers. Item takeoff is required for cooling towers with assistance from manufacturers or individuals who specialize in cooling tower construction.

Air Handling Equipment Takeoff

Air handling equipment considered by the estimator includes fans, motors, drive units, and the necessary ductwork to facilitate air flow. Additional items include accessories such as dampers, filters, louvers, and grills. Specialized air handling requirements that may be included in this portion of the estimate include sawdust collection systems, paint booths, smoke and fire dampers, fume hoods, and special ductwork linings.

Air Handlers. Air handler considerations for estimators include fan dimensions and capacities, power requirements, drive systems required for various horsepower levels, housings, accessories, and permissible manufacturers. Weather protection considerations are required for exterior applications. Fans and air handler accessories are entered into the ledger sheet, spreadsheet, or estimating software as individual items. Information is obtained from Division 15 of the specifications and the mechanical or HVAC drawings. Labor rates are based on company historical records or industry standards for installation of the various equipment.

Ductwork. Ductwork is commonly prefabricated using standard parts and fittings. Ductwork designs are shown with plan views and elevations on mechanical prints. Diameters are indicated for round ductwork.

Width and height are indicated for rectangular or square ductwork. Methods for attachment of ductwork to structural members is not shown. Hanging information is included in the specifications or general plan notes.

Diffusers are used at the ends of ductwork. A *diffuser* is an air outlet in an HVAC system that directs air in a wide pattern. Standard sizes and types are available from various manufacturers. Each diffuser is identified by size and type using symbols on mechanical plan views or on a diffuser schedule. See Figure 13-21.

Ductwork drawings may be provided for large commercial buildings or generated from the HVAC drawings. All ductwork layout drawings should be drawn to scale. Ductwork layout drawings may then be scaled to determine the lengths of various ductwork required. Stack lengths of all vertical ductwork are determined from vertical dimensions shown on the drawings plus the distance above each floor level. The number and types of various fittings required for fastening ductwork are counted by each item throughout various ductwork lengths.

Manufacturers of prefabricated ductwork provide listings of all parts and fittings together with figure or part numbers. As each size or part is taken off of the drawings, it is noted in the quantity takeoff column on a ledger sheet, spreadsheet, or estimating software.

HVAC contractors normally keep historical records for installing ductwork on previous jobs to help determine the cost of installing ductwork per foot. This allows for determination of an average labor cost for ductwork based on the number of linear feet of ductwork or the number of registers. Estimators review drawings to note any possible irregularities in ductwork runs. Ductwork may require insulation, or, because of structural conditions, other labor intensive operations. Each type of ductwork should be noted on an individual line, column, cell, or row in the estimate to indicate the number of feet of each type of ductwork to be installed.

> ▲ *HVAC systems condition air by maintaining the proper temperature, humidity, and quality of the air. HVAC systems are designed to operate at optimum energy efficiency while maintaining desired environmental conditions.*

Registers, grills, and louvers are taken off according to their size, type, and material, such as sheet metal, brass, or other decorative metal. Sizes and types are indicated on schedules or shown on elevation and detail drawings. Ventilation hoods for the collection of fumes and smoke inside buildings are shown with detail drawings. The detail drawings indicate dimensions of the hood and materials for construction. Locations and types of materials for filters within the air circulation system are shown on detail ductwork drawings. Registers, grills, louvers, and filters are taken off as individual items. Each type is separated into an individual line, column, cell, or row and accumulated to a total for the entire job.

Terminal Units. Pipe connections are made to terminal units, which transfer heat from water to the surrounding air. Mechanical and HVAC plan views show the location and type of terminal units throughout a structure. A terminal unit schedule may be included in the specifications. Specifications include terminal unit information that describes the water and air flow requirements. See Figure 13-22. Architects use various geometric shapes such as hexagons, ovals, and circles for easy identification and cross-referencing of symbols and numbers to schedules. Terminal unit detail drawings are also provided to indicate water and duct connections. Individual item takeoff is used for all terminal units.

The specifications or prints may require that ductwork be insulated to decrease hot and cold air transfer to unoccupied building spaces.

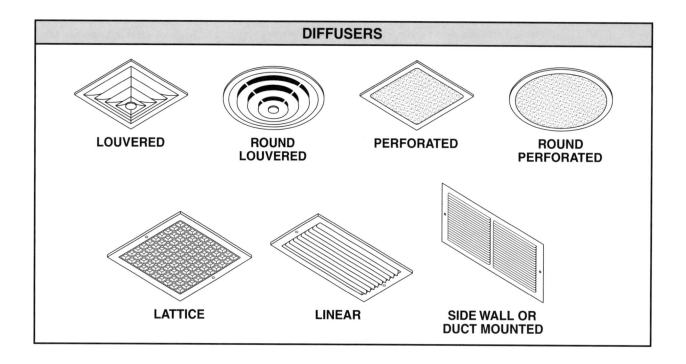

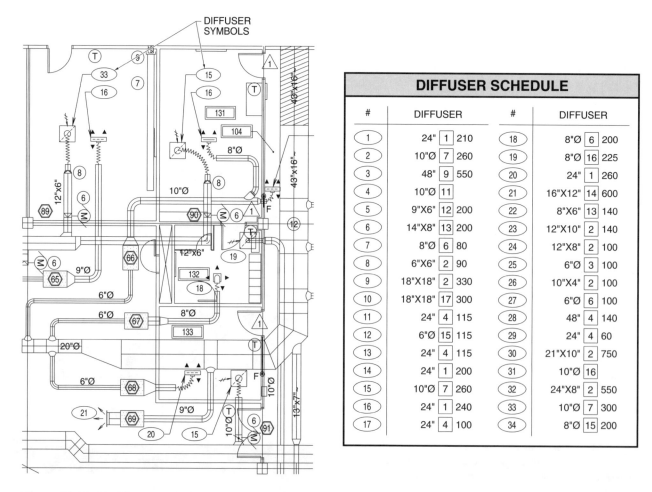

Figure 13-21. Air diffuser symbols on plan views relate to a diffuser schedule indicating diffuser size and type.

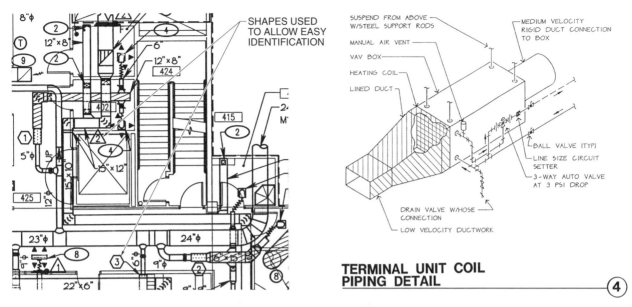

TERMINAL UNIT COIL PIPING DETAIL ④

#	INLET DIA. (IN)	MAX CFM	MIN CFM	MIN SP (IN)	NOISE CRI AT ROOM(2)	NOISE(1) CRITERIA RADIATED	EAT = 55°			HOT WATER COIL		EWT = 180°		MODEL NO.	NOTES
							LAT	LWT	GPM	WATER PD (IN)	AIR PD (IN)	# OF ROWS	MBH		
TU-1	5	180	180	0.13	37	27	99	160	0.87	0.63	-	1	-	DESV-3000	SF-4 (6)
2	9	760	760	0.19	39	27	105	158	3.8	1.05	-	2		"	"
3	6	290	110	0.17	33	25	102	159	.54	.20		1		"	" (6)
4	9	700	400	0.16	39	26	115	153	1.9	.31		2		"	"
5	5	215	80	0.11	36	27	109	156	.39	.11		1		"	" (6)
6	6	300	100	0.18	40	27	102	153	.38	.11		1		"	" (6)
7	7	400	200	0.09	35	26	94	157	.79	.53		1		"	" (6)
8	8	750	375	0.23	41	27	117	153	1.7	.25		2		"	"

(MFR: TITUS — heading above MODEL NO. column)

Figure 13-22. Estimators rely on mechanical plans, detail drawings, and schedules to take off the proper equipment for mechanical systems.

Controls. Mechanical system control considerations include the equipment being controlled and manufacturer and designer specifications. Many control devices are bundled into the cost of the equipment at the time of purchase. Other controls may require additional material and labor costs. Estimators should check the specifications for controls for each piece of equipment to ensure all control devices are included in the estimate. Other items such as thermostats are taken off as individual items.

ESTIMATING SOFTWARE – TAKEOFF USING ONE-TIME ITEMS

One-time items in Precision estimating software enable an estimator to take off items that are not currently in the database. Details can be added through the Detail Window and, if desired, one-time items for a particular estimate can be automatically saved to the database. Takeoff using one-time items is performed by following a standard procedure. See Figure 13-23.

Cleaver-Brooks

Controls are required for the operation of mechanical equipment and may be included with the equipment when purchased.

PRECISION ESTIMATING SOFTWARE – TAKEOFF USING ONE-TIME ITEMS

1. Open the One-time Item window by clicking the One-time Item (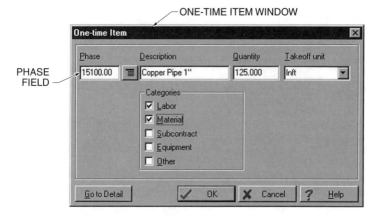) button in the toolbar.

2. Right-click in the Phase field and select Edit Phase from the shortcut menu. Click the Add button to create the Phase record. Input 15100 for Phase and Copper Pipe for description. Click OK and then click close.

3. Create the item record by filling the One-time Item fields. Input Copper Pipe 1″ for Description, 125 for Quantity, and lnft for Takeoff Unit.

4. Select the categories by clicking and placing a check in the Labor and Material Category boxes.

ONE-TIME ITEM WINDOW

PHASE FIELD

One-time Item

Phase	Description	Quantity	Takeoff unit
15100.00	Copper Pipe 1"	125.000	lnft

Categories
☑ Labor
☑ Material
☐ Subcontract
☐ Equipment
☐ Other

Go to Detail   ✓ OK   ✗ Cancel   ? Help

5. Click the Go to Detail button and enter .50 for Labor Unit Price and .98 for the Material Unit Price for the item.

6. Exit the Detail window by clicking OK and then the Close (✗) button. The one-time item displays in the spreadsheet.

Figure 13-23. One-time items enable an estimator to take off items that are not currently in the database.

Estimating

_____ 1. Division _____ of the CSI MasterFormat™ includes mechanical system specifications.

_____ 2. Type K copper pipe is color-coded _____.

_____ 3. Type L copper pipe is color-coded _____.

T   F   4. Brass pipe is available in standard and extra-strong weights.

T   F   5. Steel pipe may be seamless or welded.

_____ 6. _____ piping is piping that delivers water from the source to the point of use.

_____ 7. _____ piping is piping installed in industrial facilities for compressed air, vacuum, gas, or fuel.

_____ 8. A(n) _____ system is a system that uses water, steam, or other liquid to condition building spaces.

_____ 9. A(n) _____ pipe is a pipe that removes odors and gases from the waste piping and exhausts them away from inhabited areas.

_____ 10. A(n) _____ system is a system that uses warm or cool air to condition building spaces.

_____ 11. A(n) _____ is a closed metal container in which water is heated to produce heated water or steam.

_____ 12. A(n) _____ is the component in a hydronic air conditioning system that cools water which cools the air.

_____ 13. A(n) _____ is a temperature-actuated electric switch that controls heating and/or cooling equipment.

T   F   14. A low pressure boiler is a boiler that has a maximum allowable working pressure of up to 15 psi.

T   F   15. A high pressure boiler is a boiler that has a maximum allowable working pressure above 15 psi and over 10 boiler horsepower.

_____ 16. A(n) _____ is a heat-resistant passage in a chimney that conveys smoke or other gases of combustion.

_____ 17. A(n) _____ is a self-contained heating unit that includes a blower, burner(s), heat exchanger or electric heating elements, and controls.

_____ 18. A(n) _____ is the component in a forced-air system that cools the air.

_____ 19. A(n) _____ is an electric device that uses electromagnetism to change voltage from one level to another.

_____ **20.** A(n) _____ is an overcurrent protection device with a mechanism that automatically opens the circuit when an overload condition or short circuit occurs.

_____ **21.** A(n) _____ is a switch that disconnects electrical circuits from motors and machines.

_____ **22.** A(n) _____ is an overcurrent protection device with a fusible link that melts and opens the circuit when an overload condition or short circuit occurs.

T   F   **23.** A chiller is an air outlet in an HVAC system that directs air in a wide pattern.

T   F   **24.** A grill is a device that covers the opening of return air ductwork.

T   F   **25.** An air filter is a porous device that removes particles from the air.

Diffusers

_____ **1.** Perforated

_____ **2.** Round perforated

_____ **3.** Lattice

_____ **4.** Linear

_____ **5.** Louvered

_____ **6.** Round louvered

_____ **7.** Side wall or duct mounted

Ductwork Symbols

_____ **1.** New

_____ **2.** Round exhaust (section)

_____ **3.** Round supply (section)

_____ **4.** Rect exhaust (section)

_____ **5.** Rect supply (section)

_____ **6.** Rect return (section)

_____ **7.** Stainless steel

_____ **8.** Flexible duct

_____ **9.** Specialty exhaust

_____ **10.** Ductwork to be removed

Valve Symbols

_____   1. Safety

_____   2. Cock

_____   3. Check

_____   4. Float

_____   5. Quick open

_____   6. Diaphragm

_____   7. Motor operation globe

_____   8. Auto governor operation

_____   9. Gate

_____   10. Globe

(A)

(B)

(C)

(D)

(E)

(F)

(G)

(H)

(I)

(J)

Valves

_____   1. Pressure-reducing

_____   2. Pressure-relief

_____   3. Gate

_____   4. Globe

_____   5. Check

_____   6. Butterfly

(A)

(B)

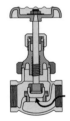
(C)

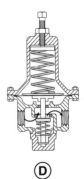
(D)

(E)

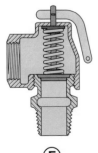
(F)

Pipe Fittings

_____	**1.** Trap
_____	**2.** Cap
_____	**3.** Wye
_____	**4.** Tee
_____	**5.** Elbows
_____	**6.** Plug
_____	**7.** Cross
_____	**8.** Union
_____	**9.** Coupling
_____	**10.** Nipple
_____	**11.** Reducer

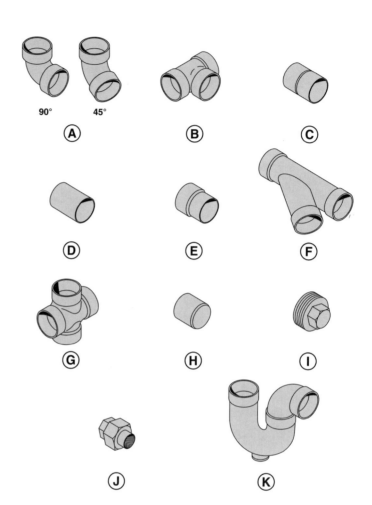

Activities
Mechanical

Name _____   Date _____

Activity 13-1 – Ledger Sheet Activity

Refer to the cost data, Print 13-1, and Estimate Summary Sheet No. 13-1. Take off the total item count for each fixture. Determine the total material and labor cost.

COST DATA			
Material	**Unit**	**Material Unit Cost***	**Labor Unit Cost***
1. Water closet, one piece, wall hung	ea	478.50	97.35
2. Water closet, bowl only, with flush valve, wall hung	ea	379.50	89.10
3. Lavatory, vitreous china, 20″ × 16″, single bowl	ea	179.30	95.70
4. Shower, built-in, 4 gpm valve	ea	74.25	71.50

* in $

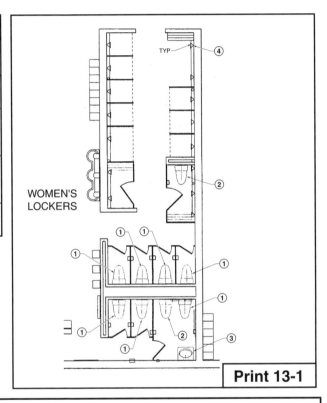

WOMEN'S LOCKERS

Print 13-1

ESTIMATE SUMMARY SHEET

Sheet No. ___13-1___
Date: _____
Checked: _____

Project: _____
Estimator: _____

No.	Description	Dimensions				Quantity		Material		Labor		Total	
							Unit	Unit Cost	Total	Unit Cost	Total	Unit Cost	Total
	Water closet, one-piece												
	Water closet, bowl only												
	Lavatory												
	Shower												
	Total												

Activity 13-2 – Spreadsheet Activity

Refer to Print 13-2 and Quantity Sheet No. 13-2 on the CD-ROM. Take off the linear feet of duct and number of supply and return air grills. Round all values up to the next whole unit. Do not make allowances for angular runs.

Activity 13-3 – Estimating Software Activity

Create a new estimate and name it Activity 13-3. Add phase 15400 Plumbing Equipment by clicking **Database** and **Phases** from the pull-down menu. Click **Add** and input *15400* for Phase Number and *Plumbing Equipment* for Description. Click **OK** and **Close**. Add a one-time item by clicking the **Add One-Time Item** button, entering *15400* for Phase Number, *water storage tank* for Description, *1* for Quantity, and *each* for Takeoff Unit. Close the One-Time Item window. Input a labor cost unit of $7882 and material price of $1500 directly into the spreadsheet. Print a standard estimate report.

Electrical

RACO, Hubbell Electrical Products

Division 16 of the CSI MasterFormat™ covers electrical service provisions from the power source into a structure, wiring and distribution throughout a structure, and controls, fixtures, switches, and power transfer equipment. Electrical prints contain information about the wiring to be installed, electrical equipment, and electrical finish materials. Electrical prints may be divided into separate sections for lighting, power supply, and signals such as fire alarms and smoke detectors. Connections to finish electrical fixtures are made after panelboards, conduit, and wiring are installed. Finish electrical fixtures include light fixtures, communication systems, and alarm systems.

POWER DISTRIBUTION SYSTEM MATERIALS AND METHODS

Electrical power distribution systems include service provisions from the power source into a structure, wiring and distribution throughout the structure, controls and switches throughout the power system, fixtures, and other equipment that transfers the power to the intended purpose. Electrical information is covered in Division 16 of the CSI MasterFormat™.

Power sources are noted on site plans and electrical prints. Locations for transformers and panelboards are shown on site plans and electrical prints along with circuit configurations. Conduit, cable, and wire placement and type are indicated with plan symbols. *Conduit* is a hollow pipe that supports and protects electrical conductors. Conduit is installed prior to the placement of poured-in-place concrete or masonry and after the placement of structural steel or wood. Exact conduit locations are not given on electrical prints. Conduit is placed according to architectural and structural plan dimensions, local building codes, and necessary connections at the ends of the conduit. During

concrete placement, the ends of conduit are capped to prevent the conduit from being filled with concrete or mortar. Conduit is fastened together with a variety of fittings.

After structural members are in place and concrete and masonry have reached their set, conductors and cables are pulled through the conduit. The types of conductors and cables are noted in the specifications and on the electrical prints. Switches, receptacles, fixtures, and other electrical equipment are installed and tested by the electrician according to local building codes and electrical specifications and drawings. A review of all portions of electrical work must be completed to develop an accurate takeoff and final bid. Electrical prints may be divided into separate sections for lighting, power supply, and signals such as fire alarms and smoke detectors. Plan views for each application are cross-referenced to ensure that all electrical needs are met.

A general contractor commonly obtains bids for electrical systems from electrical estimators. Electrical estimators are specialized in their knowledge of electrical power and wiring systems, codes, installation, labor, equipment requirements, available materials, fixtures and fittings, and the other technical aspects of electrical work.

Wiring

Electrical system wiring is located underground, placed in conduit, or left as exposed cable. Electrical system wiring may be installed in walls, above ceilings, and under floors. Electrical loads vary from low-voltage loads for items such as thermostats to high-voltage loads for welding equipment or heavy manufacturing machinery. Each wiring application is shown on the electrical specifications and prints. See Figure 14-1.

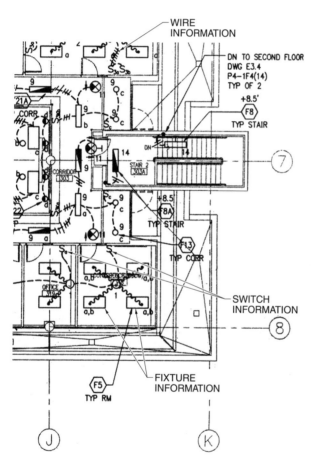

Figure 14-1. Electrical prints show wire, fixture, and switch information.

Raceways. A *raceway* is an enclosed channel for conductors. See Figure 14-2. A raceway includes all of the enclosures used for running conductors among different components of an electrical system. Raceways include intermediate metal conduit, rigid metal conduit, rigid nonmetallic conduit, electrical metallic tubing, and flexible metal conduit.

RACEWAYS			
	Abbrev	**NEC®**	**Size**
Intermediate Metal Conduit	IMC	345	½″ – 4″ in 10′ lengths
Rigid Metal Conduit	RMC	346	½″ – 6″ in 10′ lengths
Rigid Nonmetallic Conduit	RNMC	347	½″ – 6″ in 25′ lengths
Electrical Metallic Tubing	EMT	348	½″ – 4″ in 10′ lengths
Flexible Metal Conduit	FMC	350	³/₈″ – 4″ in 25′, 50′, or 100′ coils

Figure 14-2. Raceways include intermediate metal conduit, rigid metal conduit, rigid nonmetallic conduit, electrical metallic tubing, and flexible metal conduit.

Intermediate metal conduit (IMC) is a medium-weight galvanized metal conduit. IMC is available in 10′ lengths and diameters from ½″ to 4″.

Rigid metal conduit (RMC) is a heavy conduit made of metal. RMC is the universal raceway and is permitted for use in all atmospheric conditions and types of occupancies. RMC is available in 10′ lengths with a coupling on one end and in diameters ranging from ½″ to 6″.

Rigid nonmetallic conduit (RNMC) is a conduit made of materials other that metal. The most common rigid nonmetallic conduit is rigid polyvinyl chloride conduit. Rigid polyvinyl chloride (PVC) conduit is a nonconductive conduit that may replace metallic conduit in some applications. PVC conduit cannot be installed in hazardous locations or air plenums. PVC conduit is available in 10′ lengths and diameters ranging from ½″ to 6″. PVC conduit is available in thin wall (Schedule 40) and thick wall (Schedule 80). Couplings and other fittings are fastened to PVC conduit with a special adhesive. PVC conduit is also available in P&C grade for use in power and communication applications. P&C grade PVC conduit is available in 25′ lengths and diameters ranging from ½″ to 6″.

Electrical metallic tubing (EMT) is a lightweight tubular steel raceway without threads on the ends. EMT is commonly referred to as thin wall or steel tube conduit. EMT is available in 10′ lengths and diameters ranging from ½″ to 4″. EMT is galvanized or electroplated to resist corrosion. Different EMT fittings are available to facilitate connections, including compression and set screw couplings and compression and set screw connectors. A variety of hangers and supports are available to hold EMT in place by attaching it to structural members. EMT may also be encased in concrete or masonry.

Flexible metal conduit (FMC) is a raceway of metal strips that are formed into a circular cross-sectional raceway. The metal strips are helically wound, formed, and interlocked. FMC is known in the field as Greenfield. FMC is available in 25′, 50′, or 100′ coils depending on its diameter. Diameters range from ³/₈″ to 4″. A variety of connection devices are available.

IMC is available with a PVC coating to resist corrosion. This coating is 10 mils, 20 mils, or 40 mils thick. A *mil* is .001″ thick. This coating is partially removed at the ends of conduit to be joined to fittings and couplings.

Other types of conduit include fiber duct, transite duct, soapstone duct, wrought iron pipe, clay conduit, liquid-tight flexible metal conduit (LTFMC), liquid-tight flexible non-metallic conduit (LTFNMC), and flexible metallic tubing (FMT). Each are special-purpose conduits that estimators should note in specifications and electrical plans. Manufacturers and suppliers provide information about these and other specialty conduits as specified by architects and electrical engineers.

Wire and Cable. A *wire* is an individual conductor. A *conductor* is a slender rod or wire that is used to control the flow of electrons in an electrical circuit. Wires are referred to as conductors. Wire is composed of a circular, single-strand (solid) or multi-strand (stranded) conductive material (copper or aluminum). Wire may be bare or insulated. Wire is sized by a gauge number that designates the diameter of the conductive materials. The higher the gauge number, the smaller the wire diameter. See Figure 14-3. A *cable* is two or more conductors grouped together within a common protective cover and used to connect individual components. A metallic shield may also be wrapped around conductors within the outer jacket.

> ▲ When taking off conduit, estimators review the specifications and prints covering conduit and conduit installation. Notes on the prints or in the specifications may set specific minimum conduit sizes. Engineers commonly specify a minimum size conduit in the home run to the panelboard or throughout a job.

COPPER CONDUCTOR RATINGS*		
AWG	**Ampacity**	**Diameter (Mils)**
18	- - - -	40
17	- - - -	45
16	- - - -	51
15	- - - -	57
14	20	64
12	25	81
10	30	102
8	40	128
6	55	162
4	70	204
3	85	229
2	95	258
1	110	289
0	125	325
00	145	365
000	165	410

* @ 60°C

Figure 14-3. Wire is sized by gauge number.

Wire used in construction is identified by the type of insulation protecting the conductors. Wire includes:
- moisture-resistant thermoplastic
- heat-resistant thermoplastic
- moisture- and heat-resistant thermoplastic
- moisture-, heat-, and oil-resistant thermoplastic
- thermoplastic-covered fixture wire
- thermoplastic-covered fixture wire – flexible stranded
- heat-resistant thermoplastic-covered fixture wire
- asbestos-covered heat-resistant fixture wire
- underground service entrance cable – single-conductor

Cables used in construction include junior hard service cord, hard service cord, nonmetallic sheathed cable, underground feeder and branch circuit cable, service-entrance cable, aluminum wire, bare copper wire, armored cable, and high-voltage wire and cable.

Moisture-resistant thermoplastic (TW) wire is used in wet and dry applications. TW wire has no outer covering and is used with No. 14 gauge solid or stranded conductors. TW wire has a maximum operating temperature of 140°F.

Heat-resistant thermoplastic (THHN) wire has a nylon or equivalent coating and is used in dry locations. THHN wire is used with No. 14 gauge solid or stranded conductors and has a maximum operating temperature of 194°F.

Moisture- and heat-resistant thermoplastic (THW) wire is used in wet and dry locations. THW has no outer covering and is used with No. 14 gauge solid or stranded conductors. THW has a maximum operating temperature of 194°F when used with lighting equipment of 1000 V or less.

Moisture- and heat-resistant thermoplastic (XHHW) wire has a fire-resistant synthetic polymer insulation and no outer covering. XHHW wire is used in wet and dry locations, is made with No. 14 gauge solid or stranded conductors, and has a maximum operating temperature of 167°F in wet locations and 194°F in dry locations.

Moisture-, heat-, and oil-resistant thermoplastic (MTW) wire has thick (Type A) or thin (Type B) insulation. Type A insulation has no outer covering. Type B insulation has a nylon covering. MTW wire is commonly used for machine tool wiring in wet locations and has an operating temperature limit of 140°F. MTW wire is made with No. 14 gauge stranded conductors.

Wire insulation protects the conductors and is rated based on maximum operating temperature.

Thermoplastic-covered fixture (TF) wire has thermoplastic insulation with no other covering. TF wire is made with No. 16 or No. 18 gauge solid or stranded conductors and has a maximum operating temperature of 140°F. TF wire is commonly used for wiring electrical fixtures. Thermoplastic-covered fixture wire – flexible stranded (TFF) has similar characteristics to TF wire with the exception that there is no solid conductor. Heat-resistant thermoplastic covered fixture wire (TFFN) is also used for fixture wiring and has a heat resistance of 194°F.

Asbestos-covered heat-resistant fixture (AF) wire has insulation that is impregnated with asbestos to resist very high temperatures (up to 302°F). AF wire is available from No. 14 to No. 18 gauge with stranded conductors. AF wire is limited to indoor applications that are moisture-free.

Junior hard service cord (SJ) is cable made with 2, 3, or 4 conductors of No. 10 gauge or No. 18 gauge wire. SJ cable has thermoset plastic insulation and covering. SJ cable is used for hanging fixtures in damp applications.

Hard service cord may be Type S, Type SO, or Type STO. Type S has two or more conductors ranging in gauge from No. 2 to No. 18. The insulation and covering are the same as SJ cable. Type SO has the same number and type of conductors and insulation as Type S but has an outer coating of oil-resistant plastic. Type STO is similar to Type SO with the exception that the insulation and covering may be thermoplastic.

Nonmetallic sheathed (NM) cable is made with 2, 3, or 4 conductors in gauges ranging from No. 2 to No. 14. NM cable is made with or without a ground wire and may be insulated or bare. The outer covering is fire-resistant and moisture-resistant plastic.

Underground feeder and branch circuit (UF) cable is made with 2, 3, or 4 conductors with gauges ranging from No. 14 to No. 4/0. UF cable is made with or without a ground wire. Insulation thickness varies according to the gauge of the conductor wire. The outer covering is moisture- and heat-resistant.

Underground service entrance cable – single conductor (USE) wire is made with a covering that is moisture-resistant and sized from No. 12 gauge solid or stranded conductors. USE wire has a maximum operating temperature of 167°F per the NEC®.

Service-entrance (SE) cable is made with one or more conductors with gauges ranging from No. 12 to No. 4/0. The outer covering is fire-resistant and moisture-resistant.

Aluminum wire is a lightweight wire used on feeder circuits such as electrical runs from distribution switchboards to power panels. Aluminum wire is susceptible to oxidation when exposed to the atmosphere. Oxidation results in resistance to current flow that may lead to overheating at connections. Anti-oxidation materials may be applied prior to completion of the installation to guard against overheating.

Bare copper wire is available in gauges ranging from No. 14 to No. 4/0. In its uninsulated form, copper wire is most commonly used for electrical system grounding wires. Three grades of copper wire are soft drawn, medium hard drawn, and hard drawn temper.

Armored cable is made of insulated wires wrapped in a metallic flexible covering. Armored cable is commonly referred to as Type AC or BX. Armored cable is available with single or multiple conductors ranging from No. 4 to No. 14 gauge.

High-voltage wire and cable are specified by engineers for specialized applications. In some situations, detailed specifications from the electrical engineer concerning installation requirements, safe applications, and testing are required. Cable manufacturers can provide technical data and pricing for high-voltage wire and cable. Independent testing agencies provide quotations for specialized high-voltage cable testing.

A circuit run may consist of a cable or conductors within a conduit. A circuit run is often indicated on electrical plans by a solid or dashed line running between devices. The number of conductors in an electrical circuit may be indicated by slash marks across this line. The number of slash marks indicates the number of conductors in the circuit. For example, an electrical circuit line crossed by three slash marks indicates a three-conductor cable or three conductors in a conduit. Normally, a circuit line on the electrical prints without any slash marks indicates a two-conductor circuit run. See Figure 14-4. This method of identifying conductors is usually applied to circuits that require conductors no larger than No. 10 gauge. The symbol list included with the electrical prints should be consulted for information pertaining to the specific method used for designating conductor sizes and quantities in a circuit run.

> **△** *Only cable-pulling lubricant should be used when pulling wire through conduit. Oil or grease can damage the insulation.*

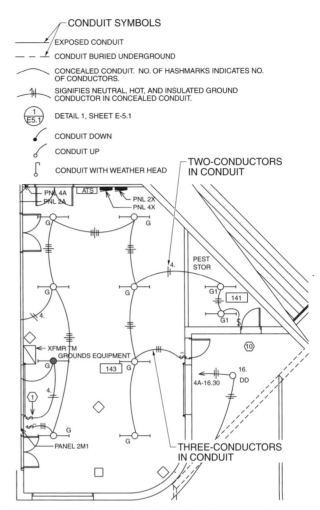

Figure 14-4. The number of slash marks on an electrical circuit run denotes the number of conductors.

Large conductors that carry electrical service from the main power source to panelboards are specified by number of wires, wire gauge, number and gauge of ground wire, and overall cable diameter. For example, a notation of 4#4, 1#10 GRD, 1¼ indicates 4 wires of No. 4 gauge, one No. 10 gauge ground wire, and an overall cable diameter of 1¼″. See Figure 14-5.

Cable Supports. Electrical cables are fastened to structural members, suspended behind walls or above ceilings, placed in cable trays, buried underground, or run within conduit. Clamps are used to fasten conduit and cables to structural steel members. Support and fastening information for cables installed in walls is not commonly provided on electrical prints. The NEC® has established standards for cable and conduit fastening requirements. Cables and conduit must be secured to wall studs or structural members within a certain distance of switches, receptacles, or other electrical fixtures.

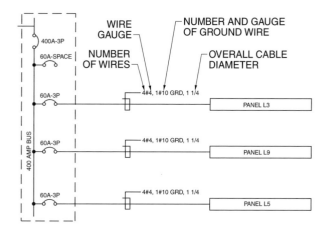

Figure 14-5. Heavy cables are specified according to the number of wires, wire gauge, number and gauge of ground wire, and the overall cable diameter.

Section views on electrical prints provide information concerning cable trays and sleeves for electrical cable and conduit. A *cable tray* is an open grid rack suspended from structural members to support a series of cables. Cable trays include ladders, troughs, and channel trays. Cable trays are commonly manufactured in 12′ sections with a variety of fittings and connectors. Cable tray locations are shown on plan views. See Figure 14-6. Support and hanging requirements vary based on the class of material used. The four classes of cable trays include Class 1 through Class 4, with Class 1 being the lightest grade and Class 4 being the heaviest. Cable trays are commonly supported by threaded rods hanging from overhead structural supports.

Boxes. A *box* is a metallic or nonmetallic electrical enclosure used for equipment, devices, and pulling or terminating conductors. Boxes protect connections, give access to wiring, and provide a method for mounting switches, receptacles, and some fixtures. Boxes are commonly made of stamped steel with punched holes for mounting and knockouts to allow for insertion of wiring and conduit. Common electrical boxes include utility (handy), square, switch, octagonal, masonry, and nonmetallic (fiberglass) boxes. See Figure 14-7.

Utility (handy) boxes are 2″ wide, 4″ high and available in depths ranging from 1½″ to 2³/₁₆″. Utility boxes are designed for surface mounting. Square boxes range in size from 4″ square to 4¹¹/₁₆″ square and in depth from 1¼″ to 2¹/₈″.

Switch boxes are sectional boxes or welded boxes. Sectional switch boxes are installed flush with the wall surface finish and can be joined to form large boxes by removing the box sides. Sectional switch boxes are 2″

wide, 4″ high and available in depths ranging from 1½″ to 3½″. Welded switch boxes are similar to sectional switch boxes except that the sides cannot be removed. Octagonal boxes are most commonly installed flush with the wall or ceiling surface and may be mounted onto a bar hanger that allows for placement between structural members. Octagonal boxes are available in 3″ or 4″ diameters with depths of 1½″ and 2¹/₈″. A masonry box is a box that has punched mounting holes, is stamped with knockouts, and has the top and bottom edges of the box turned inward. Masonry boxes are deep enough so they can be flush mounted and still contact conduit that is run through the cavity of a masonry unit. Nonmetallic (fiberglass) boxes are manufactured from PVC or fiberglass in the same common shapes and designs as metal boxes. A wide variety of covers and plates are designed to be installed on the face of the different boxes.

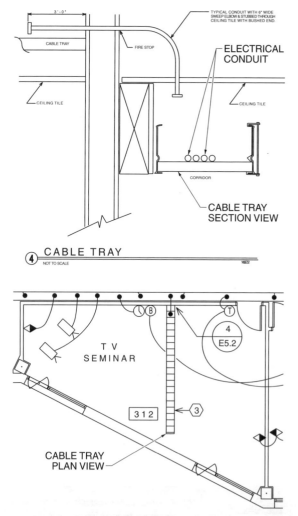

Figure 14-6. Cable tray information is given on cable tray section views and electrical plan views.

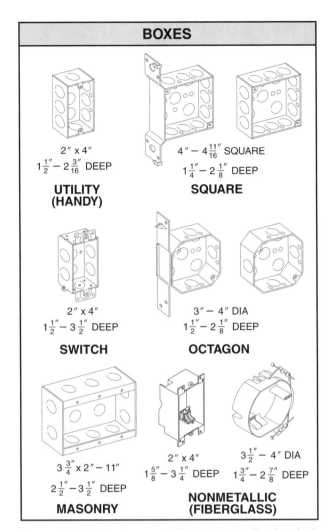

BOXES

UTILITY (HANDY)
2″ x 4″
$1\frac{1}{2}″ - 2\frac{3}{16}″$ DEEP

SQUARE
4″ – 4$\frac{11}{16}$″ SQUARE
$1\frac{1}{4}″ - 2\frac{1}{8}″$ DEEP

SWITCH
2″ x 4″
$1\frac{1}{2}″ - 3\frac{1}{2}″$ DEEP

OCTAGON
3″ – 4″ DIA
$1\frac{1}{2}″ - 2\frac{1}{8}″$ DEEP

MASONRY
3$\frac{3}{4}$″ x 2″ – 11″
$2\frac{1}{2}″ - 3\frac{1}{2}″$ DEEP

NONMETALLIC (FIBERGLASS)
2″ x 4″
$1\frac{5}{8}″ - 3\frac{1}{4}″$ DEEP
3$\frac{1}{2}$″ – 4″ DIA
$1\frac{3}{4}″ - 2\frac{7}{8}″$ DEEP

Figure 14-7. A box is a metallic or nonmetallic electrical enclosure used for equipment, devices, and pulling or terminating conductors.

Switches and Receptacles. A *switch* is a device that is used to start, stop, or redirect the flow of current in an electrical circuit. A *receptacle* is a device used to connect equipment to an electrical system with a cord and plug. The types and locations of switches and receptacles are shown on electrical plan views with various symbols and abbreviations. See Figure 14-8. See Appendix. Electrical prints for power outlet placements show receptacle locations and the type of receptacle at each location.

Switches may be classified according to their number of poles, number of closed positions, method of operation, etc. Common switches include single-pole, double-pole, two-way, three-way, four-way, key-operated, momentary-contact, maintain-contact, dimmer, photoelectric, and safety switches.

A *pole* is the number of completely isolated circuits that a switch can control. A *throw* is the number of closed contact positions per pole. Single-pole switches can carry current through only one circuit at a time. Double-pole switches can carry current through two circuits simultaneously. A *two-way switch* is a single-pole, single-throw (SPST) switch. Two-way switches allow current flow in the ON position and do not allow current flow in the OFF position. A two-way switch is used to make or break one circuit. A *three-way switch* is a single-pole, double-throw (SPDT) switch. A three-way switch is used to divert power to one of two circuit paths. A *four-way switch* is a double-pole, double-throw (DPDT) switch. A four-way switch is used to divert power from two different circuits.

A *key-operated switch* is a switch that is operated by use of a key. Key-operated switches allow for security at the switch in areas with public access. A *momentary-contact switch* is a switch that activates current for a brief moment. Momentary-contact switches control items such as a lighting contactor for a large bank of lamps. A *contactor* is a control device that uses a small control current to energize or de-energize the load connected to it. A *maintain-contact switch* is a switch that allows current flow in the up and down positions and must be manually returned to a center position to stop the flow of current.

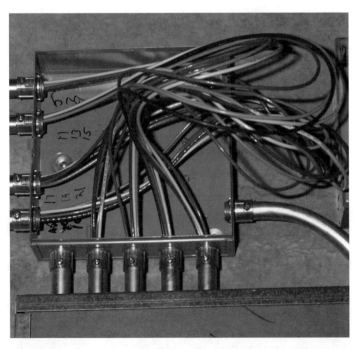

A pull box is a box used as a point to pull or feed electrical conductors into the raceway system.

ELECTRICAL SYMBOLS			
Device	**Symbol**	**Device**	**Symbol**
SINGLE RECEPTACLE OUTLET		SINGLE-POLE SWITCH	
DUPLEX RECEPTACLE OUTLET		DOUBLE-POLE SWITCH	
SPLIT-WIRED DUPLEX RECEPTACLE OUTLET		THREE-WAY SWITCH	
SINGLE SPECIAL-PURPOSE RECEPTACLE OUTLET	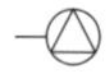	AUTOMATIC DOOR SWITCH	
DUPLEX SPECIAL-PURPOSE RECEPTACLE OUTLET		KEY-OPERATED SWITCH	
RANGE OUTLET	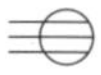 R	DIMMER	
CLOCK HANGER RECEPTACLE	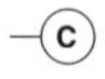	REMOTE CONTROL SWITCH	
FAN HANGER RECEPTACLE		WEATHERPROOF SWITCH	
FLOOR SINGLE RECEPTACLE OUTLET		FUSED SWITCH	
FLOOR SPECIAL-PURPOSE OUTLET		SWITCH AND SINGLE RECEPTACLE	

Figure 14-8. Symbols on electrical plan views are used to indicate various switches and receptacles.

A *dimmer switch* is a switch that changes the brightness of a lamp by changing the voltage applied to the lamp. A *photoelectric (daylight) switch* is a switch activated by light levels. A photoelectric switch is used to control dusk to dawn lighting. A *safety switch (disconnect)* is a device used periodically to remove electrical circuits from their supply source. Safety switches are used to stop current flow in an emergency situation. Safety switches may be equipped with a locking mechanism so the power cannot be activated or deactivated by accident.

A variety of receptacles are available based on design, wiring methods, grounding, use, etc. Receptacles include duplex, multioutlet assemblies, clock hanger, locking, and ground fault circuit interrupter receptacles. The most common receptacle is the duplex receptacle. A *duplex receptacle* is a receptacle that has two spaces for connecting two different plugs. A *multioutlet assembly* is a metal raceway with factory-installed conductors and attachment plug receptacles. Multioutlet assemblies provide a series of fixed placement receptacles and power sources along a continuous strip. A

clock hanger receptacle is a receptacle that has a slight recess in the face of the receptacle. A *locking receptacle* is a receptacle designed for a plug to be twisted and locked into the face of the receptacle. A *ground fault circuit interrupter (GFCI) receptacle* is a device that interrupts the flow of current to the load when a ground fault exceeding a predetermined value of current occurs. GFCIs are designed for installation in areas where exposure to water or moisture is possible. GFCIs automatically stop current flow to prevent electrical shock.

Service

A *service* is the electrical supply, in the form of conductors and equipment, that provides electrical power to a building or structure. Service components shown on electrical prints and included in this portion of the estimating process include transformers, panelboards, fuses and circuit breakers, and busways. Electrical schematic drawings indicate the various electrical loads, circuits, and demands on each leg of the electrical system. See Figure 14-9.

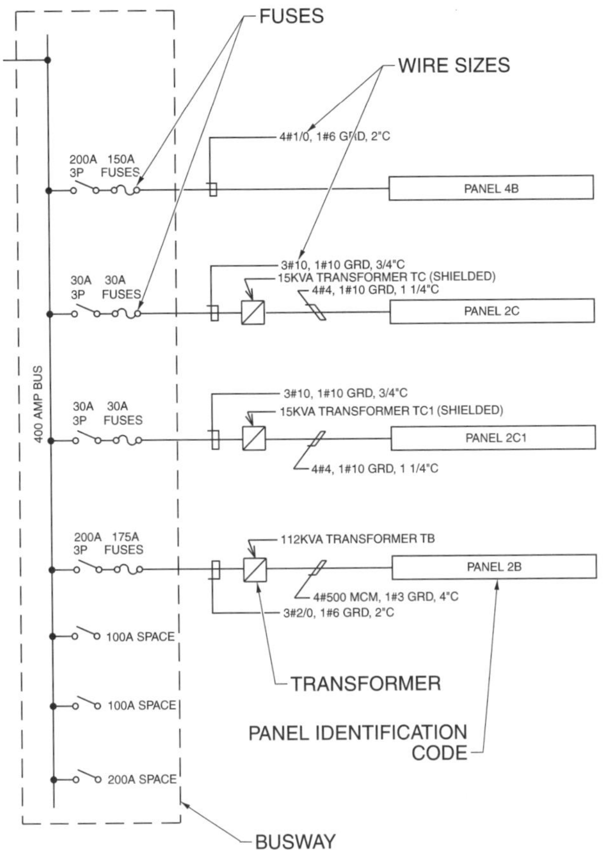

a later point in the project and installed into the enclosures at the appropriate time. See Figure 14-10. Panelboards are rated by their total load capacity as stated in amperes.

Figure 14-9. Estimators use schematic drawings to determine wire size, transformer, and fuse requirements.

Transformers. A *transformer* is an electric device that uses electromagnetism to change voltage from one level to another. Transformers are sized by the number of kilovolt-amperes (kVA) they can handle. The size of a transformer for an application is based on the amount of expected power required. The most common transformers are air-cooled. Large transformers are cooled with oil or other compounds.

Panelboards. A *panelboard* is a wall-mounted distribution cabinet containing overcurrent and short-circuit protection devices for lighting, heating, or power circuits. Panelboards consist of an enclosure (tub) and an interior on which circuit breakers or fuses are mounted and connected to busbars. Panelboard interiors are commonly assembled at a factory in accordance with specific requirements as designed by an electrical engineer. The enclosures and interiors are often shipped from the factory as separate items. This allows the enclosures to be installed on the project while the interiors are being assembled. The interiors are shipped at

Figure 14-10. Panelboards provide a connection for power distribution and circuit protection.

Fuses and Circuit Breakers. A *fuse* is an overcurrent protection device with a fusible link that melts and opens a circuit when an overload condition or short circuit occurs. A *circuit breaker* is an overcurrent protection device with a mechanism that automatically opens a circuit when an overload condition or short circuit occurs. Fuses and circuit breakers are rated by the current that can flow through the device without interrupting the circuit. The interrupting (trip) mechanism within a circuit breaker is operated thermally, magnetically, or by a thermal/magnetic combination. A thermally-sensitive component operates the trip mechanism when an electrical overload causes a gradual increase in circuit current that exceeds the current rating of the circuit breaker. A magnetically-sensitive component operates the trip mechanism when an instantaneous increase in circuit current caused by a short circuit occurs. Most circuit breakers use a combination thermal/magnetic trip mechanism.

Busways. A *busway* is a metal-enclosed distribution system of busbars available in prefabricated sections. Busways are used where conventional wiring methods are not practical or cost-effective. Busways include feeder and plug-in busways. Feeder busways are rated from 800 A to 5000 A and are used to distribute power from a large power distribution point, such as the main switchgear, to several smaller distribution points, such as panelboards. The feeder busway is tapped by use of special fittings to provide electrical power at many locations throughout the system. Feeder busway tap ratings range from 200 A to 1600 A. Plug-in busways are rated from 225 A to 600 A and are frequently installed overhead where heavy power loads are necessary and outlet locations must be flexible and movable. Plug-in busway tap ratings range from 30 A to 100 A. See Figure 14-11. Many different fittings and delivery devices are used with busways. Electrical drawings and specifications provide detailed information about busway equipment.

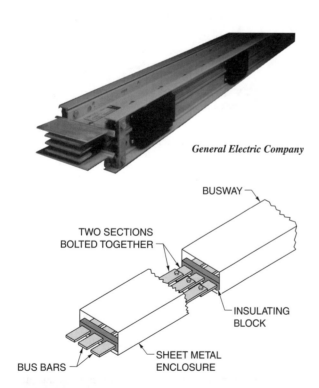

General Electric Company

Figure 14-11. Busways allow flexibility in power distribution systems where locations of electric-powered equipment change often.

FIXTURE MATERIALS AND METHODS

After panelboards, conduit, and wiring are installed, connections to finish electrical fixtures are made. Finish electrical fixtures include light fixtures, communication systems, and alarm systems. When estimating a large panelboard with many circuits and finish fixtures, ensure that proper circuits are provided for the finish fixtures. Symbols indicate devices and general locations. In a manner similar to lighting plans, the panelboard number and circuit numbers are indicated at the end of each wiring run.

Lighting

Common light fixtures include surface-mounted, ceiling-mounted, recessed, track lighting, bracket-mounted, pole-mounted, and exit lights. Common lamps include incandescent, fluorescent, and high-intensity discharge lamps. Separate plans and schedules may be provided for light fixture installation. Symbols and abbreviations indicate the light fixture installed at each location. See Figure 14-12. Specifications and electrical prints indicate lamps for each fixture.

Light Fixtures and Poles. A wide variety of designs are available for light fixtures and poles. Manufacturers and suppliers provide fixtures and poles in a broad range of designs, materials, colors, shapes, and application functions. Estimators should compare information provided by manufacturers and suppliers to the specifications and electrical plans to ensure that the size, style, and material of the fixture is in accordance with the design.

Surface-mounted light fixtures are fastened to walls, ceilings, or other structural members. Ceiling-mounted light fixtures may be surface-mounted or pendant light fixtures. Recessed light fixtures are designed for the fixture housing to be above the finished ceiling surface. Track lighting is comprised of a frame along which multiple lamp receptacles are mounted for flexibility of lighting design. Bracket-mounted fixtures allow for directional adjustment of lights and are commonly used for floodlights. Pole-mounted fixtures are mounted on steel, precast concrete, aluminum, or fiberglass poles and are commonly used for street lighting and exterior applications. Exit lights are normally surface-mounted fixtures used to indicate the direction to a safe exit route in case of an emergency.

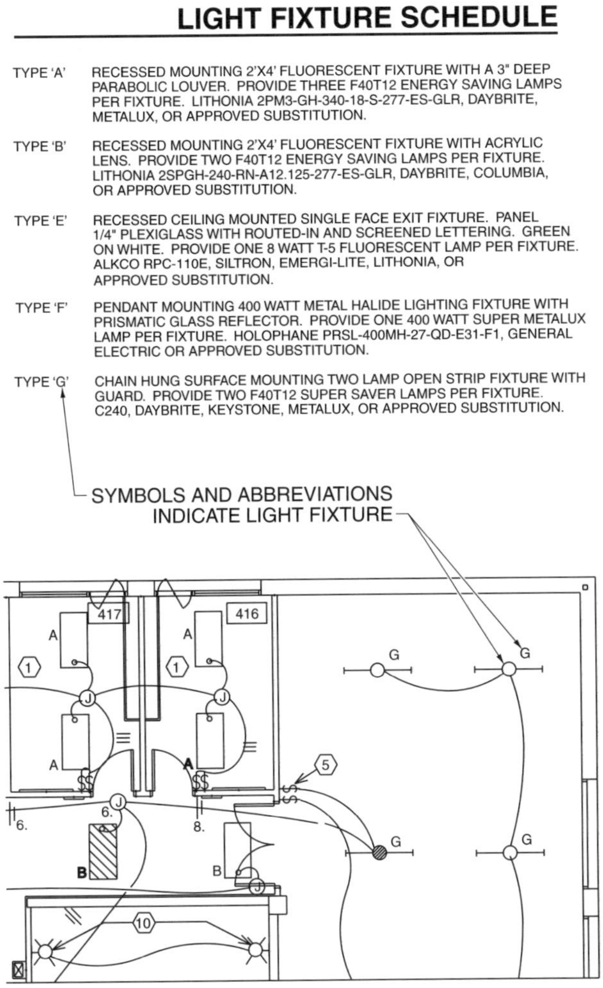

LIGHT FIXTURE SCHEDULE

TYPE 'A'   RECESSED MOUNTING 2'X4' FLUORESCENT FIXTURE WITH A 3" DEEP PARABOLIC LOUVER. PROVIDE THREE F40T12 ENERGY SAVING LAMPS PER FIXTURE. LITHONIA 2PM3-GH-340-18-S-277-ES-GLR, DAYBRITE, METALUX, OR APPROVED SUBSTITUTION.

TYPE 'B'   RECESSED MOUNTING 2'X4' FLUORESCENT FIXTURE WITH ACRYLIC LENS. PROVIDE TWO F40T12 ENERGY SAVING LAMPS PER FIXTURE. LITHONIA 2SPGH-240-RN-A12.125-277-ES-GLR, DAYBRITE, COLUMBIA, OR APPROVED SUBSTITUTION.

TYPE 'E'   RECESSED CEILING MOUNTED SINGLE FACE EXIT FIXTURE. PANEL 1/4" PLEXIGLASS WITH ROUTED-IN AND SCREENED LETTERING. GREEN ON WHITE. PROVIDE ONE 8 WATT T-5 FLUORESCENT LAMP PER FIXTURE. ALKCO RPC-110E, SILTRON, EMERGI-LITE, LITHONIA, OR APPROVED SUBSTITUTION.

TYPE 'F'   PENDANT MOUNTING 400 WATT METAL HALIDE LIGHTING FIXTURE WITH PRISMATIC GLASS REFLECTOR. PROVIDE ONE 400 WATT SUPER METALUX LAMP PER FIXTURE. HOLOPHANE PRSL-400MH-27-QD-E31-F1, GENERAL ELECTRIC OR APPROVED SUBSTITUTION.

TYPE 'G'   CHAIN HUNG SURFACE MOUNTING TWO LAMP OPEN STRIP FIXTURE WITH GUARD. PROVIDE TWO F40T12 SUPER SAVER LAMPS PER FIXTURE. C240, DAYBRITE, KEYSTONE, METALUX, OR APPROVED SUBSTITUTION.

SYMBOLS AND ABBREVIATIONS INDICATE LIGHT FIXTURE

Figure 14-12. Fixture schedules and electrical print symbols indicate the general location and light fixture required.

Lamps. An *incandescent lamp* is an electric lamp that produces light by the flow of current through a tungsten filament inside a gas-filled, sealed glass bulb. Incandescent lamps include general service, parabolic reflector, and tungsten-halogen lamps. Incandescent lamps are identified by shape. Shape designations include A for a standard lamp, G for globe-shaped, PAR for parabolic-shaped, and PS for pear-shaped. The designation ES indicates extended service lamps that are used in hard-to-reach locations. Tungsten-halogen lamps maintain a constant light output during the life of the lamp.

A *fluorescent lamp* is a low-pressure discharge lamp in which ionization of mercury vapor transforms ultraviolet energy generated by the discharge into light. Fluorescent lamps are available in standard tube design, high-output design, and special shapes such as U or circle. Fluorescent lamps are available in tube lengths from 18" to 96" and diameters from $^5/_8$" to $2^1/_8$". Energy-efficient fluorescent lamps operate on less electrical power than standard lamps.

A *high-intensity discharge (HID) lamp* is a lamp that produces light from an arc tube. An *arc tube* is the light-producing element of an HID lamp. High-intensity discharge lamps include mercury-vapor lamps, metal-halide lamps, and high-pressure sodium lamps. A *mercury-vapor lamp* is an HID lamp that produces light by an electric discharge through mercury vapor. Mercury-vapor lamps are installed where long bulb life is required and replacement of the lamp is difficult. A *metal-halide lamp* is an HID lamp that produces light by an electric discharge through mercury vapor and metal halides in the arc tube. Metal-halide lamps provide clear white light and are used in interior and exterior applications. A *high-pressure sodium lamp* is an HID lamp that produces light when current flows through sodium vapor under high pressure and high temperature. High-pressure sodium lamps are commonly used for street lamps and other exterior lighting applications.

Control and Signal Systems

Electrical control and signal systems include motor control circuits, temperature control circuits, communication circuits, and various alarm and signaling systems. Electrical control and signaling systems are low-voltage systems requiring transformers and specialized wiring. Manufacturers and suppliers provide specific information about control and signal system requirements. These systems are commonly subcontracted to specialty contractors.

Motors and Starters. A portion of the electrical work includes equipment connections to motors and starters. Motors include two-speed, variable-speed, wound-rotor, synchronous, and direct current motors. The motor used is based on the requirements of the application. Motors are normally controlled by a motor starter. Motor starters may be manual or magnetic. A *manual motor starter* is a control device that has overload protection and uses pushbuttons to energize or de-energize the load (motor) connected to it. Manual motor starters are rated in horsepower in relation to the motor to be controlled. A *magnetic motor starter* is a control device that has overload protection and uses a small control current to energize or de-energize the load (motor) connected to it. AC magnetic motor starters are the most common motor starters used for single-phase and three-phase motors. Many different magnetic motor starters are available. Estimators should consult the project specifications and equipment schedule for the starter requirements for each motor or device to be controlled.

Temperature Control. Temperature control systems are commonly installed in conjunction with the heating, ventilating, and air conditioning (HVAC) system and may not be included in Division 16 of the CSI MasterFormat™ or the specifications. However, locations for thermostats that control HVAC systems are shown on electrical plans. See Figure 14-13. Connections from the thermostat to the heating or air conditioning equipment may be shown on plan views.

Communication Systems. Communication systems include antennas, closed-circuit television, telephone equipment, intercom systems, and master clock systems. Symbols and abbreviations on electrical plans indicate connection locations for televisions, computer networks, speakers, microphones, satellite dishes, and intercoms. Each of these systems is highly specialized based on the manufacturer or equipment supplier. Estimators must work closely with the manufacturer or supplier to ensure that all components are included according to the specifications and plans.

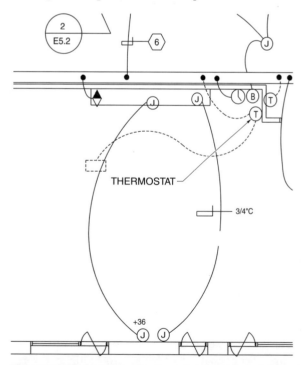

Figure 14-13. Locations for thermostats that control heating, ventilating, and air conditioning systems are shown on electrical plans.

Alarms. Configurations of wiring for smoke detection, heat detection, and security systems are also shown on electrical plans. Fire alarm and security systems are often required to be tested prior to the local government agency allowing the owner to occupy a large structure. Non-staining smoke may be released in the interior of a building to ensure that all fire warning systems are operable prior to issuance of an occupancy permit. Security system equipment includes card readers, patrol tour systems, door and window alarms, photoelectric or infrared detection systems, ultrasonic systems, microwave systems, vibration detection systems, closed circuit television systems, and audio communication systems. The security system installed depends on the time of response required, facility location, and the degree of security needed.

ELECTRICAL SYSTEM QUANTITY TAKEOFF

Electrical prints contain information about the wiring to be installed, electrical equipment, and electrical finish materials. See Figure 14-14. Specific dimensions for electrical wiring, equipment, and fixture installation are not given in all instances. Dimension information is obtained from architectural and structural plan views and elevations. Some electrical information may be shown on an electrical site plan that indicates power sources and exterior lighting.

Electrical estimators must be familiar with local building and fire protection codes pertaining to electrical work. Electrical estimators should check the specifications carefully. There may be occasions when information on architectural and electrical drawings may not be in accordance with these codes. Estimates taken only from specifications and drawings without reference to the local electrical codes could result in an inaccurate bid. Any variances between the drawings and codes should be brought to the attention of the architect prior to bid submission.

An electrical equipment circuit schedule may be included in the electrical prints. This schedule names each piece of equipment and shows the amount of electric power necessary, as well as circuit assignments. See Figure 14-15.

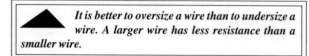

It is better to oversize a wire than to undersize a wire. A larger wire has less resistance than a smaller wire.

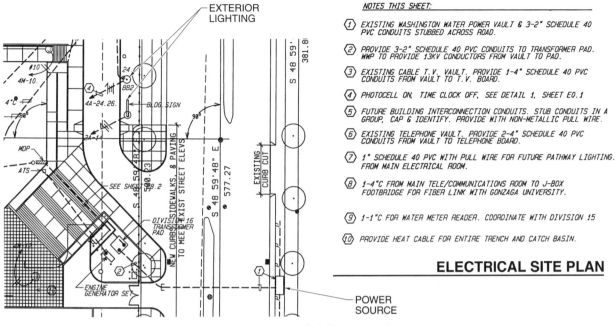

Figure 14-14. Site plans show power source and exterior lighting information.

Takeoff Methods

To ensure estimating accuracy, electrical estimators create separate worksheets for each type of electrical item needed, such as cables, light fixtures, lamps, switches, etc. The location for each item is noted on the worksheet to ensure an accurate and complete takeoff. Electrical estimators must also accurately and completely mark up electrical plans to ensure that all items are taken off. Two methods of estimating used for electrical takeoff are detailed and averaging takeoff.

Detailed. The most accurate takeoff method of estimating electrical work is making a detailed analysis of all required electrical work. Detailed takeoff is a precise method used for industrial and commercial work and wherever unique job conditions exist. A detailed analysis is made from electrical plans and specifications. If detailed electrical plans and specifications are not available, the electrical estimator prepares electrical plans (layouts) that are approved by the architect before beginning the estimate. These electrical layouts show all necessary wiring and equipment and give the size and type of electrical service and the size and location of all feeders and distribution panelboards. Electrical layouts show the size and location of all motors and controllers and the type, number, and location of all fixtures, switches, and receptacles.

Items taken off during electrical estimating include wire, wire nuts, boxes, box covers, fittings, and mounting hardware.

ELECTRICAL EQUIPMENT CIRCUIT SCHEDULE

ITEM	NAME	VOLT/PHASE	HP/AMP	SAFETY SWITCH	FUSE	CIRCUIT	CIRCUIT No.	MAG. STARTER
AC-1	AIR COMP.	480-3Ø	40 HP	3P-100 A	80 A	3#4, 1 1/4"C	4M-38.40.42.	DIV 15
	AIR DRYER	480-3Ø	40 HP	3P-30 A	50 A	3#6, 1"C	4M-43.45.47.	SIZE 1, FVN
AS-1	AIR SHOWER	460-1Ø	2 HP	3P-30 A	5.6 A	3#12, 3/4"C	4M-32.34.36.	SIZE 0
B 1	BOILER	480-3Ø	3 HP	3P-30 A	8 A	3#12, 3/4"C	4M-1.3.5	DIV 15
B 2	BOILER	480-3Ø	3 HP	3P-30 A	8 A	3#12, 3/4"C	4M-7.9.11	DIV 15
BCP-1	CIRC PUMP	480-3Ø	3 HP	3P-30 A *	8 A	3#12, 3/4"C	MCC-1A	SIZE 1, FVNR
BCP-2	CIRC PUMP	480-3Ø	3 HP	3P-30 A *	8 A	3#12, 3/4"C	MCC-1F	SIZE 1, FVNR
CAB-1	CABINET HEATER	120-1Ø	1/60 HP	DIV 15	N/A	2#12, 3/4"C	2M-9.	N/A
CAB-2	CABINET HEATER	120-1Ø	1/60 HP	DIV 15	N/A	2#12, 3/4"C	2M-9.	N/A
CAB-3	CABINET HEATER	480-3Ø	1/60 HP	3P-30 A	N/A	2#12, 3/4"C	4B-22.24.26.	N/A
CAB-4	CABINET HEATER	480-3Ø	1/60 HP	3P-30 A	N/A	2#12, 3/4"C	4B-28.30.32.	N/A
CCHWP-1	PUMP	480-3Ø	3 HP	3P-30 A *	8 A	3#12, 3/4"C	MCC-1B	SIZE 1, FVNR
CCHWP-2	PUMP	480-3Ø	1 HP	3P-30 A	2.8 A	3#12, 3/4"C	MCC-1B	SIZE 1, FVNR
CCHWP-3	PUMP	480-3Ø	1.5 HP	3P-30 A *	4 A	3#12, 3/4"C	MCC-2A	SIZE 1, FVNR
CCHWP-4	PUMP	480-3Ø	1.5 HP	3P-30 A *	4 A	3#12, 3/4"C	MCC-2A	SIZE 1, FVNR
CCHWP-5	PUMP	480-3Ø	3/4 HP	3P-30 A *	2.25 A	3#12, 3/4"C	MCC-1A	SIZE 1, FVNR
CCHWP-6	PUMP	120-1Ø	1/3 HP	1P-30 A	12 A	2#12, 3/4"C	2M-2	SIZE 00, FVNR
CH-1	CHILLER	480-3Ø	309 MCA	3P-400 A	400 A	SEE RISER	MPD	DIV 15
CH-2	FUTURE CHILLER	480-3Ø	309 MCA			SEE RISER	MPD	
CHP-1	PUMP	480-3Ø	10 HP	3P-30 A *	20 A	3#12, 3/4"C	MCC-1F	SIZE 1, FVNR
CHP-2	PUMP	480-3Ø	10 HP	-	-	3/4"C ONLY	MCC-1B	
CHWP-1	PUMP	480-3Ø	3 HP	3P-30 A *	8 A	3#12, 3/4"C	MCC-1B	DIV 16
CHWP-2	PUMP	480-3Ø	3/4 HP	3P-30 A *	2.25 A	3#12, 3/4"C	MCC-2A	DIV 16
CHWP-3	PUMP	480-3Ø	3/4 HP	3P-30 A *	2.25 A	3#12, 3/4"C	MCC-1A	DIV 16
CHWP-4	PUMP	120-1Ø	1/2 HP	3P-30 A *	15 A	2#12, 3/4"C	MCC-1E	SIZE 1, FVNR
CT-1	COOLING TOWER	480-3Ø	15 HP	3P-60 A	30 A	3#10, 3/4"C	4M-13,15,17	DIV 15
		480-3Ø	10 KW	3P-30 A	N/A	3#12, 3/4"C	4M-20,22,24	N/A
CT-2	COOLING TOWER	480-3Ø	15 HP			2)3/4"CO.		
CU-1	CONDENSER	208-1Ø	14 FLA	3P-30 A	20 A	3#10, 3/4"C	2X1-2.4.	SIZE 1, FVNR
	PAINT BOOTH	120-1Ø	15 A	N/A	N/A	2#12, 3/4"C	2A-30.	
	PAINT BOOTH	480-3Ø	3/4 HP	3P-30 A	225 A	2#12, 3/4"C	4A-23.25.27.	DIV 15
CWP-1	PUMP	480-3Ø	20 HP	3P-60 A	40 A	3#8, 1"C	4M-2,4,6	DIV 15
CWP-2	PUMP	480-1Ø	20 HP			1"C ONLY	4M	N/A

Figure 14-15. Electrical equipment circuit schedules provide information for each piece of electrical equipment required for the construction project.

In detailed takeoff, a list of the materials required is created. This list includes boxes, receptacles, conduit and fittings, demolition, equipment hookup and connection, motor controls, service and distribution equipment, switches, trenching and excavation, and specialty items such as alarms and communication systems. Each of these items is thoroughly checked by the estimator to ensure a complete and comprehensive takeoff. A ledger sheet, spreadsheet, or estimating software may be used for electrical system detailed takeoff. See Figure 14-16.

The total cost of material is obtained by consulting information from various manufacturers and suppliers. The amount of electrical labor required to install wiring and equipment is determined by applying labor units to each type of material. A *labor unit* is the average time an electrician takes to install a specific type of material. Labor units are obtained from standard estimating references or company historical data. See Figure 14-17. These labor units may be directly related to each item

in a printed chart or input into a database tied to each material unit. After the total amounts of electrical material and equipment are known, the estimator applies the appropriate labor units and calculates the total cost to complete the installation.

Variables considered during preparation of the electrical estimate include items such as the potential need for temporary service or temporary wiring for power tools during construction, job site conditions where equipment access may be restricted, and climatic conditions that affect labor production such as rain or snow. After all variables have been evaluated, an electrical estimator prepares an accurate bid for the complete electrical installation.

Averaging. Averaging takeoff is a simplified takeoff method and is useful for a non-specialist in electrical estimating. Averaging takeoff is sufficiently accurate for small takeoff jobs on basic electrical wiring systems. In averaging takeoff, costs are calculated as an average cost per outlet. Estimators determine the cost of the materials and labor required to install a typical electrical outlet. This is done for each different type of outlet in addition to service entrance conduit, meters, and panelboards. The total cost of the job is determined from these figures. Variables considered include the wiring methods used, local codes and ordinances, material costs, labor regulations, and labor costs. Due to local variations, it is not possible to quote exact cost figures for these items. Outlet, receptacle, and switch costs are commonly figured on the basis of a 15′ run of cable or conduit.

Electrical Worksheet

Project: #49864789

Estimator: MJT

No.	Desc	Duplex 15 A	Duplex 20 A	Switch SP 20 A	Receptacle Cover	EMT 1/2"
	Office 1	6	1	1	6	300
	Office 2	6	1	1	6	300
	Office 3	4	—	1	4	280

ITEMS REQUIRED

Figure 14-16. Detailed takeoff includes a listing of each item required for the electrical system.

Wiring Takeoff

Wiring, cable, and conduit takeoff begins by listing each type of wire, cable, and conduit noted on the electrical drawings. A paper or electronic template already containing this listing may be used to start the process. After all types are listed, quantity takeoff is made as to the number of linear feet for each type. From scaled drawings, the length of each circuit is measured to determine the total length for conduit and wiring. See Figure 14-18. It is common practice to determine the length of wire based on the length of conduit required. For example, for conduit with two wires, the length of conduit is multiplied by 2. Extra wire and cable amounts are added to the total linear feet to allow for connections at fixtures, junction boxes, devices, and panelboards. The wire, cable, and conduit are listed in separate sheets or pages of the takeoff. Material and labor rates for wiring and conduit are commonly based on the cost per 100′ of conduit or 1000′ of wire. Detailed costs are available from industry standard information or company historical data.

CABLE INSTALLATION LABOR

5 kV Unshielded	Unit*	Labor Hours per Unit
No. 8	100	2.3
No. 6	100	2.5
No. 4	100	2.7
No. 2	100	3.4
No. 1/0	100	3.6
No. 2/0	100	4.0
No. 3/0	100	4.2

* in lf

BUSWAY INSTALLATION LABOR

Copper Busway in Steel Case*	Unit**	Labor Hours per Unit
225	10	4.7
400	10	5.7
600	10	6.8
800	10	8.1
1000	10	9.0
1350	10	11.5
1600	10	16.9
2000	10	24.0

* rating in A
** in lf

Figure 14-17. The amount of electrical labor required to install wiring and equipment is determined by applying labor units to each type of material.

Greenlee Textron Inc.

Wire pulling equipment is included in electrical wiring takeoff.

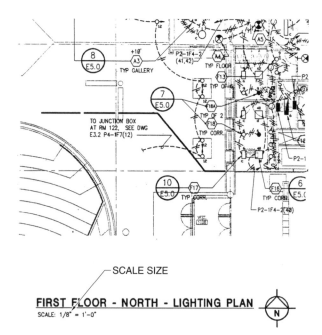

SCALE SIZE

FIRST FLOOR - NORTH - LIGHTING PLAN
SCALE: 1/8" = 1'-0"

Figure 14-18. For wiring, cable, and conduit takeoff, scaled plans are used to determine the length of each circuit run.

RIGID METAL CONDUIT ALLOWABLE FILL					
Size*	Internal Diameter*	Total Area** 100%	2 Wires 31%**	Over 2 Wires 40%**	1 Wire 53%**
½	.632	.314	.097	.125	.166
¾	.836	.549	.170	.220	.291
1	1.063	.888	.275	.355	.470
1¼	1.394	1.526	.473	.610	.809
1½	1.624	2.071	.642	.829	1.098
2	2.083	3.408	1.056	1.363	1.806
2½	2.489	4.866	1.508	1.946	2.579
3	3.090	7.499	2.325	3.000	3.975
3½	3.570	10.010	3.103	4.004	5.305
4	4.050	12.283	3.994	5.153	6.828

* in in.
** percent of fill in sq in.

Figure 14-19. Electrical conduit can only be filled to a predetermined capacity to stay within NEC® requirements.

Estimators must also include some allowance for connectors and clamps for electrical wiring, cable, and conduit. This includes clips, wire nuts, bushings, nipples, gaskets, washers, special pulling equipment required for cables, and connecting lugs. Knowledge about these specialized fittings and connectors is required to complete the wiring and conduit bid.

Conduit. Conduit size is determined based on the number of conductors or cables allowed by the NEC®. NEC® tables give information concerning the allowable percentage of conduit fill. See Figure 14-19. Estimators may allow additional space in conduit beyond the values given in the NEC® tables to keep labor costs down. Conduit with a higher percentage fill makes pulling of wire difficult. It is common to increase the conduit size after reaching approximately 80% of the allowable conduit fill. Conduit fittings are made for each connection and termination of conduit. One coupling is allowed for each 10′ length of conduit. Two additional couplings are added for each 90° elbow. Special attention is required to ensure that the proper fittings are provided in special conduit situations.

Estimators must also take wall height and fixture height into account during wiring, cable, and conduit takeoff. Additional length is added to the totals obtained from plan views based on heights above floors or below ceilings for switches, receptacles, and fixtures.

Boxes and Cable Supports. Electrical boxes are taken off as individual units. Estimators create a listing or use a template to count each box of each type on the electrical plans. Each type of box is listed and quantities are noted during plan takeoff and markup. Estimators refer to plan symbols and specification notes to ensure proper takeoff of each type of box. See Figure 14-20. Estimators must also list each cover for each box in the takeoff according to the specifications and electrical prints.

Cable tray system takeoff is based on the linear feet of tray installed and the number of each type of fitting required to complete the system. An important element to consider during cable tray system takeoff is the accessibility of the area into which the tray system is being installed. There may be potential conflicts with HVAC piping, sprinkler systems, and other overhead devices. There may also be height requirements that make hanging the cable tray difficult.

Wiring Devices. Wiring devices for takeoff include switches, receptacles, etc. As with light fixtures, these devices are taken off as the total number of each individual type of device. For example, the estimator counts the number of single-pole 20 A switches and duplex 20 A receptacles in each location. The location is entered in the left column on a ledger sheet or spreadsheet. The type of wiring device is listed in a row across the sheet. See Figure 14-21. The total of each type of device for each location of the structure is entered into the appropriate row, column, or cell. The estimator must be familiar with the various electrical symbols used by architects to depict electrical devices. Cover plates for each type of switch and receptacle are also included in this portion of the takeoff. Material and labor rates for each device are determined from manufacturer and supplier information, standard industry references, or company historical data.

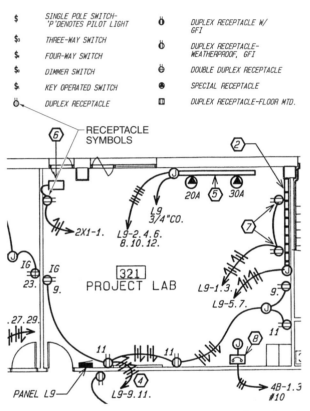

Figure 14-20. Electrical receptacles are noted on electrical drawings with standard symbols.

Lighting Takeoff

Takeoff for light fixtures is often the first step taken by the electrical estimator in the bid process. This portion of the estimate may also be completed by electrical suppliers under some conditions where there is a large fixture package on a large job. Estimators must include all portions of the lighting estimate, including fixtures, poles, lamps, fixture guards, safety wires, cable supports, suspended fixture safety clips, and other special support equipment that may not be readily apparent from the electrical prints.

Light Fixtures, Poles, and Lamps. Each type of light fixture should be listed in the first column of a ledger

sheet or spreadsheet or entered as an item into estimating software. See Figure 14-22. Each fixture is counted as an individual item. The number of each type of fixture is totaled and entered into the appropriate cell.

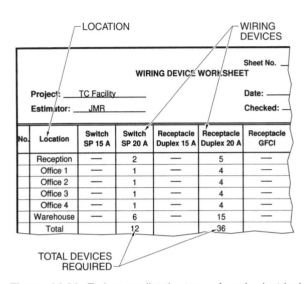

No.	Location	Switch SP 15 A	Switch SP 20 A	Receptacle Duplex 15 A	Receptacle Duplex 20 A	Receptacle GFCI
	Reception	—	2	—	5	—
	Office 1	—	1	—	4	—
	Office 2	—	1	—	4	—
	Office 3	—	1	—	4	—
	Office 4	—	1	—	4	—
	Warehouse	—	6	—	15	—
	Total		12		36	

Figure 14-21. Estimators list the type of each electrical device followed by the quantity takeoff for each device.

QUANTITY SHEET

Project: Lake Shore Facility

Estimator: AJD

No.	Description	Material			
		Quantity	Unit	Price/Unit	Total
	Light Fixture Type A	25	Ea	$70.00	$1,750.00
	Light Fixture Type B	3	Ea	$180.00	$540.00
	Light Fixture Type C	200	Ea	$77.00	$15,400.00
	Light Fixture Type D	50	Ea	$26.00	$1,300.00
	Light Fixture Type E	68	Ea	$53.00	$3,604.00
	Light Fixture Type F	63	Ea	$32.00	$2,016.00
	Light Fixture Type G	25	Ea	$127.00	$3,175.00
	Light Fixture Type H	18	Ea	$325.00	$5,850.00

Figure 14-22. Light fixtures are counted as individual items during electrical takeoff.

> ▲ *Manufacturers use footcandle levels to aid in selecting and installing lamps. Manufacturer charts list the expected footcandle rating for an area given the mounting height and spacing of the lamps.*

Light fixtures are priced in a number of different methods. Unit cost pricing is a method where each type of fixture is priced individually. Lump sum pricing is a method where a lot price is determined for all fixtures on a project. Pricing may be performed with a combination of unit cost and lump sum pricing for various types of fixtures.

Some job site conditions that may increase light fixture labor costs include a job site with excessive debris, high ceilings, inaccessible work areas due to small rooms or working around other obstructions, and the inability to move the fixtures into the final installation area.

Material and labor costs for support poles are taken off as individual units. Labor costs for poles may include equipment costs for installation and worker lifts to reach the top of the pole for fixture installation.

Lamps are taken off and totaled on a separate quantity takeoff. The number and type of lamps for each fixture are determined and multiplied by the number of fixtures to obtain a total quantity.

Light fixture support poles are taken off as individual units and require additional labor and equipment costs for installation of the fixture at the top of the pole.

Power, Control, and Signal System Takeoff

All power, control, and signal systems such as alarms, buzzers, and bells are taken off as individual units. Factors considered include consideration of the service voltage and size, single- or three-phase installation, interior or exterior installation, grounding considerations, overhead or underground service, corrosion-resistance requirements, fuses and circuit breakers, special metering equipment, utility company hookup fees, and special equipment support provisions.

Power. Panelboards, transformers, and other main service equipment are based on the total electrical load with some added capacity for potential service expansion. This portion of the takeoff includes panelboards, fuses and circuit breakers, and any metering equipment. This bid segment may be prepared by a manufacturer or supplier who specializes in service equipment. Labor unit costs are based on industry standard references or company historical data for each unit of material.

Locations of panelboards are shown on electrical plan views. The electrical plan views relate to the overall electrical service schematic and to detail drawings. See Figure 14-23. Panelboards are noted by an architectural symbol and may have a letter or number code relating to a schedule. A schedule of panelboards indicates the size of the panelboard, mounting directions, and equipment to be serviced by each circuit.

Hookup. Electrical hookup for mechanical, electrical, and kitchen equipment and control and alarm systems can be complex. Design and electrical engineers should be consulted concerning special hookup requirements for specialized systems and machinery. Items considered include the size, voltage, and possible horsepower of the equipment, motor protection requirements, disconnect and lockout switches, ground provisions, fan requirements, fire protection hookups, or any potential hazardous installation conditions. Each of these items could add to the labor hours required for hookup. Labor units for hookup of various pieces of equipment are based on standard industry references or company historical data for each piece of equipment.

▲ *Aluminum wire is larger in diameter than copper wire of equal ampacity and may require larger conduit than copper wire. In addition, aluminum expands and contracts more than copper, which can loosen a connection that has not been properly tightened.*

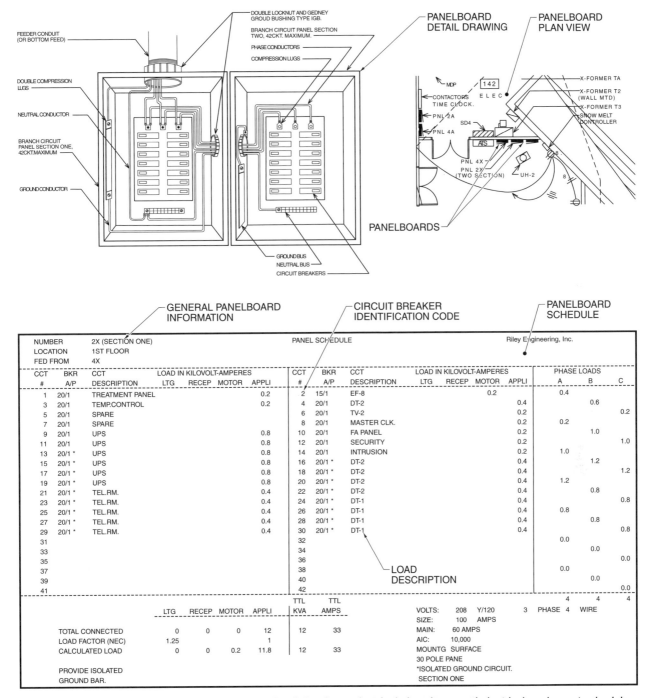

Figure 14-23. Panelboard information is given on detail drawings, electrical plan views, and electrical equipment schedules.

ESTIMATING SOFTWARE – BUILDING DATABASES USING ONE-TIME ITEMS

One-time items may be taken off and automatically added to the database in Precision estimating software by applying a standard procedure. See Figure 14-24.

Panelboards provide protection for light, heat, and/or power circuits. Panelboards are designed for placement in or on a wall or partition and must be accessible from the front. Panelboards are selected based on the required current level and type of enclosure. The enclosure may be designed for surface or flush mounting.

PRECISION ESTIMATING SOFTWARE – BUILDING DATABASES USING ONE-TIME ITEMS

1. Click the One-time Item button () on the toolbar to open the One-time Item window.

2. Create the Phase record by right-clicking in the Phase field and selecting Edit Phase from the shortcut menu. Click the Add button.

3. Input 16100 in the Phase field and Subcontractors in the Description field. Click OK and close the window.

4. Complete the One-time Item fields. Enter 16100 for Phase, Electrical Rough-In for Description, 1 for Quantity, and lsum for Takeoff unit.

5. Uncheck the Labor and Material Categories and place a check mark in the Subcontract Category box and click OK. The one-time item displays on the spreadsheet.

6. Highlight the Electrical Rough-in description and press CTRL-S to automatically jump to the Sub Amount field. Input 15000 and hit ENTER.

7. Hold down the CTRL key and hit the Left Arrow key to automatically jump to the far left column on the spreadsheet.

8. Save the one-time items associated with this estimate to the database by clicking Save One-time Items to Database from the Pricing menu. *Note:* This window presents each one-time item in the estimate to be saved or rejected. As each one-time item appears, the Item List goes to the phase specified for that item. The information shown can be accepted, new item codes entered, item details edited, different phases specified, or the item skipped altogether.

9. Precision Estimating opens the Save One-time Items to Database window. Enter 10 in the Item Field and click OK.

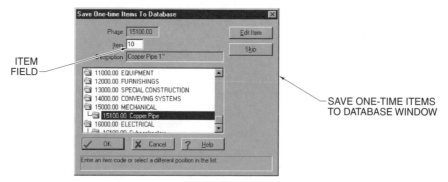

10. Scroll down and highlight Phase 16100, enter Item 10, and select OK. Once finished, click OK to return to the spreadsheet.

Note: When Precision Estimating finds a one-time item, it opens the Save One-time Items to Database window. Options include: entering the desired item code in the Item field, clicking OK to accept if an item code has been proposed, or clicking on a different item in the item list to have Precision Estimating suggest a different item code. Additional item information can be added, such as the formula, waste factor, and prices by clicking the Edit Item button and filling in the desired fields in the Adding Item window.

Figure 14-24. One-time items may be taken off and automatically added to the database.

Estimating

_____ **1.** Electrical information is covered in Division _____ of the CSI MasterFormat™.

_____ **2.** _____ is a hollow pipe that supports and protects electrical conductors.

_____ **3.** A(n) _____ is an individual conductor.

_____ **4.** A(n) _____ is a slender rod or wire that is used to control the flow of electrons in an electrical circuit.

T   F   **5.** A cable is two or more conductors grouped together within a common protective cover and used to connect individual components.

T   F   **6.** A cable tray is a closed grid rack suspended from structural members to support a series of cables.

_____ **7.** A(n) _____ is a metallic or nonmetallic electrical enclosure used for equipment, devices, and pulling or terminating conductors.

_____ **8.** A(n) _____ is a device that is used to start, stop, or redirect the flow of current in an electrical circuit.

_____ **9.** A(n) _____ is a device used to connect equipment to an electrical system with a cord and plug.

_____ **10.** A(n) _____ is the number of completely isolated circuits that a switch can control.

_____ **11.** A _____ switch is a switch activated by light levels.

A. photoelectric   C. A or B
B. daylight   D. neither A nor B

_____ **12.** A(n) _____ receptacle is a receptacle that has two spaces for connecting two different plugs.

T   F   **13.** A multioutlet assembly is a metal raceway with on-the-job installed conductors and attachment plug receptacles.

T   F   **14.** A service is the electrical supply, in the form of conductors and equipment, that provides electrical power to a building or structure.

_____ **15.** A(n) _____ is a wall-mounted distribution cabinet containing overcurrent and short-circuit protection devices for lighting, heating, or power circuits.

_____ **16.** A(n) _____ is an overcurrent protection device with a fusible link that melts and opens a circuit when an overload condition or short circuit occurs.

_____ **17.** A(n) _____ is an overcurrent protection device with a mechanism that automatically opens a circuit when an overload condition or short circuit occurs.

_____ 18. A(n) _____ is a metal-enclosed distribution system of busbars available in prefabricated sections.

_____ 19. A(n) _____ lamp is an electric lamp that produces light by the flow of current through a tungsten filament inside a gas-filled, sealed glass bulb.

_____ 20. A(n) _____ lamp is a low-pressure discharge lamp in which ionization of mercury vapor transforms ultraviolet energy generated by the discharge into light.

_____ 21. A(n) _____ lamp is a lamp that produces light from an arc tube.

_____ 22. A(n) _____ motor starter is a control device that has overload protection and uses pushbuttons to energize or de-energize the load (motor) connected to it.

_____ 23. A(n) _____ motor starter is a control device that has overload protection and uses a small control current to energize or de-energize the load (motor) connected to it.

T   F   24. Power sources are noted on site plans and electrical prints.

T   F   25. A two-way switch is a single-pole, single-throw switch.

Raceways

_____   1. Rigid metal conduit (RMC)

_____   2. Rigid nonmetallic conduit (RNMC)

_____   3. Electrical metallic tubing (EMT)

_____   4. Flexible metal conduit (FMC)

_____   5. Intermediate metal conduit

A. Metal conduit with a thicker wall than EMT, but a thinner wall than RMC

B. Lightweight tubular steel raceway without threads on the ends

C. Heavy conduit made of metal

D. Conduit made of materials other than metal

E. Raceway of metal strips that are formed into a circular cross-sectional raceway

Electrical Symbols

_____   1. Single receptacle outlet

_____   2. Single-pole switch

_____   3. Floor special purpose outlet

_____   4. Clock hanger receptacle

_____   5. Floor single receptacle outlet

_____   6. Switch and single receptacle

_____   7. Split-wired duplex receptacle outlet

_____   8. Range outlet

_____   9. Single special-purpose receptacle outlet

_____   10. Duplex receptacle outlet

Activities
Electrical

Name _____ Date _____

Activity 14-1 – Ledger Sheet Activity

Refer to the cost data, Print 14-1, and Estimate Summary Sheet No. 14-1. Take off the number of fixtures and lamps. Determine the total material and labor cost.

COST DATA			
Material	**Unit**	**Material Unit Cost***	**Labor Unit Cost***
A. Interior lighting fixture, 2′ W × 4′ L, two 40 W	ea	84.70	45.10
B. Interior lighting fixture, 2′ W × 4′ L, four 40 W	ea	100.10	52.80
C. Interior lighting fixture, acrylic lens 1′ W × 4′ L, two 40 W	ea	78.10	40.15
P. Pendant mounted	ea	47.30	49.50
Lamp, fluorescent, 4′ L, 40 W	100	247.50	312.40
Lamp, twin tube compact	100	445.50	312.40

* in $

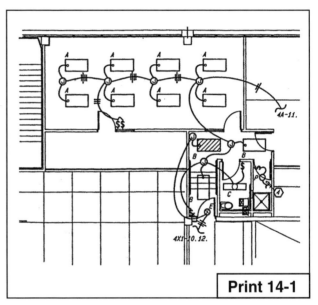

Print 14-1

ESTIMATE SUMMARY SHEET

Sheet No. _____14-1_____

Project: _____
Estimator: _____

Date: _____
Checked: _____

No.	Description	Dimensions			Quantity		Material		Labor		Total	
						Unit	Unit Cost	Total	Unit Cost	Total	Unit Cost	Total
	Lighting fixtures											
	Type A											
	Type B											
	Type C											
	Type P											
	Lamps											
	Fluorescent											
	Twin tube compact											
	Total											

Activity 14-2 – Spreadsheet Activity

Refer to the cost data, Print 14-2, and Estimate Summary Sheet No. 14-2 on the CD-ROM. Take off the number of duplex receptacles, number of smoke detectors, number of GFCI receptacles, number of fire alarms, and number of junction boxes for Suite 103. Determine the total material and labor costs.

COST DATA			
Material	**Unit**	**Material Unit Cost***	**Labor Unit Cost***
Receptacle, duplex Type NM cable	ea	5.94	19.31
Smoke detector, ceiling type	ea	82.50	45.10
GFCI receptacle, Type NM cable	ea	37.95	26.40
Fire alarm	ea	134.20	35.20
Junction box, 4″	ea	8.64	11.22

* in $

Activity 14-3 – Estimating Software Activity

Create a new estimate and name it Activity 14-3. Create a new phase record and save a new one-time item. Click the **One-Time Item** button. Right-click in the phase window. Click **Edit Phase**. Click the **Add** button. Enter *16500* for Phase and *lighting* for Description. Click **OK** and **Close**. In the one-time item window, input *Emergency Lighting* for Description, *23* for Quantity, and *each* for Takeoff Unit. Click **OK** to send the items to the spreadsheet. Enter a labor unit cost of $60 each and a material price of $30 each. Print a standard estimate report.

Specialty Items and Final Bid Preparation

Specialized applications require unique construction materials and equipment. Divisions 10 through 14 of the CSI MasterFormat™ cover specialized appliances, equipment, furnishings, construction, and conveying systems. Specialty items require highly specific estimating skills based on equipment and manufacturer recommendations and information. The final bid is the price the contractor submits to the owner to cover all expenses for materials, labor, services, overhead, and profit that are necessary to complete the job. Accurate recordkeeping is required by the estimator and construction company during the construction process to ensure accurate bidding in the future.

SPECIALTY ITEMS AND QUANTITY TAKEOFF

Unique construction materials and equipment are required for specialized applications. CSI MasterFormat™ Divisions 10, 11, 12, 13, and 14 cover specialized appliances, equipment, furnishings, construction, and conveying systems. Each of these specialty items requires highly specific estimating skills based on the equipment and manufacturer recommendations and information. Bids for these items are commonly obtained from subcontractors and specialty contractors specializing in a particular portion of this work. As with all other portions of the bid, estimators should be familiar with items required in each of these categories to ensure an accurate bid.

Division 10 – Specialties

Division 10 of the CSI MasterFormat™ includes specialty items such as partitions, access flooring, louvers, chalkboards and tackboards, toilet compartments, fireplaces, flagpoles, lockers and shelving, and pedestrian control devices. Other specialty items in this portion of the bidding documents include service walls, wall and corner guards, pest control, postal and telephone specialties, and exterior protection. Each item requires specific takeoff quantities and labor costs based on the specialty item required. Specifications for these materials commonly use manufacturer names and model numbers to identify acceptable requirements. See Figure 15-1.

Most specialty items are taken off as individual units and individual components. Individual item takeoff is used for storage units, flagpoles, signage, and lockers. Measurements in linear feet are used for takeoff of folding or sliding partitions. Interior elevations are provided to show chalkboards, projection screens, lockers, marker boards, and miscellaneous fixtures. The sizes and manufacturer codes for many of these fixtures are noted in the specifications and room finish schedule. Common specialty items include partitions, access flooring, and louvers.

SECTION 10522 - FIRE EXTINGUISHERS, CABINETS, AND ACCESSORIES

MANUFACTURERS:

Manufacturers: Subject to compliance with requirements, provide products by one of the following:

J.L. Industries
Larsen's Manufacturing Co.
Potter-Roemer, Inc.

FIRE EXTINGUISHERS:

General: Provide fire extinguishers for each cabinet and other locations indicated, in colors and finishes selected by Architect from manufacturer's standard, that comply with authorities having jurisdiction.

MANUFACTURER NAMES

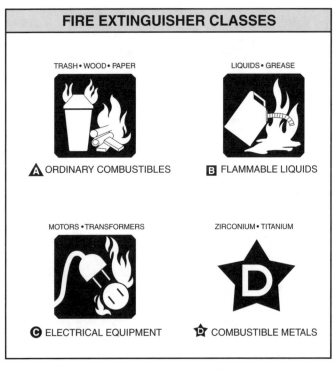

FIRE EXTINGUISHER CLASSES

TRASH • WOOD • PAPER

A ORDINARY COMBUSTIBLES

LIQUIDS • GREASE

B FLAMMABLE LIQUIDS

MOTORS • TRANSFORMERS

C ELECTRICAL EQUIPMENT

ZIRCONIUM • TITANIUM

D COMBUSTIBLE METALS

Figure 15-1. Manufacturer names are commonly used to describe specialty items such as fire extinguisher products.

Partitions. A *partition* is a vertical divider that separates one area from another. Partitions detailed in Division 10 of the CSI MasterFormat™ include operable, demountable, and wire mesh partitions, and portable partitions, screens, and panels.

An *operable partition* is a panel or partition that is moved by hand or motor along an overhead track. Operable partitions include folding (accordion), coiling, and sliding partitions. Folding, coiling, and sliding partitions are commonly supported by hanging from an overhead track supported by structural members. Folding, coiling, and sliding partitions may also be floor-supported. Folding, coiling, and sliding partitions are installed between large rooms to provide open spaces to be divided for use by multiple groups. See Figure 15-2. Panel thickness, width, height, finish, and acoustic properties vary depending on architect requirements and manufacturer designs.

Figure 15-2. Partitions allow for temporary division of large spaces into smaller units.

A *demountable partition* is a partition that is designed to be disassembled and reassembled as needed to form new room layouts. Demountable partitions are connected to permanent construction and can be moved without damage. Tracks are mounted to the finished floor and ceiling and wall panels, doors, lights, and frames are set in the tracks. Demountable partition panels may be made of gypsum drywall or other composite materials. Takeoff for demountable partitions includes the number of panels required, door units, window units, and linear feet of top and bottom track and other hardware such as stiffeners between panels.

A *wire mesh partition* is a partition fabricated of wire mesh. Portable partitions, screens, and panels include partitions, wall screens, and operable panels that are not connected to permanent construction.

Partition installation items considered by the estimator include the quality and type of partition installed and

the installation requirements. For operable and demountable partitions, the setting and fastening of the track is a key item considered in labor costs.

Access Flooring. *Access flooring* is a raised floor system supported by short posts and runners. Access flooring supports prefinished floor panels above the surface of the structural floor. See Figure 15-3. The open area below the prefinished floor panels is used for routing electrical or computer wiring and ventilation ducts. Access flooring systems may use tracks between the pedestals to support the prefinished floor panels. Prefinished floor panels are commonly manufactured in 24″ squares. Items considered by the estimator include the method of fastening the pedestals to the structural floor, track requirements, the type of finished surface on the floor panels, and the requirements for any ramps or edge treatments around the perimeter of the raised floor area.

> *Access flooring systems use square panels that are selected based on loading requirements and are supported by adjustable pedestal assemblies that positively locate, engage, and secure the panels.*

Figure 15-3. Access flooring supports prefinished floor panels above the surface of the structural floor.

Louvers. A *louver* is a ductwork cover containing horizontal slats that allow the passage of air but prevent rain from entering. Louvers allow air flow from one area to another while providing a level of weather protection, visual protection, or security. Louver styles include standard, storm-proof, operating, acoustic, and sight-proof. Louvers are commonly made of galvanized or cold-rolled steel or extruded aluminum alloy. Louver finishes may be baked enamel, clear lacquer, or anodic. Louver takeoff is performed by the item based on height, width, depth, type of louver, and finish characteristics. The number of each type of louver is entered into the proper ledger sheet, spreadsheet, or estimating software cell.

Division 11 – Equipment

Division 11 of the CSI MasterFormat™ includes food service, athletic, banking, photographic, religious, laboratory, laundry, library, and medical equipment. The most common equipment is food service equipment. Food service equipment includes a wide variety of beverage and food handling equipment. Food service equipment included in Division 11 of the CSI MasterFormat™ includes food storage, preparation, and cooking equipment. See Figure 15-4. Appliances considered include liquid dispensing systems, ice chests, coolers, fryers, kettles, steamers, deck ovens, convection ovens, refrigerators, ventilators, scales, dishwashers, and food waste disposers. Manufacturer names and model numbers are commonly given in relation to this equipment. The number of each type of food service equipment is entered into the proper ledger sheet, spreadsheet, or estimating software cell.

Athletic equipment may include basketball hoops, boxing and weightlifting equipment, nets and protective devices, and seating for grandstands and stadiums. Banking equipment includes depositories, teller windows, vaults, and drive-up banking apparatus. Photographic equipment includes darkroom items such as revolving doors, processing equipment, and transfer cabinets. Religious equipment includes pews, book racks, and other ceremonial and worship needs.

Laboratory equipment includes any specialized cabinetry or venting required for laboratory work. Laundry room equipment includes washers, dryers, exhaust equipment, and dry cleaning or ironing equipment. Library equipment includes shelving, computer stands, study carrels, map and newspaper racks, wall display units, and book stands. Medical equipment includes sterilization, examination, dental care, operating room, and radiology equipment.

QUANTITY SHEET

Project: *Restaurant*

Estimator: *RDJ*

No.	Description	Quantity	Unit	
1	Ice bin	2	ea	
2	Baked potato oven w/ stand and timer	1	ea	
3	Ice tea dispenser	2	ea	
4	Hot chocolate machine	1	ea	
5	Chicken pressure fryer assembly	1	ea	
6	Walk-in cooler/freezer	1	ea	
7	Salad bar base unit	1	ea	
8	Salad bar sneeze guard	1	ea	
9	Exhaust hood	2	ea	
10	Electric modular range top w/ stand	1	ea	

Henny Penny Corporation

Figure 15-4. Food service equipment is taken off according to the number and type of each piece of equipment required.

Division 12 – Furnishings

Division 12 of the CSI MasterFormat™ includes furnishings such as open office systems furniture, manufactured wood casework, storage units, file cabinets and filing systems, tables and chairs, and window blinds and curtains. Information determined by the estimator includes item counts based on variations of materials, finishes, hardware, fabrication, and installation. Additional information concerning wood casework is found in Division 6 of the CSI MasterFormat™.

Open Office Systems Furniture. *Open office systems furniture* is furniture that consists of panels that are connected in various configurations to provide working office space such as desktops, storage units, and lighting in a compact, flexible configuration. Open office systems furniture is designed for use in an open office plan which uses few fixed floor-to-ceiling partitions. See Figure 15-5. Estimators must be familiar with all the available components of each type of open office systems furniture. Panel finishing includes plastic laminate, fabric, wood veneer, and plastics. Common panel heights vary from 50″ to 80″, and common panel widths vary from 12″ to 48″. Items considered during open office systems furniture takeoff include panel finish, height, and width, attachments such as shelving and storage units, lighting requirements, and provisions for integrated electrical or computer wiring.

Figure 15-5. Open office systems furniture allows the integration of wall units, working surfaces, storage, and lighting.

Division 13 – Special Construction

Special construction included in Division 13 of the CSI MasterFormat™ includes seismic construction systems, fire suppression systems, air-supported structures, underground shelters, radiation, sound, and vibration control systems, walk-in coolers, swimming pools, and storage tanks. Division 13 also includes hazardous material remediation inside buildings including lead and asbestos abatement, measurement and control instrumentation, and some prefabricated struc-

tures. Additional information concerning underground shelters, outdoor hazardous material remediation, storage tanks, and swimming pools is included in Division 2 of the CSI MasterFormat™. Prefabricated metal building information is included in Division 5 of the CSI MasterFormat™. Fire suppression system information is included in Division 15 of the CSI MasterFormat™.

Seismic Construction Systems. Seismic construction concerns new construction built to specific seismic requirements or retrofitting of existing structures to improve their ability to withstand damage from earthquakes. Earthquake damage results from random vertical and horizontal vibrations. Engineering research is ongoing to determine new and improved methods of designing and retrofitting buildings to resist earthquake damage. Seismic construction systems require close adherence to design tolerances including items such as reinforcing steel placement, concrete strength, weld connections, bracing placement, and flexible pipe couplings. Estimators should check all portions of the drawings for seismic control devices and safeguards.

Fire Suppression Systems. Fire suppression systems include the pipe materials that join water supplies, sprinkler systems, and attachments to safety valves and alarms. Mechanical prints indicate locations of pipes, sprinkler head requirements, valve types, alarm box placement, and areas to be protected. See Figure 15-6.

Fire suppression systems include wet-pipe, dry-pipe, and gaseous. In wet-pipe systems, fire suppression pipes are connected to a main water supply source. Sprinkler system piping carries water throughout the structure to various locations where it is necessary to provide fire suppression. Sprinkler pipes are shown on plan views with a pipe symbol and abbreviation, such as WP for wet pipe. Isometric drawings of fire suppression piping are also provided. Sprinkler pipes are made of cast iron or any material that does not fail due to heat from a fire. Valves are placed throughout the piping system to allow periodic testing and provide maximum safety. A sprinkler head is installed at the termination point of each pipe. Additional information concerning pipe, sprinkler heads, and valves is found in the specifications. Sprinkler heads in a wet-pipe system are designed to open and release water when activated by heat. Sprinkler heads are installed after ceiling tiles are in place in suspended ceiling systems.

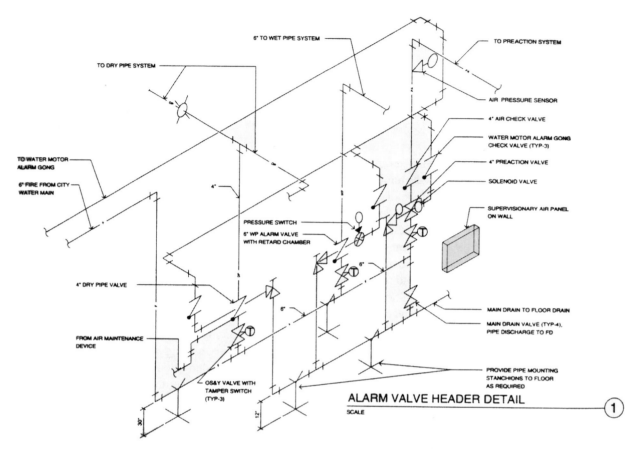

Figure 15-6. Fire suppression piping is shown on mechanical prints and included in Division 13 of the CSI MasterFormat™.

Dry-pipe system pipes are noted on plan views with the abbreviation DP for dry pipe. These pipes are connected to the same fire suppression water supply as a wet-pipe system. The portion of the system that is dry consists of pipes filled with air. When a fire occurs in an area protected with a dry-pipe system, the sprinkler heads release the air and then distribute water from the fire service main. Gaseous systems are comprised of a fire-suppressing gas such as halon piped into an area where electronic equipment such as computers are installed and cannot be exposed to water. Sprinkler pipes are connected to a tank filled with halon. Gaseous system piping is not connected to fire protection water mains.

Division 14 – Conveying Systems

Conveying systems include elevators, material handling equipment such as cranes and lifts, hoists, escalators, conveyors, and monorail systems. Division 14 of the CSI MasterFormat™ includes information concerning conveying systems. Conveying systems are specialized

according to manufacturer design and carrying capacity. Estimators review the plans and specifications for manufacturer references concerning the required conveying system. Cross-referencing with electrical, mechanical, and structural drawings may provide additional information concerning power, piping, and support requirements and tolerances.

Elevators. Elevators may be powered by electric or hydraulic power. The elevator car is guided by vertical rails fastened to the walls of the elevator shaft. The car is hoisted in the shaft by wire ropes and counterweights for an electric elevator or a post for hydraulic elevators. A machine room is commonly located at the top of the elevator shaft for electric elevators and at the bottom of the shaft for hydraulic elevators. Items considered in selection of an elevator system include the building height, use, and population, and entrance requirements. Elevator drawings and specification information required by the estimator include the net capacity, speed, travel distance, car size, and interior finish.

FINAL BID ASSEMBLY AND PRESENTATION

The final bid is assembled after computing all quantities of materials and amounts of labor and obtaining the necessary subcontractor bids. The final bid is a total sum of the costs of all items included in the construction project. It is the price that the contractor submits to the owner to cover all costs for materials, labor, services, overhead, and profit that are necessary to complete the job.

Compilation

When compiling the final bid, ensure that the most current bid documents are used. The estimator should contact the architect and owner immediately prior to the time necessary to submit the bid to ensure that the most current addenda and all changes to the project are included in the bid being submitted. The estimator compiling the bid for submission to the owner must ensure that all items are included and are priced properly.

Subcontractor Bids. Subcontractors must often wait until the final bid time is close to ensure that they are working with current bid documents and to ensure confidentiality of their bid. The estimator assembling the final bid, including the subcontractor bids, must carefully and completely check subcontractor bids even though the time requirements may be short. The lowest-price subcontractor bid may exclude items that are required in the subcontractor scope of work. The estimator assembling the final bid must ensure that all items are covered prior to developing a final cost. A manual system of tabs or file folders based on the CSI MasterFormat™ and specification sections aids in tracking subcontractor and supplier bids during the final compilation on bid day. Many bids are received at the last minute by fax, which requires planning for sufficient fax access on bid day.

Takeoff Quantities. All takeoff quantities for work being performed by the contractor assembling the bid are included. A common estimating error in bidding is unintentionally omitting one or more items in the bid. Always double-check all bidding documents to ensure that all items have been included in the final bid price.

Low Bid. In a competitive, low bid situation, contractors work hard to determine the lowest responsible bid that allows for a fair profit while ensuring all costs are covered. Estimators should quickly analyze subcontractor and supplier prices to detect pricing trends and should not take a low bid that may be inaccurate.

To ensure that the bid is complete and as low as necessary to win the work, a checklist may be used for ledger sheet bidding. A subcontractor and supplier quotation recap sheet may also be used to allow the chief estimator to quickly recognize incomplete bids or pricing trends. Estimating software may include a scan function that shows areas where blank cells or zeros appear in the final bid documents. See Figure 15-7.

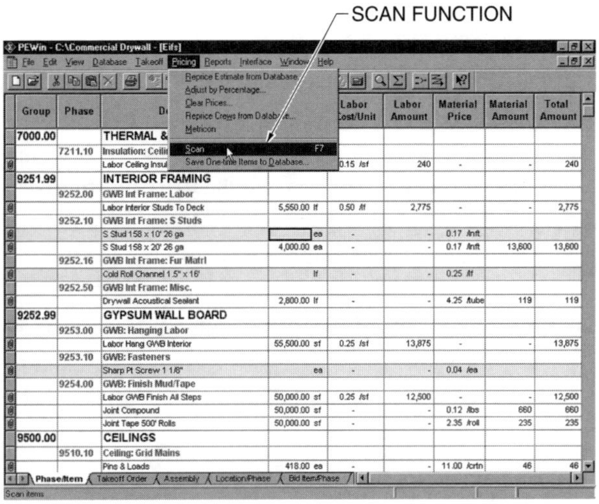

Figure 15-7. The scan function in estimating software alerts the estimator to blank cells in the bid documents.

A spreadsheet estimate recap sheet can also be formatted according to the CSI MasterFormat™ with some automatic tabulation formulas added in the proper cells. This information can include descriptions for general contractor, subcontractor, and supplier items. This sheet is maintained in a separate file from the final estimate to prevent inclusion of an incomplete price.

Some estimators use cut and add sheets to help track final changes on bid day. When subcontractors or suppliers make final changes to their pricing, cut and add sheets show the original pricing and the revised pricing for each change, including the name of the subcontractor or supplier. These cost changes are then added to the recap sheets or proper cells in estimating software.

A large bid with many items commonly requires more than one estimator to check to ensure that all costs are included. All of the costs for materials, labor, overhead, and profit are carefully analyzed to ensure that no items are missed and no items are overpriced. Estimators should allow for the possible final input of senior management in the construction firm. This can affect the final bid price and the timing of final compilation of the bid.

Negotiated Construction Work. *Negotiated construction work* is construction work for which a qualified contractor submits a bid to an owner based on the plans and specifications without a competitive bid process. In negotiated construction work, the owner and architect may review the final bid documents with the contractor to determine if there are areas for cost savings. Negotiated construction work still requires estimators to ensure that all items, materials, labor, and overhead and profit are calculated accurately and completely.

Bid Presentation. The presentation of the final bid is an item that should be included in the overhead cost that may make a significant difference in the final bid price. Some owners may require the use of a standardized bid form for bid submission. Other owners may allow estimators to submit their bids in the format determined by the estimator. One of the primary advantages of computerized bidding systems is the ability to continue to work with the bidding data and costs as close as possible to the bid date and time and still be able to produce a high-quality bid presentation. See Figure 15-8. Information is condensed and simplified prior to final presentation to the owner.

Contract Award. After an owner accepts a bid from a contractor to perform the work, a contract is agreed to and signed. Once the contract is signed, the contractor is obligated to perform the construction work as specified for the sum of money shown in the contract. The construction contract binds the contractor to furnish all the labor, materials, and subcontracted trades to completely finish the project as detailed in the specifications and print drawings.

The terms of the contract concerning work to be performed and the allocation of risk between the owner, architect, contractor, subcontractors, and suppliers should be carefully reviewed by all parties prior to the signing of the contract to avoid potential problems during construction. The contract also binds the owner to make payments based on satisfactory completion according to the agreed-upon contract terms and prices.

POST-BID TRACKING

Accurate recordkeeping is required by the estimator and construction company during the construction process to ensure accurate bidding in the future. A successful construction estimating department monitors the various costs that are incurred during a project in an attempt to keep company historical cost and labor data accurate and find efficiencies in the bidding and costing process. Estimators track materials, labor, overhead, profit, and other items that affect the final construction cost. Variances from estimates provide important information for future bidding activities.

Application of overhead costs to a construction project can have a significant effect on the success of a construction project and an accurate bid. Variances in all overhead costs in relation to the original estimate should be carefully tracked. Where costs for items such as insurance, bonding, security, taxes, or other overhead items are not consistent with the estimate, determinations should be made concerning whether these are unusual variations or items that should be calculated differently on future estimates.

Simpson Industries		Spreadsheet Report SI98-Refrigeration					Page 1 8:29 AM
Group	Description	Labor Amount	Material Amount	Sub Amount	Equip Amount	Other Amount	Total Amount
1100.00	GENERAL REQUIREMENTS	53,260	5,317	29,485	2,530	2,940	93,532
2100.00	SITEWORK	24,468	1,448	88,700	3,149		117,765
3000.00	CONCRETE	101,503	142,999	4,500			249,003
4000.00	MASONRY	79,290	32,545				111,834
5000.00	METALS	5,428	931	265,000			271,359
6000.00	WOOD & PLASTICS	3,812	853	4,400			9,066
7000.00	THERM \ MOISTURE PROTECT.	37,168	111,659	195,000			343,827
8000.00	DOORS & WINDOWS	2,834	7,029	10,000			19,863
9000.00	FINISHES	29,728	47,297	20,000			97,026

Figure 15-8. Successful estimating requires that the final presentation of the estimate be in a clear, readable, and understandable format.

Materials

All materials on the construction project should be allocated to a particular account and job site operation. Tracking of material costs compared to estimated costs show where material prices have changed over the course of the construction process. New materials or new pricing may cause variances in material costs that may become available during a construction project.

Quantities. Comparison of estimated quantities with actual material quantities indicate areas where takeoffs are excessively high or low. Cases where material takeoff quantities vary greatly from actual materials required for construction may indicate excessive waste, theft, items missed on the takeoff, inaccurate quantities being delivered, or improper waste percentages applied in the estimating formulas. Adjustments should be made to company historical data where applicable to ensure greater accuracy in material quantity takeoff on future projects.

Labor

The highest risk variable in construction is labor costs. Estimators have a variety of standard industry tables and company historical records for use in pricing labor. Labor production rates and costs must be tracked throughout the construction process to ensure minimal variance from the original labor estimate. Adjustments in scheduling or job site management may be necessary based on the continuous comparison.

Crew Performance. Work at a construction site is based on unit costs tied to the amount of production for various material units. Where labor costs and crew performances vary from the number of material units, items considered include climatic conditions, crew training, provision of proper equipment and tools, and job site safety, management, and scheduling. Accurate tracking of labor production rates for various material units can help estimators make their information more accurate for future bids based on the company crews that are performing the field work.

ESTIMATING SOFTWARE – FINALIZING ESTIMATE

In Precision estimating software, finalizing an estimate is done after takeoff is complete. These are performed by applying a standard procedure. See Figure 15-9.

PRECISION ESTIMATING SOFTWARE – FINALIZING ESTIMATE . . .

1. Ensure the estimate is maximized by pressing the maximize button (🔲) in the upper right-hand corner of the estimate.

2. Hide columns by right-clicking on the column heading and selecting Hide from the shortcut menu. Hide all columns on the spreadsheet except Group, Phase, Description, Labor Amount, Material Price, Material Amount, Sub Amount, and Total Amount. *Note:* Several columns can be hidden at once by holding down the left mouse button and dragging the highlight across the column headings.

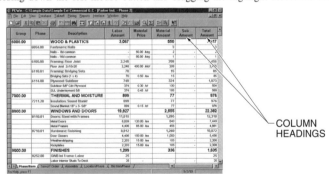

COLUMN HEADINGS

3. Make price changes by placing the cursor in the material price cell, inputting the change, and pressing ENTER. For example, a reduction in the price of metal doors is input by placing the cursor in the Material Price cell for Metal Doors, changing the price from $120/ea to $110/ea, and pressing ENTER. The Material Amount and the Total Amount columns are adjusted automatically.

4. Collapse the level of detail shown on the spreadsheet by pressing the collapse button (⊟) on the toolbar. The collapse button may be pressed again for a more summarized view. Expand the estimate by pressing the expand button (⊞) twice to show full detail.

5. Show estimate totals by category (labor, material, etc.) by pressing the totals button (Σ) on the toolbar.

6. Create an addon for profit and overhead by pressing the edit addon button (✏) in the totals window. The addon window opens.

Click the add button and set up the addon by inputting 18 in the Addon field and Profit and Overhead for Description. Check the allocatable box and Estimate Total under selection. Input a rate of 19%.

. . . PRECISION ESTIMATING SOFTWARE – FINALIZING ESTIMATE

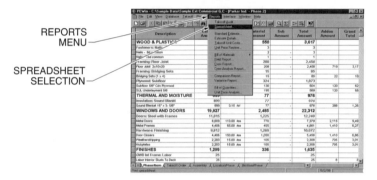

7. Click OK and Close to exit the addon window.

8. Click the insert addon button ( ) in the totals window to apply this addon to the estimate. Select addon 10 and 18 by double-clicking on each and Close to exit the insert addon window. The addon amounts are calculated automatically.

9. Close the totals window by clicking OK and Close. *Note:* Each addon has been proportionally spread across all estimate items. The addon and grand total columns can be displayed by right-clicking the Total Amount Column heading and selecting Show Hidden Columns from the shortcut menu.

10. Generate a user-defined spreadsheet report by selecting spreadsheet from the reports menu.

11. Click Report Options and Prefill from Spreadsheet buttons. Uncheck the boxes marked Print Horizontal Gridlines, Print Vertical Gridlines, and Print Cover Page and place a check in the Minimize Overline Columns box. Select Preview.

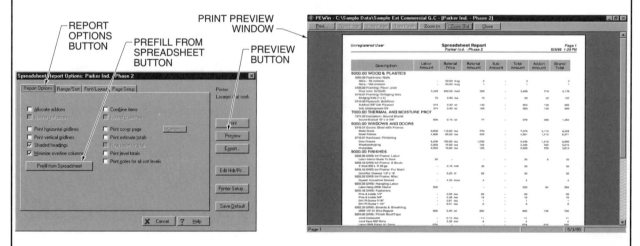

12. Click twice anywhere on the report to zoom in. Click Close in the Print Preview window and click Cancel and Close in the report options screen to get back to the spreadsheet.

Figure 15-9. In Precision estimating software, final modifications, review of the totals, and generation of user-defined spreadsheet reports can be done after takeoff is complete.

Estimating

_____ **1.** Division _____ of the CSI MasterFormat™ includes specialty items such as partitions, access flooring, louvers, chalkboards and tackboards, toilet compartments, fireplaces, flagpoles, lockers and shelving, and pedestrian control devices.

_____ **2.** A(n) _____ is a vertical divider that separates one area from another.

_____ **3.** _____ flooring is a raised floor system supported by short posts and runners.

_____ **4.** Prefinished floor panels are commonly manufactured in _____″ squares.

_____ **5.** A(n) _____ partition is a partition that is designed to be disassembled and reassembled as needed to form new room layouts.

_____ **6.** A(n) _____ partition is a panel or partition that is moved by hand or motor along an overhead track.

_____ **7.** A(n) _____ partition is a partition fabricated of wire mesh.

_____ **8.** A(n) _____ is a ductwork cover containing horizontal slats that allow the passage of air but prevent rain from entering.

_____ **9.** Division _____ of the CSI MasterFormat™ includes food service, athletic, banking, photographic, religious, laboratory, laundry, library, and medical equipment.

_____ **10.** Fire suppression system information is covered in Division _____ of the CSI MasterFormat™.

_____ **11.** _____ furniture is furniture that consists of panels that are connected in various configurations to provide working office space such as desktops, storage units, and lighting in a compact, flexible configuration.

T   F   **12.** Elevators may be powered by electric or hydraulic power.

T   F   **13.** The highest risk variable in construction is labor costs.

T   F   **14.** In estimating software, finalizing an estimate is done before takeoff is complete.

T   F   **15.** Operable partitions include folding, coiling, and sliding partitions.

_____ **16.** Division _____ of the CSI MasterFormat™ includes furnishings.

_____ **17.** _____ construction work is construction work for which a qualified contractor submits a bid to an owner based on the plans and specifications without a competitive bid process.

T   F   **18.** Work at a construction site is based on unit costs tied to the amount of production for various material units.

T   F   **19.** The preliminary bid is the price the contractor submits to the owner to cover all expenses for materials, labor, services, overhead, and profit that are necessary to complete the job.

T   F   **20.** Most specialty items are taken off as individual units.

Fire Extinguisher Classes

_____   **1.** Flammable liquids

_____   **2.** Combustible metals

_____   **3.** Ordinary combustibles

_____   **4.** Electrical equipment

Activities
Specialty Items and Final Bid Preparation

15
Chapter

Name _____ Date _____

Activity 15-1 – Ledger Sheet Activity

Refer to the cost data, equipment schedule, Print 15-1, and Estimate Summary Sheet No. 15-1. Take off the number of wire shelf units. Determine the total material and labor cost including 8% for overhead and profit.

COST DATA			
Material	**Unit**	**Material Unit Cost***	**Labor Unit Cost***
Wire shelf units, 4-tier			
18″ × 48″	ea	1210	31.35
18″ × 60″	ea	1375	31.35
24″ × 36″	ea	423.50	31.35
24″ × 42″	ea	440	31.35
24″ × 48″	ea	484	31.35
24″ × 60″	ea	654.50	31.35

* in $

EQUIPMENT SCHEDULE	
ID Number	**Equipment Description**
63	Wire shelf units – cooler
64	Wire shelf units – freezer
65	Wire shelf units – dry storage

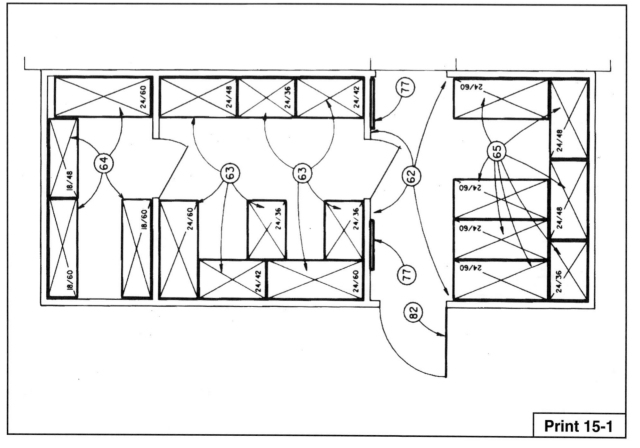

Print 15-1

341

ESTIMATE SUMMARY SHEET

Sheet No. ___15-1___

Project:_____   Date:_____
Estimator:_____   Checked: _____

No.	Description	Dimensions				Quantity		Material		Labor		Total	
							Unit	Unit Cost	Total	Unit Cost	Total	Unit Cost	Total
	Wire shelf units												
	18″ × 48″												
	18″ × 60″												
	24″ × 36″												
	24″ × 42″												
	24″ × 48″												
	24″ × 60″												
	Total												
	Total including 8% overhead and profit												

Activity 15-2 – Spreadsheet Activity

Refer to the cost data, equipment schedule, Print 15-2, and Estimate Summary Sheet No. 15-2 on the CD-ROM. Take off the number of tables and chairs. Determine the total material and labor cost including 12% for overhead and profit.

COST DATA			
Material	Unit	Material Unit Cost*	Labor Unit Cost*
Chair	ea	72.05	—
Table, two seat, 20″ × 24″	ea	456.50	28.60
Table, four seat, 30″ × 30″	ea	456.50	30.80

* in $

EQUIPMENT SCHEDULE	
ID Number	Equipment Description
105	Chairs
108	Pedestal table 20″ × 24″
109	Pedestal table 30″ × 30″

Activity 15-3 – Estimating Software Activity

Merge estimating Activity 6-3, Activity 7-3, Activity 9-3, and Activity 12-3 by selecting File from the toolbar and clicking on the merge estimate button. Name the new estimate Activity 15-3 and follow the onscreen Merge Estimate Wizard instructions. Select Activity 6-3, Activity 7-3, Activity 9-3, and Activity 12-3. Click merge estimates. Open estimate Activity 15-3. Print a bill of materials report.

Appendix

ENGLISH SYSTEM

LENGTH		Unit	Abbr	Equivalents
		mile	mi	5280′, 320 rd, 1760 yd
		rod	rd	5.50 yd, 16.5′
		yard	yd	3′, 36″
		foot	ft *or* ′	12″, .333 yd
		inch	in. *or* ″	.083′, .028 yd
AREA $A = l \times w$		square mile	sq mi *or* mi^2	640 A, 102,400 sq rd
		acre	A	4840 sq yd, 43,560 sq ft
		square rod	sq rd *or* rd^2	30.25 sq yd, .00625 A
		square yard	sq yd *or* yd^2	1296 sq in., 9 sq ft
		square foot	sq ft *or* ft^2	144 sq in., .111 sq yd
		square inch	sq in. *or* in^2	.0069 sq ft, .00077 sq yd
VOLUME $V = l \times w \times t$		cubic yard	cu yd *or* yd^3	27 cu ft, 46,656 cu in.
		cubic foot	cu ft *or* ft^3	1728 cu in., .0370 cu yd
		cubic inch	cu in. *or* in^3	.00058 cu ft, .000021 cu yd
CAPACITY WATER, FUEL, ETC.	*U.S. liquid measure*	gallon	gal.	4 qt (231 cu in.)
		quart	qt	2 pt (57.75 cu in.)
		pint	pt	4 gi (28.875 cu in.)
		gill	gi	4 fl oz (7.219 cu in.)
		fluidounce	fl oz	8 fl dr (1.805 cu in.)
		fluidram	fl dr	60 min (.226 cu in.)
		minim	min	⅛ fl dr (.003760 cu in.)
VEGETABLES, GRAIN, ETC.	*U.S. dry measure*	bushel	bu	4 pk (2150.42 cu in.)
		peck	pk	8 qt (537.605 cu in.)
		quart	qt	2 pt (67.201 cu in.)
		pint	pt	½ qt (33.600 cu in.)
DRUGS	*British imperial liquid and dry measure*	bushel	bu	4 pk (2219.36 cu in.)
		peck	pk	2 gal. (554.84 cu in.)
		gallon	gal.	4 qt (277.420 cu in.)
		quart	qt	2 pt (69.355 cu in.)
		pint	pt	4 gi (34.678 cu in.)
		gill	gi	5 fl oz (8.669 cu in.)
		fluidounce	fl oz	8 fl dr (1.7339 cu in.)
		fluidram	fl dr	60 min (.216734 cu in.)
		minim	min	1/60 fl dr (.003612 cu in.)
MASS AND WEIGHT COAL, GRAIN, ETC.	*avoirdupois*	ton	t	2000 lb
		short ton		2000 lb
		long ton		2240 lb
		pound	lb *or* #	16 oz, 7000 gr
		ounce	oz	16 dr, 437.5 gr
		dram	dr	27.344 gr, .0625 oz
		grain	gr	.037 dr, .002286 oz
GOLD, SILVER, ETC.	*troy*	pound	lb	12 oz, 240 dwt, 5760 gr
		ounce	oz	20 dwt, 480 gr
		pennyweight	dwt *or* pwt	24 gr, .05 oz
		grain	gr	.042 dwt, .002083 oz
DRUGS	*apothecaries'*	pound	lb ap	12 oz, 5760 gr
		ounce	oz ap	8 dr ap, 480 gr
		dram	dr ap	3 s ap, 60 gr
		scruple	s ap	20 gr, .333 dr ap
		grain	gr	.05 s, .002083 oz, .0166 dr ap

METRIC SYSTEM			
LENGTH	**Unit**	**Abbreviation**	**Number of Base Units**
	kilometer	km	1000
	hectometer	hm	100
	dekameter	dam	10
	meter*	m	1
	decimeter	dm	.1
	centimeter	cm	.01
	millimeter	mm	.001
AREA $A = l \times w$	square kilometer	sq km *or* km^2	1,000,000
	hectare	ha	10,000
	are	a	100
	square centimeter	sq cm *or* cm^2	.0001
VOLUME $V = l \times w \times t$	cubic centimeter	cu cm, cm^3, *or* cc	.000001
	cubic decimeter	dm^3	.001
	cubic meter*	m^3	1
CAPACITY	kiloliter	kl	1000
	hectoliter	hl	100
WATER, FUEL, ETC.	dekaliter	dal	10
	liter*	l	1
VEGETABLES, GRAIN, ETC.	cubic decimeter	dm^3	1
	deciliter	dl	.10
DRUGS	centiliter	cl	.01
	milliliter	ml	.001
MASS AND WEIGHT	metric ton	t	1,000,000
	kilogram	kg	1000
COAL, GRAIN, ETC.	hectogram	hg	100
	dekagram	dag	10
GOLD, SILVER, ETC.	gram*	g	1
	decigram	dg	.10
DRUGS	centigram	cg	.01
	milligram	mg	.001

* base units

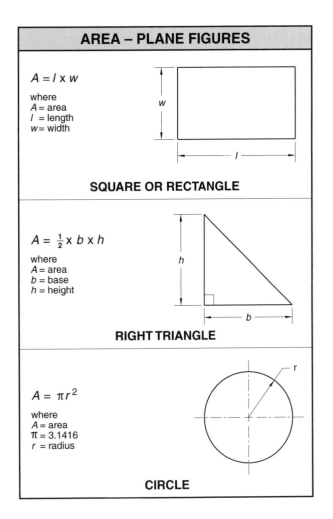

AREA – PLANE FIGURES

$A = l \times w$

where
A = area
l = length
w = width

SQUARE OR RECTANGLE

$A = \frac{1}{2} \times b \times h$

where
A = area
b = base
h = height

RIGHT TRIANGLE

$A = \pi r^2$

where
A = area
π = 3.1416
r = radius

CIRCLE

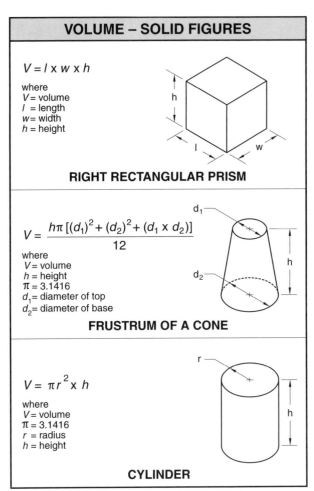

VOLUME – SOLID FIGURES

$V = l \times w \times h$

where
V = volume
l = length
w = width
h = height

RIGHT RECTANGULAR PRISM

$$V = \frac{h\pi\,[(d_1)^2 + (d_2)^2 + (d_1 \times d_2)]}{12}$$

where
V = volume
h = height
π = 3.1416
d_1 = diameter of top
d_2 = diameter of base

FRUSTRUM OF A CONE

$V = \pi r^2 \times h$

where
V = volume
π = 3.1416
r = radius
h = height

CYLINDER

DECIMAL EQUIVALENTS OF AN INCH

Fraction	Decimal	Fraction	Decimal	Fraction	Decimal	Fraction	Decimal
1/64	0.015625	17/64	0.265625	33/64	0.515625	49/64	0.765625
1/32	0.03125	9/32	0.28125	17/32	0.53125	25/32	0.78125
3/64	0.046875	19/64	0.296875	35/64	0.546875	51/64	0.796875
1/16	0.0625	5/16	0.3125	9/16	0.5625	13/16	0.8125
5/64	0.078125	21/64	0.328125	37/64	0.578125	53/64	0.828125
3/32	0.09375	11/32	0.34375	19/32	0.59375	27/32	0.84375
7/64	0.109375	23/64	0.359375	39/64	0.609375	55/64	0.859375
1/8	0.125	3/8	0.375	5/8	0.625	7/8	0.875
9/64	0.140625	25/64	0.390625	41/64	0.640625	57/64	0.890625
5/32	0.15625	13/32	0.40625	21/32	0.65625	29/32	0.90625
11/64	0.171875	27/64	0.421875	43/64	0.671875	59/64	0.921875
3/16	0.1875	7/16	0.4375	11/16	0.6875	15/16	0.9375
13/64	0.203125	29/64	0.453125	45/64	0.703125	61/64	0.953125
7/32	0.21875	15/32	0.46875	23/32	0.71875	31/32	0.96875
15/64	0.234375	31/64	0.484375	47/64	0.734375	63/64	0.984375
1/4	0.250	1/2	0.500	3/4	0.750	1	1.000

INCH — DECIMAL CONVERSION

Inch-Decimal Conversion

The conversion of inches into decimal parts of a foot and vice versa is accomplished mathematically or by memory. Mathematically, full inches and decimal equivalents of an inch are divided by twelve. The resulting answer is the decimal part of a foot. For example, 5¾″ is first changed to decimal parts of an inch (5.75″) and then divided by 12. The result of .479′ is the decimal equivalent in feet.

The memory method is referred to as field conversion. Field conversion is not as exact as the mathematical method, but is sufficient for most buildings and faster than the mathematical method. Five accuracy points throughout a foot measurement provide memory guides. The accuracy points are 0″ (.00′), 3″ (.25′), 6″ (.50′), 9″ (.75′), and 12″ (1.00′).

Field conversion assumes that each one-eighth of an inch equals one hundredth of a foot ($\frac{1}{8}$″ = .01′). Mathematically, $\frac{1}{8}$″ equals approximately .0104′. Also, each full inch is assumed to equal eight hundredths of a foot (1″ = .08′). Mathematically, 1″ = .0833′.

The addition or subtraction of .01′ for each $\frac{1}{8}$″ and .08′ for each 1″ from the closest of the five accuracy points gives an accurate dimension anywhere throughout the length of a foot by performing simple mathematical calculations. For example, to convert 4¼″ from inches to decimal parts of a foot, select the closest accuracy point. The closest accuracy point to four inches is three inches. Three inches (3″) equals twenty-five hundredths of a foot (.25′). Add .08′ for each 1″ and the total is .25′ + .08′ = .33′ (.33′ is equal to 4″). Two hundredths of a foot (.02′) equals two-eights (¼″). Therefore, .33′ (4″) + .02′ (¼″) = .35′ (4¼″).

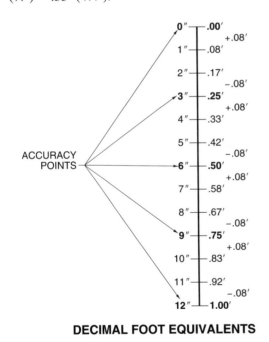

DECIMAL FOOT EQUIVALENTS

$9 \frac{5}{8}$″ = .xx′

1. 9″ = .75′

2. $\frac{5}{8}$″ = .05′

3. ADD ⟶ .75′ + .05′ = .80′

FIELD CONVERSION

$5 \frac{3}{4}$″ = .xx′

1. CHANGE $5 \frac{3}{4}$″ TO DECIMAL $5 \frac{3}{4}$″ = 5.75″

2. DIVIDE BY 12

$$\begin{array}{r} .479' \\ 12\overline{)5.75''} \\ \underline{4\,8} \\ 95 \\ \underline{84} \\ 110 \\ \underline{108} \end{array}$$

MATHEMATICALLY

ALPHABET OF LINES

NAME AND USE	CONVENTIONAL REPRESENTATION	EXAMPLE	
OBJECT LINE Define shape. Outline and detail objects.	THICK	OBJECT LINE	
HIDDEN LINE Show hidden features.	$\frac{1}{8}''$ (3 mm) THIN $\frac{1}{32}''$ (0.75 mm)	HIDDEN LINE	
CENTER LINE Locate centerpoints of arcs and circles.	$\frac{1}{16}''$ (1.5 mm) THIN $\frac{1}{8}''$ (3 mm) $\frac{3}{4}''$ (18 mm) TO $1\frac{1}{2}''$ (36 mm)	CENTER LINE CENTERPOINT	
DIMENSION LINE Show size or location. **EXTENSION LINE** Define size or location.	DIMENSION LINE DIMENSION THIN 2'-6" EXTENSION LINE	DIMENSION LINE $1\frac{3}{4}$ EXTENSION LINE	
LEADER Call out specific features.	OPEN ARROWHEAD THIN X CLOSED ARROWHEAD 3X	$1\frac{1}{2}$ DRILL LEADER	
CUTTING PLANE Show internal features.	THICK $\frac{1}{8}''$ (3 mm) $\frac{1}{16}''$ (1.5 mm) A $\frac{3}{4}''$ (18 mm) TO $1\frac{1}{2}''$ (36 mm)	LETTER IDENTIFIES SECTION VIEW CUTTING PLANE LINE	
SECTION LINE Identify internal features.	$\frac{1}{16}''$ (1.5 mm) THIN	SECTION LINES	
BREAK LINE Show long breaks. **BREAK LINE** Show short breaks.	$\frac{3}{4}''$ (18 mm) TO $1\frac{1}{2}''$ (36 mm) THIN FREEHAND THICK	LONG BREAK LINE SHORT BREAK LINE	

ABBREVIATIONS OF TERMS. . .

A

acoustic	ACST
acoustical tile	AT or ACT
adhesive	ADH
aileron	AIL
air conditioner	AIR COND
alloy	ALY
alternating current	AC
ambient	AMB
American Wire Gauge	AWG
ampere	AMP
anchor bolt	AB
architecture	ARCH
area	A
area drain	AD
asphalt	ASPH
asphalt tile	AT
as required	AR
astragal	A
automatic	AUTO
automatic sprinkler	AS
auxiliary	AUX
azimuth	AZ

B

bathroom	B
beam	BM
bearing	BRG
bearing plate	BPL or BRG PL
bench mark	BM
board	BD
board foot	BF
brick	BRK
bridge	BRDG
building	BLDG
building line	BL
built-up roofing	BUR

C

cased opening	CO
casement	CSMT
cast iron	CI
cast-iron pipe	CIP
cast steel	CS
Celsius	°C
cement	CEM
cement floor	CF
center	CTR
centerline	CL
center-to-center	C to C
ceramic	CER
ceramic tile	CT
ceramic-tile floor	CTF
circuit breaker	CB
circuit interrupter	CI
cleanout	CO
coarse	CRS
coaxial	COAX
cold air	CA
cold water	CW
column	COL
compacted	COMP
composition	CMPSN
compressed air	COMPA
concrete	CONC
concrete block	CCB
concrete floor	CCF
concrete pipe	CP
condenser	COND
conductor	CNDCT
conduit	CND
construction joint	CJ
contour	CTR
control joint	CLJ
cubic	CU
cubic foot per minute	CFM
cubic foot per second	CFS
cubic inch	CU IN
cubic yard	CU YD
culvert	CULV
cutoff valve	COV
cutout valve	COV
cylinder lock	CYLL

D

dead load	DL
decibel	DB
deck	DK
demolition	DML
diagonal	DIAG
dimension	DIM
direct current	DC
discharge	DISCH
disconnect switch	DS
dishwater	DW
distribution panel	DPNL
division	DIV
double-acting	DBL ACT
double-pole double-throw	DPDT
double-pole double-throw switch	DPDT SW
double-pole single-throw	DPST
double-pole single-throw switch	DPST SW
double-pole switch	DP SW
downspout	DS
drain	DR
drain tile	DT
drinking fountain	DF
drywall	DW
duplex	DX
duty cycle	DTY CY

E

each	EA
east	E
electric panel	EP
electromechanical	ELMCH
electronic	ELEK
elevation	EL
elevator	ELEV
entrance	ENTR
equipment	EQPT
equivalent	EQUIV
estimate	EST
expanded metal	EM
exterior	EXT
exterior grade	EXT GR

F

face brick	FB
Fahrenheit	°F
fastener	FSTNR
fiberboard	FBRBD
fiberboard, corrugated	FBDC
fiberboard, double wall	FDWL
fiberboard, solid	FBDS
figure	FIG
finish	FNSH
finish all over	FAO
finished floor	FNSH FL
finish grade	FG
finish one side	F1S
finish two sides	F2S
firebrick	FBCK
fire door	FDR
fire extinguisher	FEXT
fire hydrant	FHY
fireproof	FPRF
fire wall	FW
fixed window	FX WDW
flagstone	FLGSTN
flammable	FLMB
flashing	FL
floor	FL
floor drain	FD
fluorescent	FLUOR
footing	FTG
foot per minute	FPM
foot per second	FPS
foundation	FDN
four-pole	4P
four-pole double-throw switch	4PDT SW

four-pole single-throw switch	4PST SW
four-pole switch	4PSW
front view	FV
furring	FUR
fuse	FU
fuse block	FB
fuse box	FUBX
fuse holder	FUHLR
fusible	FSBL

G

gauge	GA
gallon per hour	GPH
gallon per minute	GPM
garage	GAR
glass	GL
glass block	GLB
glaze	GLZ
grade	GR
gravel	GVL
gross vehicle weight	GVW
ground	GND
grounded (outlet)	G
ground-fault circuit interrupter	GFCI
gypsum	GYP

H

hazardous	HAZ
heating, ventilating, and air conditioning	HVAC
heavy-duty	HD
hertz	Hz
highway	HWY
horizontal	HOR
hot water	HW
hydraulic	HYDR

I

inch	IN
inch per second	IPS
inch-pound	IN LB
infrared	IR
interior	INTR
iron pipe	IP

J

jamb	JB or JMB
joist	J
junction	JCT

K

keyway	KWY
kiln-dried	KD
knife switch	KN SW
knockout	KO

L

lavatory	LAV
left	L
left hand	LH
lighting	LTG
linoleum	LINOL
lintel	LNTL
live load	LL

M

main	MN
manhole	MH
masonry	MSNRY
maximum working pressure	MWP
medicine cabinet	MC
metal anchor	MA
metal door	METD
metal flashing	METF
metal lath and plaster	MLP
metal threshold	MT
mezzanine	MEZZ
mile per gallon	MPG
mile per hour	MPH
miscellaneous	MISC
monolithic	ML
mortar	MOR

. . .ABBREVIATIONS OF TERMS

N

nameplate	NPL
National Electrical Code	NEC
net weight	NTWT
north	N
nosing	NOS

O

on center	OC
opaque	OPA
open web joist	OJ or OWJ
outlet	OUT

P

panel	PNL
parallel	PRL
peak-to-peak	P-P
perpendicular	PERP
phase	PH
piling	PLG
pitch	P
plank	PLK
plate glass	PLGL
plumbing	PLMB
plywood	PLYWD
pneumatic	PNEU
point of beginning	POB
pole	P
porcelain	PORC
pound	LB
pound-foot	LB FT
precast	PRCST
prefabricated	PREFAB
prefinished	PFN
property line	PL
pull box	PB
pull switch	PS

Q

quadrant	QDRNT
quarry tile	QT
quart	QT
quick-acting	QA

R

recess	REC
reference	REF

reinforced concrete	RC
reinforcing steel	RST
reverse-acting	RACT
revolution per minute	RPM
revolution per second	RPS
ribbed	RIB
right	R
right hand	RH
rigid	RGD
riser	R
roll roofing	RR
roofing	RFG
rotor	RTR
rough sawn	RS

S

sanitary	SAN
schedule	SCHED
section	SECT
sheet	SH
sheet metal	SM
sill cock	SC
single-phase	1PH
single-pole	SP
single-pole double-throw	SPDT
single-pole double-throw switch	SPDT SW
single-pole single-throw	SPST
single-pole single-throw switch	SPST SW
single-pole switch	SP SW
skylight	SLT
sliding door	SLD
slope	SLP
solenoid	SOL
south	S
specification	SPEC
square	SQ
square foot	SQ FT
square inch	SQ IN
square yard	SQ YD
stainless steel	SST
structural glass	SG
surfaced or dress four sides	S4S
surfaced or dressed one side	S1S
switch	SW

T

telephone	TEL

temperature	TEMP
three-phase	3PH
three-pole	3P
three-pole double-throw	3PDT
three-pole single-throw	3PST
three-way	3WAY
three-wire	3W
threshold	TH
timber	TMBR
total	TOT
triple-pole double-throw	3PDT
triple-pole double-throw switch	3PDT SW
triple-pole single-throw	3PST
triple-pole single-throw switch	3PST SW
triple-pole switch	3P SW
truss	TR
two-phase	2PH
two-pole	DP
two-pole double-throw	DPDT
two-pole single-throw	DPST

V

valley	VAL
valve	V
vent	V
vertical	VERT
voltage	V
voltage drop	VD
volume	VOL

W

walk in closet	WIC
waste pipe	WP
waste stack	WS
water	WTR
water closet	WC
water heater	WH
water meter	WM
watt	W
weatherproof	WTHPRF
welded wire fabric	WWF
west	W
without	W/O
wood	WD

ARCHITECTURAL SYMBOLS. . .

Material	Elevation	Plan	Section
EARTH			
BRICK	WITH NOTE INDICATING TYPE OF BRICK (COMMON, FACE, ETC.)	COMMON OR FACE / FIREBRICK	SAME AS PLAN VIEWS
CONCRETE		LIGHTWEIGHT / STRUCTURAL	SAME AS PLAN VIEWS
CONCRETE BLOCK		OR	OR
STONE	CUT STONE / RUBBLE	CUT STONE / RUBBLE / CAST STONE (CONCRETE)	CUT STONE / CAST STONE (CONCRETE) / RUBBLE OR CUT STONE
WOOD	SIDING / PANEL	WOOD STUD / REMODELING / DISPLAY	ROUGH MEMBERS / FINISHED MEMBERS / PLYWOOD
PLASTER		WOOD STUD, LATH, AND PLASTER / METAL LATH AND PLASTER / SOLID PLASTER	LATH AND PLASTER
ROOFING	SHINGLES	SAME AS ELEVATION VIEW	
GLASS	OR / GLASS BLOCK	GLASS / GLASS BLOCK	SMALL SCALE / LARGE SCALE

...ARCHITECTURAL SYMBOLS

Material	Elevation	Plan	Section
FACING TILE	CERAMIC TILE	FLOOR TILE	CERAMIC TILE LARGE SCALE / CERAMIC TILE SMALL SCALE
STRUCTURAL CLAY TILE			SAME AS PLAN VIEW
INSULATION		LOOSE FILL OR BATTS / RIGID / SPRAY FOAM	SAME AS PLAN VIEWS
SHEET METAL FLASHING		OCCASIONALLY INDICATED BY NOTE	
METALS OTHER THAN FLASHING	INDICATED BY NOTE OR DRAWN TO SCALE	SAME AS ELEVATION	SMALL SCALE / STEEL / CAST IRON / ALUMINUM / BRONZE OR BRASS
STRUCTURAL STEEL	INDICATED BY NOTE OR DRAWN TO SCALE	OR	REBARS / SMALL SCALE / LARGE SCALE / L-ANGLES, S-BEAMS, ETC.

PLOT PLAN SYMBOLS

N	NORTH	●	FIRE HYDRANT		WALK	E OR	ELECTRIC SERVICE
	POINT OF BEGINNING (POB)	⊠	MAILBOX		IMPROVED ROAD	G OR	NATURAL GAS LINE
▲	UTILITY METER OR VALVE	○	MANHOLE		UNIMPROVED ROAD	W OR	WATER LINE
	POWER POLE AND GUY	⊕	TREE	BL	BUILDING LINE	T OR	TELEPHONE LINE
	LIGHT STANDARD		BUSH	PL	PROPERTY LINE		NATURAL GRADE
	TRAFFIC SIGNAL		HEDGE ROW		PROPERTY LINE		FINISH GRADE
	STREET SIGN		FENCE		TOWNSHIP LINE	+ XX.00′	EXISTING ELEVATION

ELECTRICAL SYMBOLS. . .

LIGHTING OUTLETS

OUTLET BOX AND INCANDESCENT LIGHTING FIXTURE — CEILING / WALL

INCANDESCENT TRACK LIGHTING

BLANKED OUTLET — B / B

DROP CORD — D

EXIT LIGHT AND OUTLET BOX. SHADED AREAS DENOTE FACES.

OUTDOOR POLE-MOUNTED FIXTURES

JUNCTION BOX — J / J

LAMPHOLDER WITH PULL SWITCH — L PS / L PS

MULTIPLE FLOODLIGHT ASSEMBLY

EMERGENCY BATTERY PACK WITH CHARGER — B

INDIVIDUAL FLUORESCENT FIXTURE

OUTLET BOX AND FLUORESCENT LIGHTING TRACK FIXTURE

CONTINUOUS FLUORESCENT FIXTURE

SURFACE-MOUNTED FLUORESCENT FIXTURE

PANELBOARDS

FLUSH-MOUNTED PANELBOARD AND CABINET

SURFACE-MOUNTED PANELBOARD AND CABINET

CONVENIENCE OUTLETS

SINGLE RECEPTACLE OUTLET

DUPLEX RECEPTACLE OUTLET

TRIPLEX RECEPTACLE OUTLET

SPLIT-WIRED DUPLEX RECEPTACLE OUTLET

SPLIT-WIRED TRIPLEX RECEPTACLE OUTLET

SINGLE SPECIAL-PURPOSE RECEPTACLE OUTLET

DUPLEX SPECIAL-PURPOSE RECEPTACLE OUTLET

RANGE OUTLET — R

SPECIAL-PURPOSE CONNECTION — DW

CLOSED-CIRCUIT TELEVISION CAMERA

CLOCK HANGER RECEPTACLE — C

FAN HANGER RECEPTACLE — F

FLOOR SINGLE RECEPTACLE OUTLET

FLOOR DUPLEX RECEPTACLE OUTLET

FLOOR SPECIAL-PURPOSE OUTLET

UNDERFLOOR DUCT AND JUNCTION BOX FOR TRIPLE, DOUBLE, OR SINGLE DUCT SYSTEM AS INDICATED BY NUMBER OF PARALLEL LINES

BUSDUCTS AND WIREWAYS

SERVICE, FEEDER, OR PLUG-IN BUSWAY — B B B

CABLE THROUGH LADDER OR CHANNEL — C C C

WIREWAY — W W W

SWITCH OUTLETS

SINGLE-POLE SWITCH — S

DOUBLE-POLE SWITCH — S$_2$

THREE-WAY SWITCH — S$_3$

FOUR-WAY SWITCH — S$_4$

AUTOMATIC DOOR SWITCH — S$_D$

KEY-OPERATED SWITCH — S$_K$

CIRCUIT BREAKER — S$_{CB}$

WEATHERPROOF CIRCUIT BREAKER — S$_{WCB}$

DIMMER — S$_{DM}$

REMOTE CONTROL SWITCH — S$_{RC}$

WEATHERPROOF SWITCH — S$_{WP}$

FUSED SWITCH — S$_F$

WEATHERPROOF FUSED SWITCH — S$_{WF}$

TIME SWITCH — S$_T$

CEILING PULL SWITCH — S

SWITCH AND SINGLE RECEPTACLE — S

SWITCH AND DOUBLE RECEPTACLE — S

A STANDARD SYMBOL WITH AN ADDED LOWERCASE SUBSCRIPT LETTER IS USED TO DESIGNATE A VARIATION IN STANDARD EQUIPMENT — a.b / a.b / S$_{a.b}$

. . .ELECTRICAL SYMBOLS

COMMERCIAL AND INDUSTRIAL SYSTEMS

PAGING SYSTEM DEVICE

FIRE ALARM SYSTEM DEVICE

COMPUTER DATA SYSTEM DEVICE

PRIVATE TELEPHONE SYSTEM DEVICE

SOUND SYSTEM

FIRE ALARM CONTROL PANEL — FACP

SIGNALING SYSTEM OUTLETS FOR RESIDENTIAL SYSTEMS

PUSHBUTTON

BUZZER

BELL

BELL AND BUZZER COMBINATION

COMPUTER DATA OUTLET

BELL RINGING TRANSFORMER — BT

ELECTRIC DOOR OPENER — D

CHIME — CH

TELEVISION OUTLET — TV

THERMOSTAT — T

UNDERGROUND ELECTRICAL DISTRIBUTION OR ELECTRICAL LIGHTING SYSTEMS

MANHOLE — M

HANDHOLE — H

TRANSFORMER-MANHOLE OR VAULT — TM

TRANSFORMER PAD — TP

UNDERGROUND DIRECT BURIAL CABLE

UNDERGROUND DUCT LINE

STREET LIGHT STANDARD FED FROM UNDERGROUND CIRCUIT

ABOVE-GROUND ELECTRICAL DISTRIBUTION OR LIGHTING SYSTEMS

POLE

STREET LIGHT AND BRACKET

PRIMARY CIRCUIT

SECONDARY CIRCUIT

DOWN GUY

HEAD GUY

SIDEWALK GUY

SERVICE WEATHERHEAD

PANEL CIRCUITS AND MISCELLANEOUS

LIGHTING PANEL

POWER PANEL

WIRING – CONCEALED IN CEILING OR WALL

WIRING – CONCEALED IN FLOOR

WIRING EXPOSED

HOME RUN TO PANEL BOARD
Indicate number of circuits by number of arrows. Any circuit without such designation indicates a two-wire circuit. For a greater number of wires indicate as follows: —///— (3 wires) —////— (4 wires), etc.

FEEDERS
Use heavy lines and designate by number corresponding to listing in feeder schedule

WIRING TURNED UP

WIRING TURNED DOWN

GENERATOR — G

MOTOR — M

INSTRUMENT (SPECIFY) — I

TRANSFORMER — T

CONTROLLER

EXTERNALLY-OPERATED DISCONNECT SWITCH

PULL BOX

PLUMBING SYMBOLS. . .

FIXTURES...	...FIXTURES	...PIPING
STANDARD BATHTUB	LAUNDRY TRAY	CHILLED DRINKING WATER SUPPLY — DWS —
OVAL BATHTUB	BUILT-IN SINK	CHILLED DRINKING WATER RETURN — DWR —
WHIRLPOOL BATH	DOUBLE OR TRIPLE BUILT-IN SINK	HOT WATER
SHOWER STALL	COMMERCIAL KITCHEN SINK	HOT WATER RETURN
SHOWER HEAD	SERVICE SINK SS	SANITIZING HOT WATER SUPPLY (180 °F)
TANK-TYPE WATER CLOSET	CLINIC SERVICE SINK	SANITIZING HOT WATER RETURN (180°F)
WALL-MOUNTED WATER CLOSET	FLOOR-MOUNTED SERVICE SINK	DRY STANDPIPE — DSP —
FLOOR-MOUNTED WATER CLOSET	DRINKING FOUNTAIN DF	COMBINATION STANDPIPE — CSP —
LOW-PROFILE WATER CLOSET	WATER COOLER	MAIN SUPPLIES SPRINKLER — S —
BIDET	HOT WATER TANK HWT	BRANCH AND HEAD SPRINKLER
WALL-MOUNTED URINAL	WATER HEATER WH	GAS – LOW PRESSURE — G — G —
FLOOR-MOUNTED URINAL	METER M	GAS – MEDIUM PRESSURE — MG —
TROUGH-TYPE URINAL	HOSE BIBB HB	GAS – HIGH PRESSURE — HG —
WALL-MOUNTED LAVATORY	GAS OUTLET G	COMPRESSED AIR — A —
PEDESTAL LAVATORY	GREASE SEPARATOR G	OXYGEN — O —
BUILT-IN LAVATORY	GARAGE DRAIN	NITROGEN — N —
WHEELCHAIR LAVATORY	FLOOR DRAIN WITH BACKWATER VALVE	HYDROGEN — H —
CORNER LAVATORY	**PIPING...**	HELIUM — HE —
FLOOR DRAIN	SOIL, WASTE, OR LEADER – ABOVE GRADE	ARGON — AR —
FLOOR SINK	SOIL, WASTE, OR LEADER – BELOW GRADE	LIQUID PETROLEUM GAS — LPG —
	VENT	INDUSTRIAL WASTE — INW —
	COMBINATION WASTE AND VENT — SV —	CAST IRON — CI —
	STORM DRAIN — SD —	CULVERT PIPE — CP —
	COLD WATER	CLAY TILE — CT —
		DUCTILE IRON — DI —
		REINFORCED CONCRETE — RCP —
		DRAIN – OPEN TILE OR AGRICULTURAL TILE

...PLUMBING SYMBOLS

PIPE FITTING AND VALVE SYMBOLS

	FLANGED	SCREWED	BELL & SPIGOT		FLANGED	SCREWED	BELL & SPIGOT		FLANGED	SCREWED	BELL & SPIGOT
BUSHING				REDUCING FLANGE				AUTOMATIC BY-PASS VALVE			
CAP				BULL PLUG				AUTOMATIC REDUCING VALVE			
REDUCING CROSS				PIPE PLUG				STRAIGHT CHECK VALVE			
STRAIGHT-SIZE CROSS				CONCENTRIC REDUCER				COCK			
CROSSOVER				ECCENTRIC REDUCER				DIAPHRAGM VALVE			
45° ELBOW				SLEEVE				FLOAT VALVE			
90° ELBOW				STRAIGHT-SIZE TEE				GATE VALVE			
ELBOW – TURNED DOWN				TEE – OUTLET UP				MOTOR-OPERATED GATE VALVE			
ELBOW – TURNED UP				TEE – OUTLET DOWN				GLOBE VALVE			
BASE ELBOW				DOUBLE-SWEEP TEE				MOTOR-OPERATED GLOBE VALVE			
DOUBLE-BRANCH ELBOW				REDUCING TEE				ANGLE HOSE VALVE			
LONG-RADIUS ELBOW				SINGLE-SWEEP TEE				GATE HOSE VALVE			
REDUCING ELBOW				SIDE OUTLET TEE – OUTLET DOWN				GLOBE HOSE VALVE			
SIDE OUTLET ELBOW – OUTLET DOWN				SIDE OUTLET TEE – OUTLET UP				LOCKSHIELD VALVE			
SIDE OUTLET ELBOW – OUTLET UP				UNION				QUICK-OPENING VALVE			
STREET ELBOW				ANGLE CHECK VALVE				SAFETY VALVE			
CONNECTING PIPE JOINT				ANGLE GATE VALVE – ELEVATION				GOVERNOR-OPERATED AUTOMATIC VALVE			
EXPANSION JOINT				ANGLE GATE VALVE – PLAN							
LATERAL				ANGLE GLOBE VALVE – ELEVATION							
ORIFICE FLANGE				ANGLE GLOBE VALVE – PLAN							

HVAC SYMBOLS

EQUIPMENT SYMBOLS	DUCTWORK	HEATING PIPING

EQUIPMENT SYMBOLS

EXPOSED RADIATOR

RECESSED RADIATOR

FLUSH ENCLOSED RADIATOR

PROJECTING ENCLOSED RADIATOR

UNIT HEATER (PROPELLER) – PLAN

UNIT HEATER (CENTRIFUGAL) – PLAN

UNIT VENTILATOR – PLAN

STEAM

DUPLEX STRAINER

PRESSURE-REDUCING VALVE

AIR LINE VALVE

STRAINER

THERMOMETER

PRESSURE GAUGE AND COCK

RELIEF VALVE

AUTOMATIC 3-WAY VALVE

AUTOMATIC 2-WAY VALVE

SOLENOID VALVE

DUCTWORK

DUCT (1ST FIGURE, WIDTH; 2ND FIGURE, DEPTH) — 12 X 20

DIRECTION OF FLOW

FLEXIBLE CONNECTION

DUCTWORK WITH ACOUSTICAL LINING

FIRE DAMPER WITH ACCESS DOOR — FD | AD

MANUAL VOLUME DAMPER — VD

AUTOMATIC VOLUME DAMPER

EXHAUST, RETURN OR OUTSIDE AIR DUCT – SECTION — 20 X 12

SUPPLY DUCT – SECTION — 20 X 12

CEILING DIFFUSER SUPPLY OUTLET — 20" DIA CD 1000 CFM

CEILING DIFFUSER SUPPLY OUTLET — 20 X 12 CD 700 CFM

LINEAR DIFFUSER — 96 X 6-LD 400 CFM

FLOOR REGISTER — 20 X 12 FR 700 CFM

TURNING VANES

FAN AND MOTOR WITH BELT GUARD

LOUVER OPENING — 20 X 12-L 700 CFM

HEATING PIPING

HIGH-PRESSURE STEAM	—— HPS ——
MEDIUM-PRESSURE STEAM	—— MPS ——
LOW-PRESSURE STEAM	—— LPS ——
HIGH-PRESSURE RETURN	—— HPR ——
MEDIUM-PRESSURE RETURN	—— MPR ——
LOW-PRESSURE RETURN	—— LPR ——
BOILER BLOW OFF	—— BD ——
CONDENSATE OR VACUUM PUMP DISCHARGE	—— VPD ——
FEEDWATER PUMP DISCHARGE	—— PPD ——
MAKEUP WATER	—— MU ——
AIR RELIEF LINE	—— V ——
FUEL OIL SUCTION	—— FOS ——
FUEL OIL RETURN	—— FOR ——
FUEL OIL VENT	—— FOV ——
COMPRESSED AIR	—— A ——
HOT WATER HEATING SUPPLY	—— HW ——
HOT WATER HEATING RETURN	—— HWR ——

AIR CONDITIONING PIPING

REFRIGERANT LIQUID	—— RL ——
REFRIGERANT DISCHARGE	—— RD ——
REFRIGERANT SUCTION	—— RS ——
CONDENSER WATER SUPPLY	—— CWS ——
CONDENSER WATER RETURN	—— CWR ——
CHILLED WATER SUPPLY	—— CHWS ——
CHILLED WATER RETURN	—— CHWR ——
MAKEUP WATER	—— MU ——
HUMIDIFICATION LINE	—— H ——
DRAIN	—— D ——

REFRIGERATION SYMBOLS

GAUGE		PRESSURE SWITCH		DRYER	
SIGHT GLASS		HAND EXPANSION VALVE		FILTER AND STRAINER	
HIGH SIDE FLOAT VALVE		AUTOMATIC EXPANSION VALVE		COMBINATION STRAINER AND DRYER	
LOW SIDE FLOAT VALVE		THERMOSTATIC EXPANSION VALVE		EVAPORATIVE CONDENSOR	
IMMERSION COOLING UNIT		CONSTANT PRESSURE VALVE, SUCTION		HEAT EXCHANGER	
COOLING TOWER		THERMAL BULB		AIR-COOLED CONDENSING UNIT	
NATURAL CONVECTION, FINNED TYPE EVAPORATOR		SCALE TRAP		WATER-COOLED CONDENSING UNIT	
FORCED CONVECTION EVAPORATOR		SELF-CONTAINED THERMOSTAT			

FIRE DOOR OPENING CLASSIFICATIONS

Class	Time*	Application
A	3	Walls separating buildings or dividing one building into separate fire areas
B	1 – 1½	Enclosures of vertical egress in a structure such as stairs and elevators
C	¾	Hallways and room partitions
D	1½	Exterior walls subject to fire exposure from outside the structure
E	¾	Exterior walls subject to moderate to light fire exposure

* in hr

QUANTITY SHEET

Project: _____

Estimator: _____

Sheet No. _____

Date: _____

Checked: _____

No.	Description	Dimensions			Unit		Unit		Unit		Unit		Unit	

ESTIMATE SUMMARY SHEET

Project: _____
Estimator: _____

Sheet No. _____
Date: _____
Checked: _____

No.	Description	Dimensions	Quantity	Material			Labor			Equipment		
				Unit	Unit Cost	Total	Unit Cost	Total	Unit Cost	Total	Unit Cost	Total
Total												

Glossary

The terms in this glossary are defined as they relate to the building trades.

abbreviation: A letter or group of letters representing a term or phrase.

absorption refrigeration: A refrigeration process that uses the absorption of one chemical by another chemical and heat transfer to produce a refrigeration effect.

abutment: The supporting structure at the end of a bridge, arch, or vault.

access door: A door used to enclose an area that houses concealed equipment.

access flooring: A raised floor system supported by short posts and runners.

access road: A temporary road installed by the general contractor to facilitate transportation of material and craftworkers to and from a job site.

addendum: A change to the originally-issued contract documents.

aggregate: Granular material such as gravel, sand, vermiculite, or perlite that is added to cement to form concrete, mortar, or plaster.

air conditioner: The component in a forced-air system that cools the air.

air filter: A porous device that removes particles from air.

air handler: A device used to distribute conditioned air to building spaces.

alkyd primer: A thermoplastic or synthetic resin-based primer.

angle of repose: The slope that a material maintains without sliding or caving in.

aquastat: A temperature-actuated electric switch used to limit the temperature of boiler water.

ashlar: Masonry composed of squared building stones of various sizes.

asphalt: Dark-colored pitch primarily composed of crushed stone and bituminous materials.

assembly: A collection of items needed to complete a particular unit of work.

backfilling: The replacing of soil around outside foundation walls after the walls have been completed.

backing: The material that supports carpet pile yarns.

balloon framing: Construction in which one-piece studs extend from the first floor line or sill to the top plate of the upper story.

baluster: The upright member that runs between the handrail and the treads.

bar joist: An open web joist with steel angles at the top and bottom of the joist and bars for the intermediate members.

base flashing: A sheet metal strip installed at the lowest meeting point of a vertical and horizontal surface.

base molding: The molding at the bottom of a wall at the intersection with the floor.

batten: A narrow strip of wood used to cover the joints between boards.

beam: A structural member installed horizontally to support loads over an opening.

beam and column construction: Construction consisting of bays of framed structural steel which are repeated to create large structures.

bevel siding: Siding that has a tapered cross-section.

bid: An offer to perform a construction project at a stated price.

bituminous roofing: Roofing comprised of layers (plies) of asphalt-impregnated felt or fiberglass material that is fastened to the roof deck and mopped with hot coal tar to create a waterproof surface.

blocking: A piece of wood fastened between structural members to strengthen the joint, provide structural support, or block air passage.

blockout: A frame set in a concrete form to create a void in the finished concrete structure.

blower control: A temperature-actuated switch that controls the blower motor in a furnace.

board: A wood member that has been sawn or milled having a nominal size from $1'' \times 4''$ up to $1'' \times 12''$.

board foot: A unit of measure for lumber based on the volume of a piece $12''$ square and $1''$ thick.

boiler: A closed metal container (vessel) in which water is heated to produce heated water or steam.

boiler horsepower (BHP): The evaporation of 34.5 lb of water per hour at a feedwater temperature of 212°F.

bollard: A stone guard attached to a corner or a freestanding stone post to protect it against damage from vehicular traffic.

bond: 1. An insurance agreement for financial loss caused by the act or default of an individual or by some contingency over which the individual has no control. **2.** The arrangement of masonry units in a wall by lapping them one upon another to provide a sturdy structure.

bond breaker: A chemical compound applied to the surface of concrete forming material or the slab that comes in contact with the concrete as it is placed.

bored pile: A foundation support formed by drilling into the ground to a predetermined depth and diameter and filling the hole with concrete.

boulder: A rock that has been transported by some geological action from its site of formation to its current location.

box: A metallic or nonmetallic electrical enclosure used for equipment, devices, and pulling or terminating conductors.

brace: A structural piece, either permanent or temporary, designed to resist weights or pressures of loads.

brick: A rectangular block used for building or paving purposes.

building paper: Felt material saturated with tar to form a waterproof sheet.

busway: A metal-enclosed distribution system of busbars available in prefabricated sections.

butterfly valve: A valve that controls fluid flow by a square, rectangular, or round disc mounted on a shaft that seats against a resilient housing.

cabinet: An enclosure fitted with any combination of shelves, drawers, and doors usually used for storage.

cable: Two or more conductors grouped together within a common protective cover and used to connect individual components.

cable tray: An open grid rack suspended from structural members to support a series of cables.

CAD compatibility: The ability of an estimating software package to directly transfer quantity takeoff information from a CAD-generated set of plans.

caisson: A poured-in-place concrete piling of large diameter created by boring a hole and filling it with cast-in-place concrete.

camber: The slight upward curve in a structural member designed to compensate for deflection of the member under load.

cap flashing: A sheet metal member wrapped around the top of a wall or other vertical projection to provide moisture protection.

carpet: A floor covering composed of natural or synthetic fibers woven through a backing material.

carpet face weight: The total weight of pile yarns measured in ounces per square yard.

casting bed: The forms and support used for forming precast concrete members.

cavity wall: A masonry wall with at least a 2″ void between faces.

cellular insulation: Insulation composed of small individual cells separated from each other.

channel: Light-gauge metal members used for framing and supporting lath or drywall.

check valve: A valve that allows flow in only one direction.

chemical cross-linking: A curing process in which the curing is dependent on molecular action between the coating and the hardener (curing compound) as opposed to evaporation.

chiller: The component in a hydronic air conditioning system that cools water, which cools the air.

circuit breaker: An overcurrent protection device with a mechanism that automatically opens the circuit when an overload condition or short circuit occurs.

cladding: 1. Wall surface material attached to a structural steel frame to span between supporting members and provide closure to the structure. **2.** A layer of material greater than .04″ thick used to improve corrosion resistance or other properties.

clapboard siding: Siding that has a consistent thickness and square edges.

clay: A natural mineral material that is compact and brittle when dry but plastic when wet.

clear and grub: The removal of vegetation and other obstacles from a construction site in preparation for new construction.

clock hanger receptacle: A receptacle that has a slight recess in the face to the receptacle.

closed stringer: A finish member installed at the meeting of a staircase and wall.

cofferdam: A watertight enclosure used to allow construction or repairs to be performed below the surface of water.

coiling door: A door that moves vertically to coil around a steel rod.

column: A vertical structural member used to support axial compressive loads.

completion bond: A short-term insurance policy insuring the owner that a construction project will be fully completed and free from encumbrances and liens when turned over to the owner.

composition shingle: An asphalt or fiberglass shingle coated with a layer of fine mineral gravel.

compressor: A mechanical device that compresses refrigerant.

concealed flashing: Flashing hidden from the interior and exterior of a building.

concrete: A material consisting of portland cement, aggregate, and water, which solidifies through chemical reaction into a hard, strong mass.

concrete construction: Construction in which concrete and reinforcing steel are placed in forming materials to create a high-strength finished structure.

concrete masonry unit: A precast hollow or solid masonry unit made of portland cement and fine aggregate, with or without admixtures or pigments.

condenser: A heat exchanger that removes heat from high-pressure refrigerant vapor.

conductor: A slender rod or wire that is used to control the flow of electrons in an electrical circuit.

conduit: A hollow pipe that supports and protects electrical conductors.

Construction Specifications Institute (CSI): An organization that developed standardized construction specifications.

construction system: A method used in the design and construction of a structure, including materials, construction sequence, structural design, and finish materials.

contactor: A control device that uses a small control current to energize or de-energize the load connected to it.

contour line: A dashed or solid line on a plot plan used to show elevations of the surface.

contraction joint: A groove in a vertical or horizontal concrete surface to create a weakened plane and control the location of cracking due to imposed loads.

contractor: An individual or company responsible for performing construction work in accordance with the terms of a contract.

control joint: A thin strip of perforated metal applied on lath and plaster surfaces to relieve stress resulting from expansion and contraction in large ceiling and wall surfaces.

control zone: Any part of a building that is controlled by one controlling device.

core drilling: The process of making holes in the ground to retrieve soil samples for analysis.

corner bead: A light-gauge, L-shaped galvanized metal device used to cover outside corner joints in plaster walls.

cost allocation: The designation of costs to particular equipment.

cost index: A compilation of a number of cost items from various sources combined into a common table for reference use.

cost plus basis: A method of determining profit in which costs are submitted for payment for labor, materials, equipment, and overhead, and an amount is added automatically for profit.

course: A continuous horizontal layer of masonry units bonded with mortar.

cove molding: The trim material often used to cover the joint between the tread and the stringer and the joint between the tread and the top of the riser.

cradling: The temporary supporting of existing utility lines in or around an excavated area to properly protect the lines.

crew-based estimating: An estimating practice in which a range of quantities and costs are included into a single calculation based on a specific quantity of construction put in place.

cripple: A short wall stud spanning from a plate to an opening sill or header.

CSI MasterFormat™: A master list of numbers and titles for organizing information about construction requirements, products, and activities into a standard sequence.

curtain wall: A non-bearing prefabricated panel suspended on or fastened to structural members.

cutting plane: A line that cuts through a part of a structure on a drawing.

damper: A movable plate that controls and balances air flow in a forced-air system.

dampproofing: The treatment of a material so it is moisture resistant.

database assembly: An assembly that is stored in the database and can be used in any estimate based on that database.

database interactivity: The direct transfer of stored information by a software program from a database into a spreadsheet.

dead load: A permanent, stationary load composed of all building material and fixtures and equipment permanently attached to a structure.

decking: Light-gauge metal sheets used to construct a floor or deck form or used as floor or deck members.

demolition: The organized destruction of a structure.

demountable partition: A partition that is designed to be disassembled and reassembled as needed to form new room layouts.

denier: The unit of fineness of yarn.

depreciation: The accounting practice of reducing the value of equipment by a standard amount each year.

design build (schematic) estimating: An estimating practice in which the development of an estimate is based on the building function or the functional area of a building.

detail drawing: A drawing showing a small part of a plan, elevation, or section view at an enlarged scale.

detail takeoff: An estimating practice in which a takeoff of each individual construction component is done.

diffuser: An air outlet in an HVAC system that directs air in a wide pattern.

digitizer: An electronic input device that converts analog data to digital form.

dimension lumber: A wood member precut to a particular size for the building industry, normally having a nominal size from 2″ × 4″ up to 4″ × 6″.

dimmer switch: A switch that changes the brightness of a lamp by changing the voltage applied to the lamp.

disconnect: A switch that disconnects electrical circuits from motors and machines.

door closer: A device that closes a door and controls the speed and closing action of the door.

driven pile: A steel or wood member that is driven into the ground with a pile-driving hammer.

dry film thickness: The thickness of the resultant mass after the VOCs have evaporated.

ductwork: The distribution system for forced-air heating or cooling systems.

duplex receptacle: A receptacle that has two spaces for connecting two different plugs.

elastomeric roofing: Roofing made of a pliable synthetic polymer.

electrical metallic tubing (EMT): A lightweight tubular steel raceway without threads on the ends.

electronic estimating method: An estimating method in which takeoff and pricing is accomplished using computer spreadsheets and integrated database systems.

elevation drawing: An orthographic view of a vertical surface without allowance for perspective.

enamel: A paint with a large amount of varnish.

entrance door: The primary means of egress of most pedestrian traffic for a commercial building such as retail, office, health care, and manufacturing facilities.

Environmental Protection Agency (EPA): A federal government agency established in 1970 to control and abate pollution in the areas of air, water, solid waste, pesticides, radiation, and toxic substances.

equipment leasing: The use of equipment based on a contract that stipulates a set amount of time and a set cost for the equipment use.

equipment rental: The use of equipment for a fixed cost per a given unit of time.

escutcheon: A protective cover placed around a door knob.

estimating: The computation of construction costs of a project.

estimating method: The approach used by an estimator to perform project analysis, takeoff, and pricing in a consistent and organized manner.

estimating practice: The system used to integrate all parts of the estimating process in a cohesive, consistent, reliable manner to ensure an accurate final bid.

evaporator: A heat exchanger that absorbs heat from the surrounding air to evaporate refrigerant.

excavation: Any construction cut, cavity, trench, or depression in the earth's surface formed by earth removal.

expansion joint: A separation between adjoining sections of a concrete slab to allow movement caused by expansion and contraction.

expansion tank: A tank that allows the water in a hydronic heating system to expand without raising the water pressure to dangerous levels.

expansion valve: A device that reduces the pressure on liquid refrigerant by allowing the refrigerant to expand.

exposed flashing: Flashing open to view and most commonly made of metal.

exposure: The amount of a shingle or shake that is seen after installation.

exterior insulation and finish system (EIFS): An exterior siding material composed of a layer of exterior sheathing, insulation board, reinforcing mesh, a base coat, and a textured finish.

face brick: Brick commonly used for an exposed surface.

fiberboard: A wood panel product manufactured from fine wood fibers mixed with binders.

fibrous insulation: Insulation composed of small diameter fibers.

final grading: The process of smoothing and sloping a site for paving and landscaping.

finish wood construction: The use of wood members for finish applications.

finish wood member: Any decorative and nonstructural wood in a structure.

fire break: The space or fire-resistant materials between structures or groups of structures to prevent fire from spreading to adjacent areas.

firebrick: Brick manufactured from clay containing properties that resist crumbling and cracking at high temperatures.

fire clay: Clay that is highly resistant to heat and does not deform.

fire door: A fire-resistant door and assembly, including the frame and hardware, commonly equipped with an automatic door closer.

flashing: Any material installed at the seam or joint of two building components to inhibit air, water, or fire passage.

flexible metal conduit (FMC): A raceway of metal strips that are formed into a circular cross-sectional raceway.

float glass: High-quality glass used in architectural and specialty applications.

flue: A heat-resistant passage in a chimney that conveys smoke or other gases of combustion.

fluorescent lamp: A low-pressure discharge lamp in which ionization of mercury vapor transforms ultraviolet energy generated by the discharge into light.

folding door: A door formed with panel sections joined with hinges along their vertical edges and supported by rollers in a horizontal upper track.

footing: The section of a foundation that supports and distributes structural loads directly to the soil.

forced-air system: A system that uses warm or cool air to condition building spaces.

form: A temporary structure or mold used to retain and support concrete while it sets and hardens.

form tie: A metal bar, strap, or wire used to hold concrete forms together and resist the pressure of wet concrete.

formwork: The system of support for freshly-placed concrete.

foundation: The primary support for a structure through which the imposed load is transmitted to the footing or earth.

four-way switch: A double-pole, double-throw (DPDT) switch.

frame construction: Construction in which a structure is built primarily of wood structural members.

furnace: A self-contained heating unit that includes a blower, burner(s), heat exchanger or electric heating elements, and controls.

furring: Wood or metal strips fastened to a structural surface to provide a base for fastening finish material.

fuse: An overcurrent protection device with a fusible link that melts and opens the circuit when an overload condition or short circuit occurs.

gang form: A large form constructed by joining small panels.

gate valve: A valve that has an internal gate that slides over the opening through which fluid flows.

general conditions: The written agreements describing building components and various construction procedures included in the specifications for a construction project.

general contractor: Contractor who has overall responsibility for a construction project.

geotextile: A sheet of material that stabilizes and retains soil or earth in position on slopes or other unstable conditions.

girder: A large horizontal structural member that supports loads at isolated points along its length.

girt: A horizontal bracing member placed around the perimeter of a structure.

glass block: Hollow opaque or transparent block made of glass.

glazing: The glass material installed in an opening or the job of setting and installing glass into an opening.

globe valve: A valve that controls the flow of water by means of a circular disc.

glulam (glue-and-laminated): A structural member constructed by bonding several layers of lumber with adhesive.

gooseneck: A curved or bent section of a handrail.

grading: 1. The classification of various pieces of wood according to their quality and structural integrity. **2.** The process of lowering high spots and filling in low spots of earth at a construction site.

granite: Extremely hard natural rock consisting of quartz, feldspar, and other minerals produced under intense heat and pressure.

granular insulation: Insulation composed of small nodules which contain voids or hollow spaces.

gravel: Pieces of rock smaller than boulders and larger than sand.

gridding: The division of a topographical site map for large areas to be excavated or graded into small squares or grids.

grill: A device that covers the opening of return air ductwork.

ground fault circuit interrupt (GFCI) receptacle: A device that interrupts the flow of current to the load when a ground fault exceeding a predetermined value of current occurs.

groundwater: Water present in subsurface material.

grout: A thin, fluid mortar made of a mixture of portland cement, fine aggregate, lime, and water.

handrail: A rail that is grasped by the hand for support in using a stairway.

hardpan: A hard, compacted layer of soil, clay, or gravel.

hardwood: Wood produced from broad leaf, deciduous trees such as ash, birch, maple, oak, and walnut.

hazardous material: A material capable of posing a risk to health, safety, or property.

header: A horizontal framing member at the top of a window or door opening.

hearth: The fireproof floor immediately surrounding a fireplace.

heavy soil: Soil that can be loosened by picks, but is hard to loosen with shovels.

high-intensity discharge (HID) lamp: A lamp that produces light from an arc tube.

high pressure boiler: A boiler that has a maximum allowable working pressure above 15 psi and over 6 BHP.

high-pressure sodium lamp: An HID lamp that produces light when current flows through sodium vapor under high pressure and high temperature.

hinge: Pivoting hardware that joins two surfaces or objects and allows them to swing around the pivot.

hydronic system: A system that uses water, steam, or other fluid to condition building spaces.

hydrostatic pressure: Pressure exerted by a fluid at rest.

I beam: A structural steel member with a cross-sectional area resembling the capital letter I.

igneous rock: Rock formed from the solidification of molten lava.

incandescent lamp: An electric lamp that produces light by the flow of current through a tungsten filament inside a gas-filled, sealed glass bulb.

infill material: Material including sand, gravel, and crushed rock installed to increase the stability of the soil.

intermediate metal conduit (IMC): A medium-weight, galvanized metal conduit.

Internet capability: The ability of estimating software to link and transfer information from the world wide web into estimating software.

isolation joint: A separation between adjoining sections of concrete used to allow movement of the sections.

isometric drawing: A drawing in which all horizontal lines are drawn at a 30° angle from horizontal and all vertical lines are drawn at a 90° angle from horizontal.

job cost analysis: The study of the final costs of building a project as compared to the original estimate.

joint sealant: Any material installed between two members to seal the seam between the members.

joist: A horizontal structural member that supports the load of a floor or ceiling.

Keene's cement: A hard, dense finish plaster.

key-operated switch: A switch that is operated by use of a key.

labor unit: The average time an electrician takes to install a specific type of material.

lagging: Vertical planks used to support earth on the side of an excavation.

landing: The platform that breaks the stair flight from one floor to another.

landing newel post: The main post supporting the handrail at a landing.

latex primer: A water-based primer that dries by the evaporation of water.

lath: The backing fastened to structural members onto which plaster is applied.

ledger sheet: A grid consisting of rows and columns on a sheet of paper into which item descriptions, locations, code numbers, quantities, and costs are entered.

license: A privilege granted by a state or government body to an individual or company to perform work or provide professional service.

lift slab construction: A method of concrete construction in which horizontal slabs are cast on top of each other, jacked into position, and secured to columns at the desired elevation.

light: A pane of glass or translucent material in a door or window.

light soil: A granular substance that is readily shoveled by hand without the aid of machines.

limestone: Sedimentary rock consisting of calcium carbonate.

limit switch: A switch that contains a bimetal element, which senses the temperature of the surrounding air.

liquidated damages: Penalties assessed against the contractor or subcontractor based on failure to complete work within a specific time period.

live load: The predetermined load a structure is capable of supporting.

load-bearing framing: A system of medium-gauge steel members designed by structural engineers to support live and dead loads.

locking receptacle: A receptacle designed for a plug to be twisted and locked into the face of the receptacle.

lockset: The complete assembly of a bolt, knobs, escutcheon, and all mechanical components for securing a door and providing means for opening.

long span construction: Construction consisting of large horizontal steel members, such as girders and trusses, that are fastened together to create large girders and trusses.

louver: A ductwork cover containing horizontal slats that allow the passage of air but prevent rain from entering.

low pressure boiler: A boiler that has a maximum allowable working pressure of up to 15 pounds per square inch (psi).

low-temperature limit control: A temperature-actuated electric switch that energizes the damper motor and shuts the damper if the ventilation air temperature drops below a setpoint.

lumber: Sawn and sized lengths of wood used in construction.

magnetic motor starter: A control device that has overload protection and uses a small control current to energize or de-energize the load (motor) connected to it.

maintain-contact switch: A switch that allows current flow in the up and down positions and must be manually returned to a center position to stop the flow of current.

makeup air: Air that is used to replace air that is lost to exhaust.

manual motor starter: A control device that has overload protection and uses pushbuttons to energize or de-energize the load (motor) connected to it.

marble: Crystallized limestone.

masonry construction: Construction in which masonry members such as clay bricks or tiles, concrete bricks or blocks, and natural or artificial stone are set in mortar.

masonry control joint: A continuous vertical joint in a masonry wall without mortar in which rubber, plastic, or caulking is installed.

mason's rule: A rule graduated in increments equal to various mortar joint thicknesses.

mechanical compression refrigeration: A refrigeration process that produces a refrigeration effect with mechanical equipment.

medium soil: Soil that can be loosened by picks, shovels, and scrapers.

mercury-vapor lamp: An HID lamp that produces light by an electric discharge through mercury vapor.

metal-halide lamp: An HID lamp that produces light by an electric discharge through mercury vapor and metal halides in the arc tube.

metal roofing: Roofing made of steel, aluminum, copper, and various metal alloys.

metamorphic rock: Sedimentary or igneous rock that has been changed in composition or texture by extreme heat, pressure, or chemicals.

mil: A unit of measure .001″ thick.

millwork: Finished wood materials or parts, such as moldings, jambs, and frames completed in a mill or manufacturing plant.

model estimating: An estimating practice in which an estimate is produced by inputting fundamental structure parameters such as building function, size, height, location, and construction system.

momentary-contact switch: A switch that activates current for a brief moment.

mortar: A bonding mixture consisting of lime, cement, sand, and water.

mullion: A vertical dividing member between two window units.

multioutlet assembly: A metal raceway with factory-installed conductors and attachment plug receptacles.

negotiated construction work: Construction work for which a qualified contractor submits a bid to an owner based on the plans and specifications without a competitive bid process.

nominal size: A designated or theoretical size that may vary from the actual value.

nosing: The projection of the tread beyond the face of the riser.

oblique drawing: A drawing in which one face of an object is shown in a flat plane with the receding lines projecting back from the face.

Occupational Safety and Health Administration (OSHA): A federal government agency established under the Occupational Safety and Health Act of 1970, which requires all employers to provide a safe environment for their employees.

Office of Federal Contract Compliance Programs (OFCCP): A federal agency whose mandate is to promote affirmative action and equal employment opportunity on behalf of minorities, women, the disabled, and Vietnam veterans.

one-time assembly: An assembly that is stored with the estimate and cannot be used in other estimates.

open office systems furniture: Furniture that consists of panels that are connected in various configurations to provide working office space such as desktops, storage units, and lighting in a compact, flexible configuration.

open stringer: A finish member that has been cut out to support the open side of a stairway.

open web joist: A framing member constructed with steel angles that are used as chords with steel bars extending between the chords at an angle.

operable partition: A panel or partition that is moved by hand or motor along an overhead track.

oriented strand board (OSB): A wood panel in which wood strands are mechanically oriented and bonded with resin under heat and pressure.

orthographic projection: A drawing in which each face of an object is projected onto flat planes at 90° to one another.

overdig: The amount of excavation required beyond the dimensions of a building to provide an area for construction activities.

overhead: The cost of doing business that is not related to a specific job.

overhead costs: Expenses that are not attributable to any one specific operation.

overhead door: A door which, when opened, is suspended in a horizontal track above the opening.

paint: A mixture of minute solid particles (pigment) suspended in a liquid medium (vehicle).

panelboard: A wall-mounted distribution cabinet containing overcurrent and short-circuit protection devices for lighting, heating, or power circuits.

particleboard: A wood panel constructed of wood particles and flakes that are bonded together with a synthetic resin.

partition: A vertical divider that separates one area from another.

payline: A measurement used by excavation contractors to compute the amount of necessary excavation as stated in the specifications.

percentage basis: A method of determining profit in which a percentage is added to the final estimated cost.

performance bond: A short-term insurance agreement that guarantees that a contractor will execute a construction job in the manner described in the contract documents.

permanent soil stabilization: The process of providing long-term stabilization of the ground at a construction site through mechanical compaction or the use of infill materials.

permit: Written permission granted by the authority having jurisdiction.

perspective drawing: The representation of an object as it appears when viewed from a given point.

photoelectric (daylight) switch: A switch activated by light levels.

pile: 1. A vertical structural member installed in the ground to provide vertical and/or horizontal support. **2.** The upright yarns that are the wearing surface of a carpet.

pile cap: A large unit of concrete placed on top of a pile or group of piles.

pilot safety control: A safety control that determines if the pilot light is burning.

pipe fitting: A device fastened to ends of pipes to terminate or connect individual pipes.

plan markup: The color coding or marking off of items during the takeoff process.

planting: The landscaping process of excavating, treating and preparing soil, and installing plant material in specified locations.

plan view: A drawing of an object as it appears looking down from above.

plaster: A mixture of portland cement, water, and sand used as an interior and exterior wall finish.

plastic laminate: A sheet material composed of multiple layers of paper material and resins bonded together under high heat and pressure.

plate: A horizontal support member in a frame wall, usually 2″ material.

plate glass: Polished glass manufactured in large sheets.

platform (western) framing: Construction in which single or multistory buildings are built one story at a time with work proceeding in consecutive layers or platforms.

ply: The number of single ends of yarn twisted to create a large yarn.

plywood: A wood product manufactured from thin layers of lumber.

pole: The number of completely isolated circuits that a switch can control.

post: A vertical support member such as a column or pillar.

post-tensioned beam: A beam containing steel reinforcing cables that are tensioned after the concrete has set.

poured-in-place concrete: Plastic (uncured) concrete that is poured (placed) in wood or metal forming material that is set to a specific shape and acts as a mold for the concrete.

prebid meeting: A meeting in which all interested parties in a construction project review the project, question the architect or owner concerning methods for accomplishing the work, and share information necessary to understand the entire scope of the work.

precast concrete: Concrete components that are formed, placed, and cured to a specific strength at a location other than their final installed location.

prefabricated unit: A stand-alone module composed of many different building components.

pressure control: A pressure-actuated mercury switch that controls the boiler burner(s) by starting and stopping the boiler burner(s) based on the pressure inside the boiler.

pressure-reducing valve: A valve that limits the maximum pressure at its outlet, regardless of the inlet pressure.

pressure-relief valve: A valve that sets a maximum operating pressure level for a circuit to protect the circuit from overpressure.

pressure switch: An electric switch operated by the amount of pressure acting on a diaphragm or bellows element.

prestressed beam: A beam containing steel reinforcing cables that are put in tension before the concrete is placed.

primer: A coating applied to new and old surfaces to improve the adhesion and final quality of the finish coating.

process piping: Piping installed in industrial facilities for compressed air, vacuum, gas, or fuel.

profit: The amount of money in excess of project and company expenses.

purlin: A horizontal support that spans across adjacent rafters.

quick takeoff: An estimating software function that enables the estimator to enter quantities directly into the estimating software spreadsheet.

rabbet: A square or rectangular groove cut in the edge or side of a board or plank.

raceway: An enclosed channel for conductors.

rail: A horizontal member of a door, window frame, or cabinet face frame.

receptacle: A device used to connect equipment to an electrical system with a cord and plug.

reflected ceiling plan: A plan view with the view point located beneath the object.

refractory material: A material that can withstand high temperatures without structural failure.

register: A device that covers the opening of supply air ductwork.

reinforced concrete: Concrete to which reinforcement has been added in the form of steel rods, bars, or mesh to increase its strength and resistance to cracking.

reinforcement: Material embedded in another material to provide additional support or strength.

remediation: The act or process of correcting.

retaining wall: A wall constructed to hold back earth.

rigid metal conduit (RMC): A heavy conduit made of metal.

rigid nonmetallic conduit (RNMC): A conduit made of materials other that metal.

riser: The piece forming the vertical face of a step.

rock wool insulation: Lightweight heat and sound insulating material made by blowing steam through molten rock or slag.

roof: The covering for the top exterior surface of a building or structure.

roofing tile: Clay, slate, or lightweight concrete members installed as a weather-protective finish on a roof.

roof paver: A flat, precast concrete block that is set in place on a roof deck to provide a walking surface and to protect waterproofing materials.

runner: A U-shaped channel fastened to the floor and ceiling in a metal framing system.

R value: A unit of measure of the resistance to heat flow.

safety switch (disconnect): A device used periodically to remove electrical circuits from their supply source.

safety valve: A valve that prevents the boiler from exceeding its maximum allowable working pressure.

sand: A loose granular material consisting of particles smaller than gravel but coarser that silt.

sandstone: Sedimentary rock consisting of quartz held together by silica, iron oxide, and/or calcium carbonate.

sanitary sewer: A piping system designed to collect liquid waste from a building and channel it to the appropriate removal or treatment system.

sealant: A liquid or semiliquid material that forms an air- or waterproof joint or coating.

sealer: A coating applied to a surface to close off the pores and prevent penetration of liquids to the surface.

sedimentary rock: Rock formed from sedimentary materials such as sand, silt, and rock and shell fragments.

service: The electrical supply, in the form of conductors and equipment, that provides electrical power to a building or structure.

shake: A hand-split wood roofing material.

shaking out: The process of unloading steel members in a planned manner to minimize the moving of pieces during the erection process.

she bolt: A special bolt that threads into the ends of a tie rod.

sheet glass: Clear or opaque glass manufactured in continuous, long, flat pieces that are cut to desired sizes and shapes.

sheet pile: Interlocking vertical support driven into the ground with a pile driver to form a continuous structure.

shellac: A coating made from special resins dissolved in alcohol.

shingle: A thin piece of wood, asphalt-saturated felt, fiberglass, or other material applied to the surface of a roof or wall to provide a waterproof covering.

shiplap siding: Siding that has rabbeted edges that allow for a fitted overlap between edges.

shoring: Wood or metal members used to temporarily support formwork or other structural components.

sill plate: A support member (usually a treated 2″ × 4″ or 2″ × 6″) laid flat and fastened to the top of a foundation wall.

silt: Fine granular material produced from the disintegration of rock.

skylight: A roof opening covered with glass or plastic designed to let in light.

slab: A horizontal or nearly horizontal layer of concrete.

slab-on-grade: A concrete slab that is placed directly on the ground.

slate: Fine-grain metamorphic rock that is easily split into thin sheets.

sliding door: A horizontal-moving door suspended on rollers that travel in a track that is fixed at the top of the opening or rollers mounted in the bottom of the door.

slope cut: An inclined or sloped wall excavation.

slurry wall: A wall built around a foundation to hold back ground water.

smoke shelf: A section of a chimney directly over a fireplace and below the flue.

snap tie: A concrete form tie that is snapped off when the concrete is set.

softwood: Wood produced from a conifer (evergreen) tree.

soil amendment: A material added to existing job site soil to increase its compaction and ability to maintain the desired slope at the completion of construction.

solvent: The volatile portion of the vehicle, such as turpentine and mineral spirits, that evaporates during the drying process.

spandrel beam: A beam in the perimeter of a structure that spans columns and commonly supports a floor or a roof.

specifications: Written information included with a set of prints clarifying and supplying additional data.

spline: A thin strip of wood inserted into grooves in edges of adjoining members to align and reinforce the joint.

spread footing: A foundation footing with a wide base designed to add support and spread the load over a wide area.

spreadsheet: Computer program that uses cells organized into rows and columns that perform various mathematical calculations when formulas and numbers are entered.

stack switch: A mechanical combustion safety control that contains a bimetal element that senses flue gas temperature and converts it to mechanical motion.

stain: A pigmented wood finish used to seal wood while allowing the wood grain to be exposed.

starting newel post: The main post supporting the handrail of a stair at the bottom of a stairway.

stile: A vertical member of a door, window frame, or cabinet face frame.

storm sewer: A piping system designed to collect storm water and channel it to a retention pond or other means of removal.

street bond: A bond that guarantees the repair of streets adjacent to the construction project if they are damaged during construction.

stringer (carriage): A support for stairs or steps.

strongback: A vertical support attached to concrete formwork behind the horizontal walers to provide additional strength against deflection during concrete placement.

structural clay tile: A hollow or solid clay building unit.

structural plastic: Plastic used for exterior, light bearing applications.

structural steel construction: Construction in which a series of horizontal beams and trusses and vertical columns are joined to create large structures with open areas.

structural wood construction: The use of wood members as the means of support for all imposed loads of a structure.

structural wood member: A wood component that provides support for live and dead loads to a structure.

stud: A vertical framing member in a wall.

subcontractor: A contractor who works under the direct control of the general contractor.

substrate: The underlying surface or material.

subsurface investigation: The analysis of all materials below the surface of a construction site to a predetermined depth.

supply piping: Piping that delivers water from the source to the point of use.

suspended ceiling: A ceiling hung on wires from structural members.

swell: The volume growth in soil after it is excavated.

switch: A device that is used to start, stop, or redirect the flow of current in an electrical circuit.

symbol: A pictorial representation of a structural or material component used on prints.

takeoff: The practice of reviewing contract documents to determine quantities of materials that are included in a bid.

temporary soil stabilization: The process of holding existing surface and subsurface materials in place during the construction process.

terminal unit: A device that transfers heat or coolness from the water or steam in a piping system to the air in building spaces.

terrazzo: A mixture of cement and water with colored stone, marble chips, or other decorative aggregate embedded in the surface.

thermostat: A temperature-actuated electric switch that controls heating and/or cooling equipment.

three-way switch: A single-pole, double-throw (SPDT) switch.

throw: The number of closed contact positions per pole.

tie: A device used to secure two or more members together.

tie rod: A steel rod that runs between forms to hold them together after concrete is placed.

tile: A thin building material made of cement, fired clay, glass, plastic, or stone.

tilt-up construction: A method of concrete construction in which concrete members are cast horizontally at a location close to their final position and tilted in place after removal of the forms.

tilt-up panel: A panel that is precast and lifted into place at the job site.

timber: A heavy wood member that has a nominal size of 5″ × 5″ or larger.

timber framing: Construction in which large, laminated wood members (glulam) or one-piece timbers are used to form large open areas.

time and material basis: A method of determining profit in which the owner pays a fixed rate for time spent on a project and pays for all materials.

traditional estimating method: An estimating method in which quantities are calculated and entered onto ledger sheets for pricing.

transformer: An electric device that uses electromagnetism to change voltage from one level to another.

traprock: Fine-grained, dark-colored igneous rock.

tread: The horizontal surface of a step.

tremie: A pipe or tube through which concrete is placed into vertical formwork or under water.

trench box: A reinforced wood or metal assembly used to shore the sides of a trench.

trencher: A machine with a rotating device that digs a narrow, long hole into the earth.

truss: A manufactured roof or floor support member.

tuft: The cut or uncut loop comprising the face of a carpet.

two-way switch: A single-pole, single-throw (SPST) switch.

value engineering: A process in which construction professionals such as project managers, estimators, and engineers employed by the firm developing the bid are allowed to suggest changes to the contract documents.

valve: A device that controls the pressure, direction, or rate of fluid flow.

varnish: A clear or tinted liquid consisting of resin dissolved in alcohol or oil that dries to a clear, protective finish.

vent pipe: A pipe that removes odors and gases from the waste piping and exhausts them away from inhabited areas.

volatile organic compounds (VOCs): The portion of a paint mixture that easily evaporates after application.

waler: A horizontal member used to align and brace concrete forms or piles.

wall bearing construction: Construction integrating horizontal steel beams and joists into other construction methods such as masonry and reinforced concrete.

wallpaper: An interior patterned-paper or vinyl sheet wall covering.

waterproofing: The treatment of a material so it is impervious to water.

weave: The pattern or method used to join surface and backing carpet yarns.

well point: A pipe with a perforated point that is driven into the ground to allow groundwater to drain.

wide-flange beam: A structural steel member with parallel flanges that are joined with a perpendicular web.

wire: An individual conductor.

wire mesh partition: A partition fabricated of wire mesh.

zero line: A line connecting points on a topographical map where existing and planned elevations are equal.

zone valve: A valve that regulates the flow of water in a control zone or terminal unit of a building.

Before removing the CD-ROM, please note that the textbook cannot be returned if the CD-ROM sleeve seal is broken.

Installing Precision Estimating – Extended Edition Trial Version

The software must be installed using the Setup command. The application files cannot be copied directly onto the computer.

1. Insert the CD-ROM into the appropriate drive on the computer.

2. Select **Run** from the **Start** menu.

3. Enter the letter of the source drive followed by a colon (:) and the setup command. For example: d:setup. Click **OK**.

4. Follow the on-screen instructions. When prompted for the application components to be installed, make sure that both the Precision Estimating and Sample Databases boxes are checked.

5. Accept the default destination folder (c:\Program Files\Timberline\Precision).

6. Once the setup has finished, select **Yes** to restart the computer.

Starting the Demo

1. From the Start menu, choose Programs, Precision Estimating, Precision Estimating - Extended Edition.

2. The first time Precision Extended is started, a Demo Activation window appears indicating that 180 days remain to use the software. Click **OK**.

3. The next window to appear states "You must have a database open to work with Precision Estimating. Would you like to open a database now?" Select **Yes**.

4. Once the database window appears, double-click on Sample Ext Commercial G.C. and then double-click on the pei.dat file. *Note:* You are now connected to the Sample Ext Commercial G.C. database and ready to create an estimate.

Precision Estimating – Extended Edition is designed to work on Microsoft® Windows 95, 98, and Windows NT™. Requirements include at least a 486 or Pentium™ computer with 16 MB of memory, 80 MB of available hard disk space, a VGA monitor, a CD drive, and a mouse.

The frequently asked questions link on the CD-ROM is the first resource used if questions arise during the use of this product. This link provides product information posted on the American Tech web site. Technical support may also be obtained by sending an e-mail message to support_bte@americantech.org or by calling (708) 957-1241 Monday through Friday from 8:00 AM to 4:00 PM Central time and requesting Building Trades Estimating technical support. For more information on American Tech training materials, visit the American Technical Publishers, Inc. web site at www.americantech.org or call 1-800-323-3471.

For more information on Timberline software products, visit the Timberline Software Corporation web site at www.timberline.com or call 1-800-628-6583.